Principles of Molecular Probe Design and Applications

Wellington Pham

Principles of Molecular Probe Design and Applications

 Springer

Wellington Pham
Department of Radiology and Radiological
Sciences
Vanderbilt University School of Medicine
Nashville, TN, USA

ISBN 978-981-19-5738-3   ISBN 978-981-19-5739-0 (eBook)
https://doi.org/10.1007/978-981-19-5739-0

This Springer imprint is published by the registered company Springer Nature Singapore Pte Ltd.
The registered company address is: 152 Beach Road, #21-01/04 Gateway East, Singapore 189721,
Singapore

Preface

Molecular imaging emerged as a key player in the pursue of precision and personalized medicine. The development of molecular probes indeed serves as a key component in this journey. This book is modified with more updates on the original lecturing materials for the probe design class I taught at Vanderbilt School of Medicine. I remember spending a tremendous amount of time collecting information to cover the design and synthesis of imaging agents. Molecular imaging is a burgeoning field; the pace of literature outputs grows so fast beyond manageable comprehension. In this book, I try to classify a great deal of information about molecular imaging probes into groups and how they relate to each other, like a periodic table of molecular probes. This helps to manage literature effectively. On the other hand, it facilitates the creative designing of new probes. If we know what category the probe belongs to, we can certainly predict its physical property.

The materials present herein comprised a collection of classical and contemporary designs of molecular probes spanning several decades. Aside from providing practical methods, the other idea of this book focuses on inspirational work from esteemed pioneers in the field. It is worthwhile to understand the background behind every invention. By understanding the conceptual background behind these peculiar cases, I hope that young scientists will be motivated and translate the knowledge into their future inventions.

Although this book is designed for graduate school, it is also suitable for senior students who have completed a year of organic chemistry and biology. Due to the limited space, some topics were briefly discussed. However, the instructors can use these sections as guidance for more elaboration in their lectures. Further, this book is also suitable as a reference in the laboratory to assist scientists with their practical laboratory work. While writing this book, I always imagine that I am lecturing, so I hope the materials present vibrant discussions.

I want to take this opportunity to thank my mentor Steven Peseckis who guided me through the art of synthetic chemistry. After graduation from his laboratory, I pursued my career in molecular imaging with many pioneers in this field. I am honored and humbled for the chance to work in their laboratories over the past

22 years. This book is not possible without mentioning Tatsushi Toyokuni, Ralph Weissleder, Ching-Hsuan Tung, Bruce Rosen, Anna Moore, and John Gore.

I also thank many authors who supported me during the writing by providing permission/materials for the book, including Anna Wu, Jeff Bulte, Bengt Langstrom, Angelique Louie, Robert Pooley, Marcos Serrou do Amaral, Nicholas Long, John Gore, Johannes Broichhagen, David Hodson, Joanna Fowler, Marc Vandrell, Veronique Gouverneur, Victor Pike, Kenneth Dahl, Carolyn Anderson, Joy Chong, Jan Kotek, James Eills, Yuanpei Li, Michael Smith and Philip Low.

Last but not least, I want to acknowledge the support and encouragement from my parents and my brother Albert. Writing a book amid the pandemic is not an easy task; it means fewer family outings, leisure, and gatherings. For that, I sincerely thank my wife Michiyo and Haruka for their everyday support. More than that, many thanks to Haruka for training her dad how to draw cartoons digitally. She demonstrated that a ten-year-old girl can accomplish a serious job if given that opportunity.

Thank you all,

Nashville, TN, USA   Wellington Pham
May 2022

The original version of the book was updated. Index list was included in the backmatter.

Contents

Chapter 1
Overview of the Fundamentals of Chemistry for Molecular Probe Design

1.1 Valence Bonding

This chapter will review some of the fundamental knowledge in organic chemistry as a framework before delving more into the chemical design of imaging probes in the following chapters. If you examine the molecules in this universe, they are simply formed by binding individual elements together, and this is called valence bonding. The valence number is used to describe the unique characteristics of these elements to bind to atoms through the chemical bonds. The concepts of valence, no doubt, are the main principles that help to explain the chemical bonds, molecular orbitals, and chemical structures. The valence bonding theory is essential for providing a rational understanding of how molecules are formed at the atomic levels. For example, one would wonder why some elements could form more covalent bondings than others and so on.

All started with Sir Isaac Newton, who first described the initial explanation of chemical bonds in 1704 in his book entitled *Opticks* (Cantor 1984). He explained that atoms attach to each other by some type of "force." However, until 1916, Gilbert Lewis formally defined the concept of chemical bonds in the journal of the American Chemical Society, where he described how chemical bonding results from the sharing electrons between two participating atoms (Shaik 2007). Let us examine the simplest molecule, such as hydrogen gas, which has the simplest covalent bond in organic chemistry. The H:H constitutes the overlapping of 2 electrons donated by each hydrogen element in the region between two nuclei. The diatomic molecule is formed in an apparent satisfaction of the "octet rule" since each hydrogen atom contains only one electron in its outer shell, and thus, it needs to combine to fill the first orbital. As the result, the molecule is more stable compared to a single hydrogen atom, which has only one electron.

In general, an atom is the smallest particle of an element that has the chemical genetic of that element indestructible throughout the chemical and physical encounters. In general, atoms that possess a lone pair of electrons in the outermost energy level would bind with other atoms that have unpaired electrons to achieve their full

W. Pham, *Principles of Molecular Probe Design and Applications*,
https://doi.org/10.1007/978-981-19-5739-0_1

complement of electrons. The associated atoms share their valence electrons in order to form a covalent bonding, and thus, a molecule is formed. So, a molecule is an electrically neutral group of two or more atoms held together by chemical bonds (Kemp et al. 2003). According to Dalton's Atomic Theory, all atoms of an element have identical properties, which differ from those of other elements (Whitten et al. 1988). Further, it also stated that atoms of different elements can combine with each other in minimal whole-number ratios to form chemical compounds. For example, hydrogen and oxygen atoms form covalent bonds between themselves to form a water molecule, which is a liquid.

This theory could be expanded beyond the formation of compounds made of two or more different elements. Let us expand that concept a bit more to explain common organic molecules, which we see in the laboratory or at least in our everyday lives. For example, the formation of molecules from 2 identical atoms is called a diatomic molecule. Hydrogen (H_2), nitrogen (N_2), fluorine (F_2), chlorine (Cl_2), and bromine (Br_2) belong to this group. Since these diatomic structures are made of two atoms of the same element, they are also referred to as homonuclear molecules. Diatomic molecules made up of two different atoms are called heteronuclear molecules. For example, carbon monoxide (CO), nitric oxide (NO), hydrogen chloride (HCl), and hydrogen fluoride (HF) are heteronuclear compounds.

When three or more atoms are held together by covalent bonds to form a molecule, these are called polyatomic structures. Homonuclear polyatomic molecules include ozone (O_3), sulfur (S_8), and phosphorus (P_4), while the heteronuclear polyatomic molecules are ammonia (NH_3), ethanol (CH_2OH), and methane (CH_4).

Aside from covalent bonds, another typical chemical association is called ionic bondings, which are formed by the complete transfer of one or more electrons to create a positive cation and negative anion, usually depicted as A^+ or: B^-. Typically, metallic elements on the left side of the periodic table form ionic bonds with non-metallic ions on the right side of the periodic table by transferring the electrons to the non-metal partners and becoming a positively charged ion (cation). The electron-accepted non-metals become negatively charged ion (anion). For example, sodium chloride (NaCl), magnesium chloride ($MgCl_2$), magnesium sulfate ($MgSO_4$), potassium cyanide (KCN), sodium iodide (NaI), manganese oxide (MnO), and zinc oxide (ZnO) belong to this group.

According to the quantum theory, electrons of an atom constantly travel around the nucleus in a way analogous to the satellite orbiting around the earth, and thus generating a so-called orbital. As a result, there is a continual trade-off between the potential and kinetic energy as it varies in distance from the nucleus, but the total remains exactly the same since the energy of an electron in an atom is restricted in a fixed value, or in other word, it is quantized (Stowell 1988). The lowest energy state of an electron in an atom is termed the ground state. Any other higher energy levels are called excited states. According to the Heisenberg Uncertainty Principle, which states that there is no way to determine both the precise values related to the position and momentum of an electron simultaneously. Therefore, it is likely true that an exact direction of an electron cannot be determined. The electrons can travel randomly around the nucleus while maintaining fixed energy. Nevertheless, the atomic orbital, the region of space next to the atomic nucleus, has the highest probability of finding an electron. The orbitals have different energy levels; for example, the s orbital has low energy since it is the closest to the nucleus, while the 2s and 2p orbitals are farther from the nucleus and resulted in higher energy level, with those of 2p is higher than those of 2s. And the energy levels continue to rise with other distancing orbitals, such as d and f orbitals. The wave function values of s orbitals have a spherical shape around the nucleus. In contrast, p orbitals have two lobes, like an 8-shape where high electron clouds could be found at two opposite lobes away from the nucleus, and it diminishes gradually to zero at the nodal plan of the nucleus. Different from the spherical s orbitals, the two-lobed p, four-lobed d, and diffused shape f orbitals are directional.

According to the Pauli Exclusion Principle, no more than two electrons can occupy an atomic orbital. The two electrons must have opposite spins; meaning that if one has an up-spin (+1/2), the other must have a down-spin ($-1/2$). The spin states of an electron are defined as [2(electron spin number) + 1]. In regard to the energy levels in an atomic model, the principal energy levels are related to the space that contains the electrons. The energy levels are quantized as 1, 2, 3, and so on away from the nucleus. Each energy level can accommodate $2n^2$ electrons, where n is the number of the level. For instance, the first level can hold a maximum of $2(1)^2 = 2$ electrons, and the second level contains $2(2)^2 = 8$ electrons. The process of filling the electrons in the energy levels should be started from the lowest energy level all the way up. According to the Aufbau Principle, also called the "building-up principle," it states that atomic orbitals (AOs) should be filled in the order of increasing energy. The electrons fill the orbitals starting with the lowest energy levels before moving up to the higher energy states. The lower the number of the principal energy levels, the closer the electron is to the nucleus, and thus, it is more difficult to remove this electron from the atom.

Let us examine the elements in row 1 of the periodic table. This row represents the first energy level; the atoms in this row have only 1 AO, which can hold up to 2 electrons. For example, hydrogen has only one electron, while helium (He), a noble gas, fills up the AO fully with two electrons (Fig. 1.1). From this information, one would predict that helium is very stable since there are no free electrons to share, so it is an inert gas.

While in row 2, the electrons of the atoms are filled in the second energy level. Among them, neon (Ne) fills all the AO completely; thus, no free electrons to share, and thus, Ne is also a very stable and inert gas.

It is worthwhile to remind that Hund's Rule, also known as Rule of Maximum Multiplicity, should be observed when filling the electrons in the orbitals of a given sublevel (Fig. 1.2). It postulates that all the orbitals in a sublevel must be filled with singlet electrons before the double occupancy occurs, and these unpaired electrons

Fig. 1.1 The atomic orbitals of the atoms in row 1 of the periodic table

should have the same spin. This electronic configuration is very stable because there is less electrostatic repulsion between electrons in different orbitals than between paired electrons in the same orbital.

So now, let us discuss what it means to have a covalent bond of a molecule after reviewing the AO theory. Basically, a covalent bond is the overlap of two AOs of the participating atoms to produce the molecular orbitals (MOs) enclosing both atoms. For example, the MOs of the simplest molecule, such as H_2, are formed from the contribution of the $1s^1$ electron configuration of two individual hydrogen atoms. In those two generated MOs, one by reinforcement resulted in lower energy and a second by cancellation, which resulted in high energy (Fig. 1.3).

Most of the covalent bonds involve one, two, or three pairs of electrons. As shown in the case of H_2, a single covalent bond occurs when two hydrogen atoms share a pair of electrons. While a double covalent bond involves the sharing of two electron

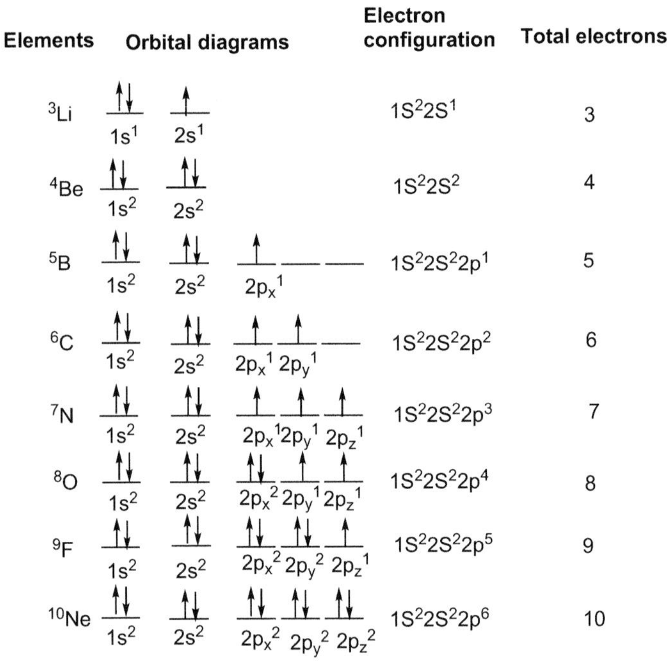

Elements	Orbital diagrams	Electron configuration	Total electrons
^{3}Li	$1s^1$ $2s^1$	$1S^22S^1$	3
^{4}Be	$1s^2$ $2s^2$	$1S^22S^2$	4
^{5}B	$1s^2$ $2s^2$ $2p_x^{\ 1}$	$1S^22S^22p^1$	5
^{6}C	$1s^2$ $2s^2$ $2p_x^{\ 1}\ 2p_y^{\ 1}$	$1S^22S^22p^2$	6
^{7}N	$1s^2$ $2s^2$ $2p_x^{\ 1}2p_y^{\ 1}\ 2p_z^{\ 1}$	$1S^22S^22p^3$	7
^{8}O	$1s^2$ $2s^2$ $2p_x^{\ 2}2p_y^{\ 1}2p_z^{\ 1}$	$1S^22S^22p^4$	8
^{9}F	$1s^2$ $2s^2$ $2p_x^{\ 2}2p_y^{\ 2}\ 2p_z^{\ 1}$	$1S^22S^22p^5$	9
^{10}Ne	$1s^2$ $2s^2$ $2p_x^{\ 2}\ 2p_y^{\ 2}\ 2p_z^{\ 2}$	$1S^22S^22p^6$	10

Fig. 1.2 Hund's rule for filling the electrons in the orbitals

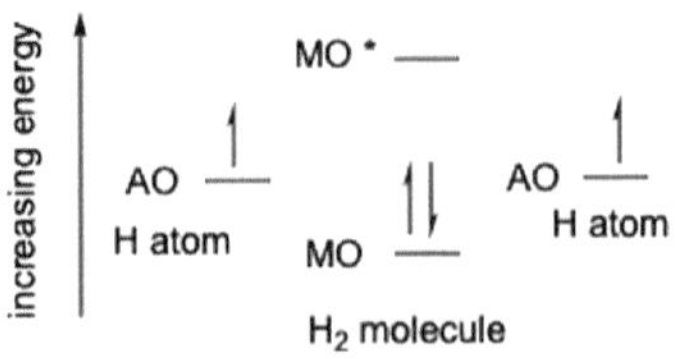

Fig. 1.3 Hydrogen molecular orbital electron configuration energy diagram

pairs, a triple covalent bond involves three electron pairs. In short, we usually term these associations as single, double, and triple bonds. It is apparent that the electron density is highest along the axis of these bonds.

Is there a meaningful way to determine how many bonds an atom has? The answer is yes. After arranging the number of electrons of an element with appropriate AOs, the unshared electrons will be revealed, and thus, the valency can be considered as being equal to the number of unpaired electrons present in that atom. For example, lithium has one unshared electron, and its valency should be one; while nitrogen has three unshared electrons in its outermost shells; thus, its valency is three. So, nitrogen can share three bonds with other elements, such as hydrogens, to make ammonia (NH_3), which is a gas.

As said, this rule is not always infallible, as a reminder, we are trying to generalize our observation to explain how things are formed in this universe, and as a matter of fact, not every atom behaves the same, and thus, new methods always evolve for explaining rare phenomena, such as in the case of carbon. Theoretically, carbon has a valency of 2, and thus, CH_2 molecule does exist, but it is not very stable. So, something is amiss in here given a carbon atom has four covalent bonds. To explain this phenomenon, one would need to borrow the orbital hybridization theory to explain this observation. Probably, the diagram in Fig. 1.4 explains it better; for instance, carbon has an electron configuration of $1s^2\,2s^2\,2p^2$. This arrangement means that only two valence electrons in carbon's outermost shell can form covalent bonds, the 2p orbital electrons. However, the close proximity between the energy gap of the 2s and 2p orbitals enables the hybridization of 2s with 2p orbitals to form the new sp^3 hybrids, which now can accommodate four singlet electrons.

It turns out that orbital hybridization is indeed a compelling theory used to explain many irregular cases. Let us examine another case, like boron, which has a ground state electronic configuration of $1s^2 2s^2 2p^1$. Based on this setup, only one covalent bond can be formed with the half-filled orbital, but that cannot explain the reality that boron has 3 covalent bonds. Using the hybridization concept, one would recognize the promotion of 2s with three of the 2p orbitals to form the sp^2 hybrid orbitals. With 3 half-filled orbitals, boron can form three covalent bonds, leaving one empty 2pz orbital. Taken altogether, this orbital hybridization is very useful for explaining how many bonds an atom can have and its subsequent electron pair geometry. One of the simplest ways to figure out the geometry of a molecule is by counting and adding up the order of the shells. For example, sp hybridization has one s and one p

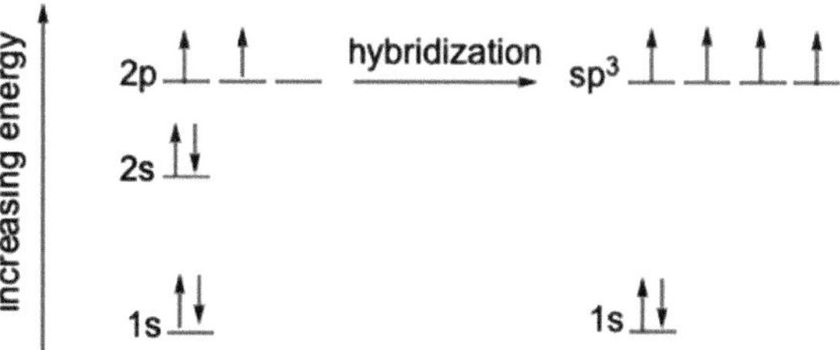

Fig. 1.4 Orbital hybridization to explain the electronic configuration of carbon

shell; each has an order of one, so the number of attachments linked to the central atom is 2. Then, sp^2 is 3, sp^3 4, sp^3d 5, sp^3d^2 6, sp^3d^3 7, and so on. The spatial distribution of 2 groups in sp is 180-degree bond angle linear electron pair ($BeCl_2$, $HgBr_2$, BeF_2, CHN); 3 groups in sp^2 hybridization, 120-degree bond angles trigonal planar (BCl_3, BH_3, BF_3, $CH_2 = O$, $CH_2 = NH$); 4 groups in sp^3, 109-degree bond angles tetrahedral (CH_4, CCl_4, C_2H_6, NH_4^+, CH_3-NH_2); 5 groups in sp^3d, 120- and 90-degree bond angles trigonal bipyramidal (PF_5); 6 groups of sp^3d^2 90-degree bond angles octahedral geometry (SF_6); and 7 groups of sp^3d^3, 72- and 90-degree bond angles pentagonal bipyramidal geometry (IF_7).

1.2 Chemical Bonds of Molecular Probes in Biological System

This section focuses on the intermolecular interactions between molecules in a biological milieu. Although this book focuses on imaging probes, all interactions of the probes at target sites are identical to those of drug molecules. However, it is worthwhile to mention that imaging probes differ from drugs. Besides specific recognition of targeted sites, the probes convey the process by emitting signals for monitoring purposes. The ideal probes with the combined sensitivity and specificity render the use of low doses, so significant that no pharmacological responses are involved. As a general rule, prior knowledge about the target site would facilitate rational design on how and where the probe should be targeting with a well-defined backbone (Silverman 1992). For instance, the probe–target interaction relies on a few atoms that are stabilized by a number of biophysical parameters, such as size, shape, molecular weight, covalent bonds, hydrogen bonds, hydrophobic interactions, van der Waals forces, and so forth (Bhinge et al. 2004).

1.2.1 Covalent Bonding

In most cases, the interaction between molecular probes with the intended targets is a reversible process. The probes should have appropriate pharmacokinetics and binding affinity to report the target's mode of action. Then, washing out and clearance should be next after doing what they are designed to do. Only a small number of probes targeting catalytic enzymes are reported to modulate the enzyme activity via covalent bonds, in an approach similar to the development of suicide inhibitors in medicinal chemistry (Silverman 1992). Different from small drug molecules, extending the presence of molecular probes in the biological systems may have unprecedented implications, including profound toxicity and long-term adversary effects because they may carry radioisotopes, transition metals, and chelators. Particularly, as we will discuss more detail in Chap. 4, the dissociation of metals from the chelators has two negative consequences; one is the toxic caused by the metals. And the other is the available chelators that will scavenge metals in the biochemical pathways, such as magnesium, calcium, potassium, and more, resulted in toxicity. In this regard, the "suicide" probes should be screened with these issues in mind before in vivo applications. Figure 1.5 showed a "suicide" fluorescence resonance energy transfer (FRET)-based probe, in which a hydroxyl moiety of a serine residue from a protease enzyme's active site would form a stable covalent linkage with the coumarin reporter dye via opening a lactam ring, tethering between the FRET pair (Mizukami et al. 2009).

The covalent bond is very stable, with the bond energy somewhere between 40 and 110 kcal/mol (Silverman 1992). Certainly, no biological or chemical compositions of a living system can disintegrate this conjugation. The covalent binding is generated by the nucleophilic attack from the enzyme to the highly strained and reactive lactam ring of the probe, resulting in an irreversible binding.

Besides the β-lactams shown in this probe, other electrophilic centers can also be used for the design of "suicide" probes, including the α, β-unsaturated carbons, epoxides, hydrazides, boronates, propargylamines, aldehydes, alkyl halides, disulfides, sulfoxides, acetates, acrylamides, cyanamides, sulfones, sulfonamides, nitriles, and thiourea groups. When introduced into a biological system, specific binding between the probe and the target is achieved via the covalent bonds, generated by active groups from the enzyme, including serines, lysines, cysteines, and histidines.

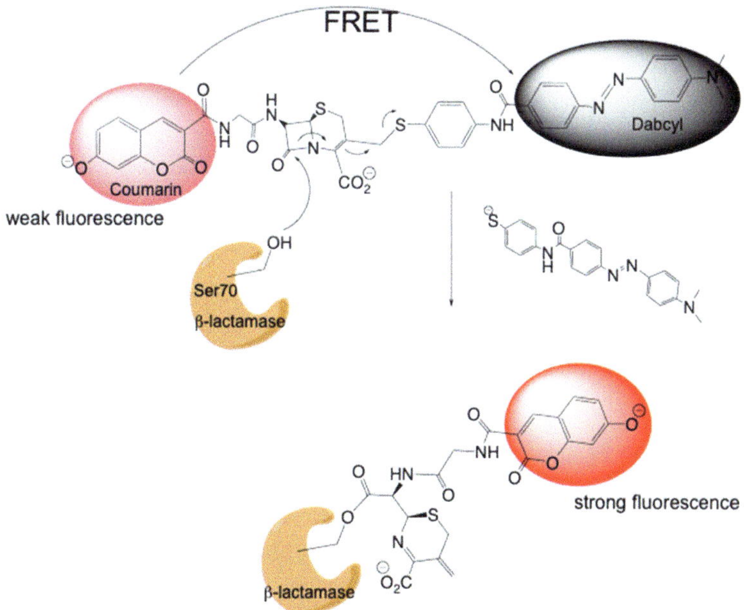

Fig. 1.5 "Suicide" FRET probe for imaging enzymes. Data obtained from Mizukami et al. (2009) with permission from the American Chemical Society

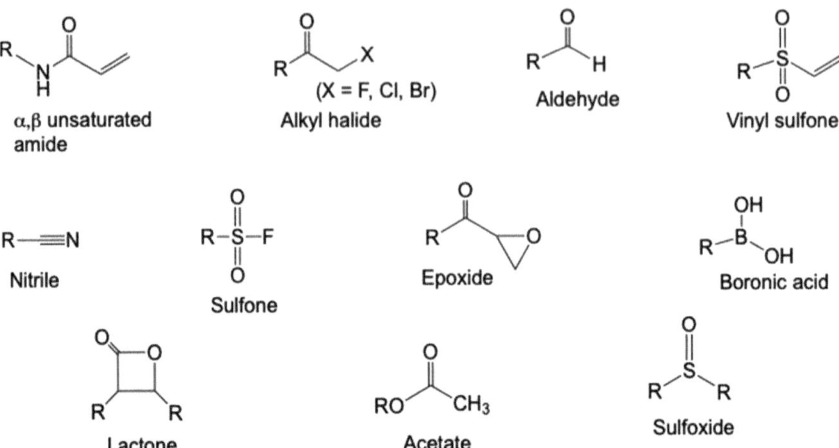

Fig. 1.6 Electrophilic handles of the probes for designing interaction with enzyme via covalent bonding

Particularly, α, β-unsaturated carbonyls are the most popular analogs for this type of reaction, given their ease to develop, and the intermediates are stable. The Michael addition reaction through thiol groups available in the proteins/enzyme's binding site is a neat and spontaneous reaction. At neutral physiological pH, the nucleophilicity of the thiol residue is more robust than any other nucleophiles. The order of strength starts first with cysteine (1) > histidine (10^{-2}) > methionine (10^{-3}) > lysine/serine (10^{-5}) > threonine/tyrosine (10^{-6}) (Way 2000). According to Pearson's theory, both thiol and α, β-unsaturated carbonyls are soft molecules, which are more polarizable than hard molecules; thus, their unions are favored with a large negative value for ΔH and very stable complexes (Pearson 1990). Here are the chemical structures of useful electrophiles for "suicide" probe design (Fig. 1.6).

Enhanced binding at the target site using irreversible interaction is a powerful approach for improved signal detection and selectivity. Nevertheless, using probes with the irreversible binding is the associated toxicity and implications of off-target modifications, for instance, by reacting with proteins other than the intended target, particularly those with hyper-reactive amino acids, like cysteine. To mitigate these issues, very dedicated chemistry had been reported for targeting enzymes/proteins with reversible covalent binding. The method was initially intended for drug discovery but can be applicable to probe development. The idea of this construct was to maintain covalent targeting since it is crucial for enhancing selectivity but without concerns about "suicide" interaction leading to irreversible adduct formation (Serafimova et al. 2012). This chemistry was based on a past observation that thiol groups reacted with 2-cyanoacrylates instantaneously without a catalyst under normal physiological conditions. However, no product could be isolated, and it is postulated that the Michael addition reaction might go through a quick equilibrium reaction. Further analysis suggested that the electron-withdrawing nitrile group

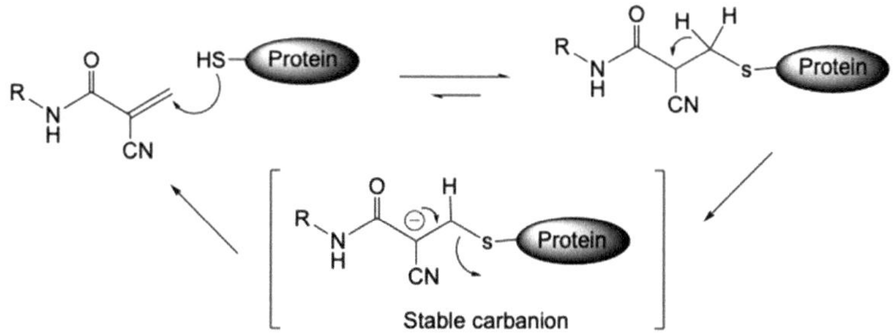

Fig. 1.7 Reducing toxicity of the probe using a reversible binding mechanism. Derived from Serafimova et al. (2012)

creates electron-deficient olefins, which change the course of the supposed to be irreversible to reversible Michael reaction. The addition of the nitrile group to the α-carbon increases the susceptibility of the β-carbon to nucleophilic attack but also stabilizes the resultant carbanion (Fig. 1.7). This leads to not only reversible Michael reaction with protein cysteinyl thiols, but the reaction is also very rapid.

Active carbonyl structures are also good candidates, serving as electrophiles for the design of covalently reversible probes. For example, aldehyde or ketone may undergo reversible addition reactions with alcohols or thiols to create hemiacetals/hemiketals or hemithioacetals/hemithioketals, respectively. However, due to the nature of unspecificity and highly reactive agents, aldehyde can cause off-target interactions, and thus, concomitant toxicity is inevitable. Another approach to improve the specificity is the use of boronates, which act as an electrophile mimicking the carbonyl carbon. They also form slow off-rate binding like aldehydes to form covalent tetrahedral adducts analogous to those formed by active carbonyl compounds. But unlike aldehydes and Michael acceptors, boronates do not interact with thiols, so they are specific for only amines and hydroxyl moieties usually found in histidine and serine, respectively. Boronates react with amino groups in a kinetic reaction, but if there is an available hydroxyl group in the vicinity, it will displace the amino group in a thermodynamically favorable reaction (Deadman et al. 1995) (Fig. 1.8). Another advantageous application of boronates in vivo is that the materials are relatively stable under physiological conditions.

Among the imaging targets, cathepsin enzymes are probably one of the most suitable candidates to be targeted by the covalent-binding probes. Besides acting as degrading proteases, a large body of research showed the role of cysteine cathepsin in tumor progression. The increased levels of these peptidases were correlated with poor prognosis and high tumor grade in different tumor types (Vasiljeva et al. 2019). Further, aberrant cathepsin activity also implicates other diseases, including arthritis, neurodegenerative diseases, cardiovascular disease, obesity, and cystic fibrosis (Olson and Joyce 2015). In general, proteases recognize their substrate target as amino acid sequences around their cleavage site. In most cases, the recognition is up to 4 amino acids on each side of the scissile bond (Tsvirkun et al. 2018). In this probe design for in vivo imaging applications, a known peptide-based cathepsin

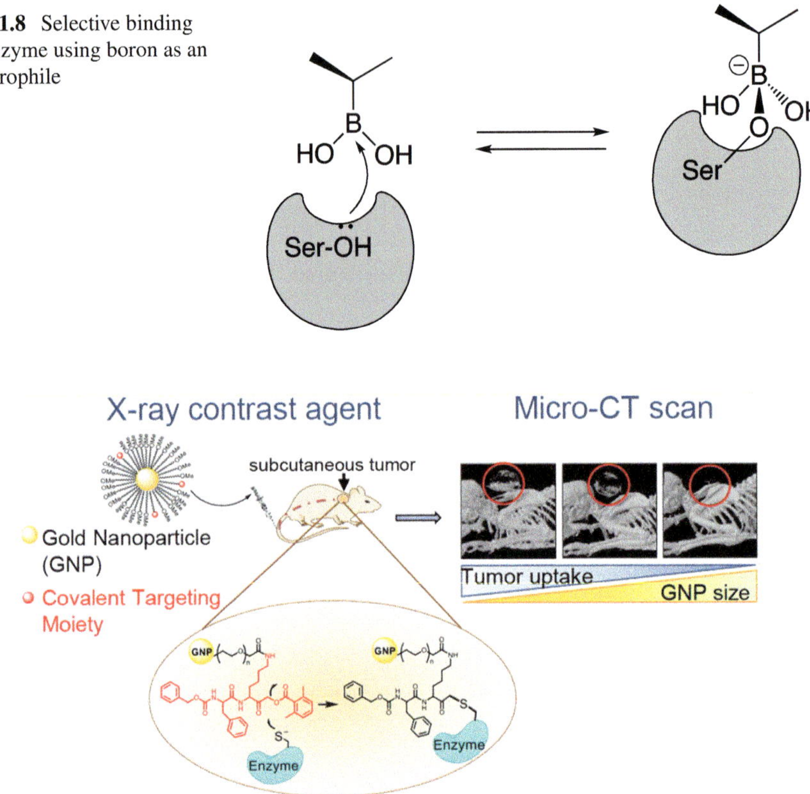

Fig. 1.8 Selective binding to enzyme using boron as an electrophile

Fig. 1.9 Enzyme-catalyzed conversion of a pro-probe into a probe. Data obtained from Tsvirkun et al. (2018) with permission from the American Chemical Society

inhibitor was modified on the ε-amino group of lysine with a linker for bioconjugation (Fig. 1.9). The other end of the linker possesses a thiol group, which is used to attach gold nanoparticles via the Au–S bond. The C-terminus of the peptide is derivatized with a 2,6-dimethylbenzoic ester, which serves as a leaving group. Upon exposure to the enzyme binding domain, the thiol group from cysteine will promptly bind the probe, displacing the 2,6-dimethylbenzoic ester with a thioether bond. This gold nanoparticle cathepsin-targeted activity-based probe is ideal for functional computed tomography (CT) imaging of elevated cathepsin activity within cancerous tissue.

1.2.2 Hydrogen Bonding

This is the strongest bond in biological milieus after covalent bonds. The typical energy of hydrogen bonds is about 5–6 kcal/mol (Sheu et al. 2003). A hydrogen bond is formed in polar molecules, either by the intramolecular or intermolecular association between hydrogen and the highly electronegative elements, which carry an unshared pair of electrons, such as N, O, or F. The resulting bond is facilitated by the dipole–dipole attraction between the partial positive charge of H and the partial negative charge of another atom, including halides. The order of electronegative atoms that can participate in the hydrogen bonding is $F > O > N >>> Cl$. Due to the exceptional force of interaction, a hydrogen bond (Fig. 1.10) is the main mechanism behind the surprisingly high boiling point of small molecules like water, ethanol, methanol, or ammonia, compared to similar compounds with the same molecular weight.

There are several polar head groups in a biological system that have a partial charge that can form hydrogen bonds with probes in the target site, such as hydroxyl, carboxyl, and amino groups. Basically, all amino acids can form hydrogen bonds, acting either as donors or acceptors. It is noteworthy that hydrogen bonds are the most unique among other bonds in that they can have a positive charge in physiological pH while still forming the bond. This is particularly true for histidine. It is found that peptides containing protonated histidine residues form hydrogen bonds better than those with unprotonated histidines (Patronov et al. 2014). Furthermore, hydrogen bonds play an essential role in maintaining the structural integrity of proteins through the α-helix and β-sheet conformation and double helix of DNA (Stryer 1995). In these macromolecules, the intramolecular H-bonding helps them fold in a specific domain to exert their individually unique physiological and biochemical functions. But most of all, aside from electrostatic interactions, hydrogen bonds are the primary determinants to dictate the specific binding between biological macromolecules and their probes (Wade et al. 1993). For the rational design of probes with available three-dimensional molecular structures of the local environment and binding motifs, carefully navigating the distribution of the probe with associated hydrogen bonds would enhance binding affinity and specificity. As a donor, the ligand mostly interacts with amino acid sequence containing leucine, glutamic acid, and histidine, while the ligands tend to interact with glycine and leucine as an acceptor (Panigrahi 2008), Before actual receptor/probe binding happens, the receptor and probe force the neighboring water molecules into an ordered shell. Then, the binding of the probe to the receptor releases some of the ordered water, and the resulting increase in entropy provides a thermodynamic push toward the formation of the probe–receptor complex (Panigrahi 2008).

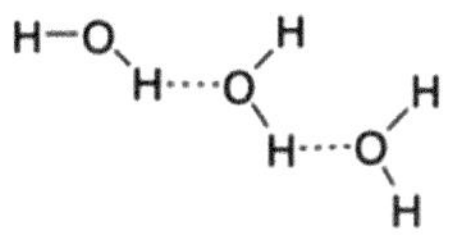

Fig. 1.10 Hydrogen
bonding

1.2.3 Hydrophobic Interactions

The specific association between the probes at the binding pocket of the target site is more complicated. It involves other critical encounters, than just covalent bonding or ionic interaction. The hydrophobic interaction represents one such example. It is described as the interaction between hydrophobic moieties with each other in the presence of water. We can use the term "like dissolves like" to explain this interesting, phenomenal interaction between lipophilic groups in an aqueous environment. This effect has important implications in maintaining the structural integrity of protein and especially biological membranes, as well as for a plausible explanation of drug/probe associated with receptors. In Chap. 2, we will discuss how hydrophobic interaction affects the fluorescence property of fluorescent dyes. The hydrophobic moieties are vital players in molecular recognition and binding. When two hydrophobic molecules encounter each other as they are in a neighborhood brought about by a random movement, they tend to berth closer; as such, the surrounding water molecules become sequestered and associate with each other and, as a result, increase in the entropy. Therefore, there is a spontaneous decrease in the free energy that favors probe/drug–target association. This phenomenon lends support for the explanation of hydrophobic ligands that bind to the respective receptors (Davis and Teague 1999). Probably, the family of COX-2 inhibitors shown in Fig. 1.11 best exemplifies the significant role of probe/drugs' hydrophobic structures in the enhanced recognition of the receptor's hydrophobic surface. For instance, these two potent COX-2 inhibitors, Flurbiprofen and SC-558 share the same binding position inside the COX-2 enzyme's binding pocket.

The distal ring on flurbiprofen overlaps with the bromophenyl ring of SC-558. The bromophenyl ring is bound in a hydrophobic cavity formed by Phe 381, Leu 384, Tyr 385, Trp 387, Phe 513, and Ser 530; with the contribution from the backbone atoms of Gly 526 and Ala 527, the pyrazole of compound SC-558 superimposes with the fluorophenyl ring of flurbiprofen. Furthermore, the carboxylate of the flurbiprofen and trifluoromethyl group of SC-558 bind in the same enzyme cavity (Kurumbail et al. 1996). Aside from hydrophobic work, increasing the size of the hydrophobic ring also improves binding affinity. This approach has been reported using a very large hydrophobic ring structure, such as the 7-member ring of azulene, demonstrating that it has better IC_{50} values compared to Indomethacin. A $[^{18}F]$COX-2 PET radioligand derived from this work has also been reported for specific imaging of COX-2 in a preclinical animal model of cancer (Nolting et al. 2013).

1.2.4 Van Der Waals Forces

This biological interaction is also referred to as dipole–dipole interaction. It is the weakest non-covalent interaction so far in the definition of association between two molecules. This force is mostly applicable for molecules in the condensed phase and

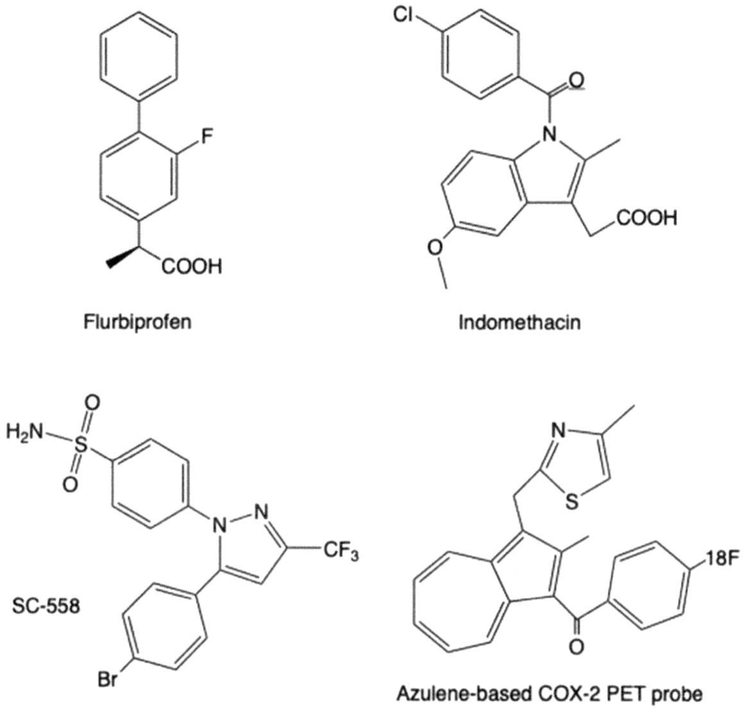

Fig. 1.11 Fitting the COX-2 inhibitors inside the enzyme binding pockets

becomes weak in small molecules. The interaction is significant only when examined with tiny distance $(1/d^7)$. Van der Waals' force of attraction depends on the overall size and shape of molecules and their molecular weights. For example, van der Waals' forces are stronger for linear molecules than for branch molecules, given the former has more surface contact. Thus, the van der Waals interaction serves best for describing the fact why the boiling point of 1-bromopropan (PrBr, BP: 71 °C) is much higher than that of isopropyl bromide (i-PrBr, BP: 59 °C). The same holds true for propanol (BP: 97 °C) versus isopropanol (BP: 82.5 °C) or butanol (BP: 118 °C) versus isobutanol (BP: 108 °C). Overall, van der Waals interaction is a unique example of the many diverse forms of attracted forces that evolved among the biological proteins. Each has its own might, although the attraction is weak for the van der Waals bond, but significant enough to withhold the protein in a unique conformation since proteins possess several permanent dipoles, such as carbonyl or amide groups.

1.3   Terminology and Types of Reactions and Mechanisms

1.3.1   Terminology

Chemists recognize organic molecules by chemical structures designed based on chemical bonds and valence rules. Although there are few major elements in organic synthesis, the arrangement of these elements in several combinations generates a large number of different compounds. Therefore, it is necessary to group them into general categories for classification to improve communication, share, and search in the database. One of the first steps involves learning how to name organic molecules or sometimes called terminology. This is a non-trivial task, and it requires a lot of practice and a general understanding of the rules. In the electronic era, chemical names can indeed be readily available in several domains and search engines, such as ChemDraw or SciFinder. Nevertheless, understanding the general rule of naming chemical nomenclature is advantageous; it provides knowledge about the chemical composition of the molecules and informs the spatial arrangement. Particularly, assigning appropriate positions of the carbons and their substituted moieties enables discussion without ambiguity. Before we delve further, let us examine some of the chemical structures to demonstrate this notion. For example, how to draw an appropriate chemical structure of a compound called (2E, 11Z)-ethyl tetradeca-2,11-dienoate? First of all, the ending with "oate" suggests this is an ester. The main structure has fourteen carbons (tetradeca), in which the carbon at positions two and eleven has double bonds (2,11-diene). In addition, the nomenclature also implies that the two double bonds are located on opposite sides. As the rule of thumb, if the heaviest atomic number atom attached at the double bond lies across the double bond from each other, the conformation should be designated as E (Entgegen, in German, means "opposite"). If they lie on the same side of the double bond, it is Z (Zusammen, means "together"). Finally, we can perceive the chemical structure of (2E, 11Z)-ethyl tetradeca-2,11-dienoate as shown in Fig. 1.12.

Let us examine another chemical structure, for instance, menthol ((1R,2S,5R)-2-isopropyl-5-methylcyclohexan-1-ol). Overall, this suggests a six-member ring (cyclohexan) with a hydroxyl group at position one (1-ol). The name also suggests an isopropyl and a methyl group located at positions two and five, respectively (2-isopropyl-5-methyl). Further, three stereocenters occupy positions one, two, and five with the assignment of 1R, 2S, and 5R, respectively (Fig. 1.13).

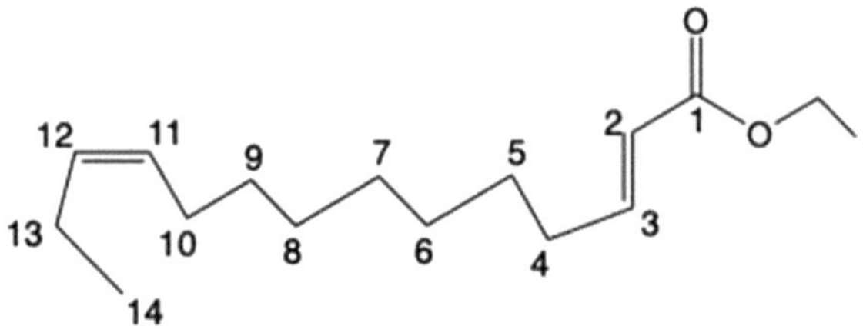

Fig. 1.12 Structure of (2E, 11Z)-ethyl tetradeca-2,11-dienoate

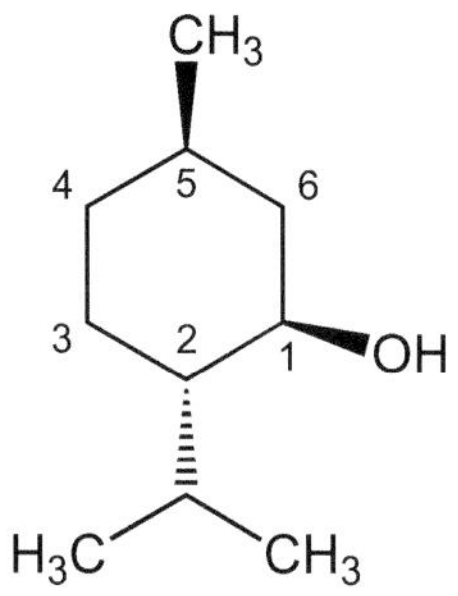

Fig. 1.13 The chemical
structure of menthol

1.3.1.1 Polycyclic Structures

Polycyclic structures comprised a family of compounds that have a number of fused rings together linked through the bridge(s). These aromatic hydrocarbons are crucial intermediates for developing fluorescent dyes since they help extend the absorbance wavelength and serve as an electron sink. More discussion about this design can be found in Chap. 2. With a large repertoire of diverse structures available from commercial sources, diffused rings also serve as great building blocks for generating other types of imaging probes. They are perfect models for plug-and-try to enhance binding specificity. If the binding site geometry is known, diffused polycyclic molecules can be designed to precisely fit that space.

Now, let us get back to learning how to name these polycyclic molecules. As referred in Fig. 1.14, the azulene and the octa-tetraene compounds contain a bridge on each structure linking two distinguished 5-, 7 and 4-, 6-membered rings, respectively. Therefore, these structures are considered a bicyclic system. All the names of the bicyclic systems must have three numbers in the bracket to indicate the ring sizes and the position of the bridge where the rings fuse. So for these molecules' compound, the technical term should be bicyclo[5.3.0] azulene or bicyclo[4.2.0]octa-1,3,5,7-tetraene. The number zero means no carbon in the middle of that bridge.

When working with fused ring systems, it is important to remember how to label the numerical positions on the rings. Usually, the counting should be started with the outermost ring located on the right side. The numbers continue to increase with the direction dictated by the priority of the next carbon atom, usually starting with the most hydrogen-bearing carbon. In other words, the carbon with the most number of hydrogen is numbered as low as possible. For instance, in the case of 1,5-dihydropyrene, the C1 carbon is adjacent to the hydrogen-bearing carbon to the right; thus, the counting should start in a clockwise manner. A similar rationale explains 1H-phenalene and 4,9-dihydropyrene should be counted in a counterclockwise direction (Fig. 1.14). For a substituted fused system, the substituted carbons should be numbered so the sum should be as small as possible. For example, the sum of C2 and C5 carbons for the methyl groups of 2,5-dimethyl azulene (Fig. 1.14) is seven for a clockwise assignment. If the direction turns the other way around, the sum of these two substituted carbons will be higher, thus unacceptable.

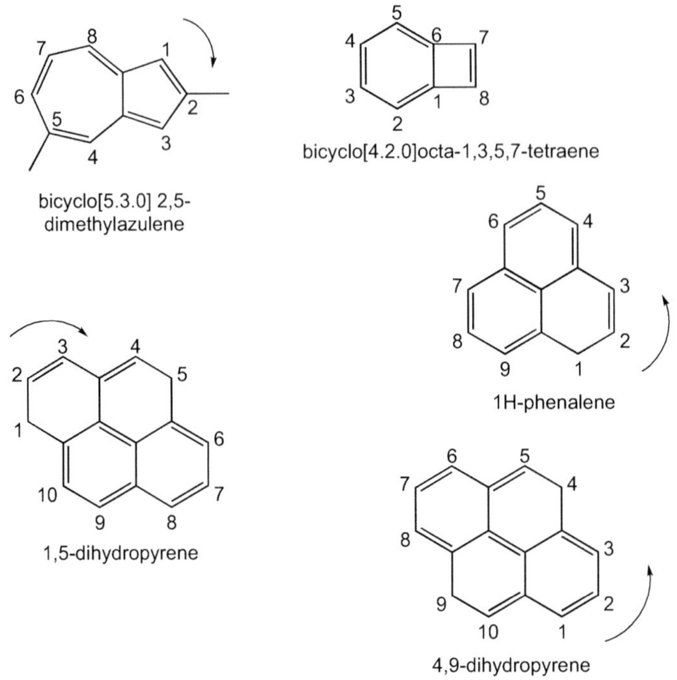

bicyclo[5.3.0] 2,5-
dimethylazulene

bicyclo[4.2.0]octa-1,3,5,7-tetraene

1H-phenalene

1,5-dihydropyrene

4,9-dihydropyrene

Fig. 1.14 Numbering the position of carbons in a continuous order in a ring system

In addition to the positions at the carbons, the sides of the ring should also be labeled in the same direction of the numbering system, starting with letters in the alphabetical order, as shown in the indole 1,1,2-trimethyl-1H-indene, Fig. 1.15. When the indole is attached with a 5-membered ring, the compound is now called 3,4,4-trimethyl-2,4-dihydro-1H-cyclopenta[cd]indene. The sides c and d in the square brackets indicate where the new ring was attached to the indole.

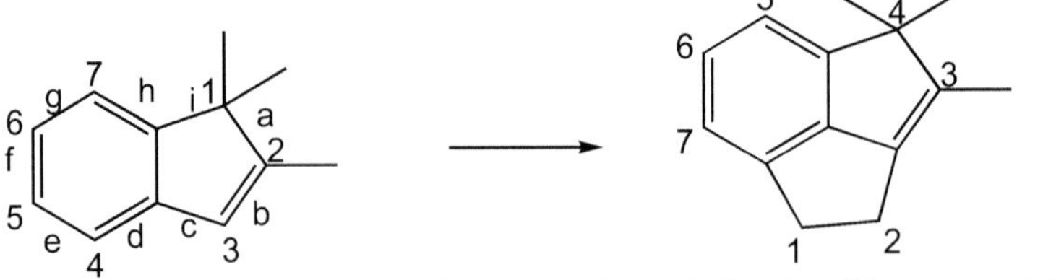

1,1,2-trimethyl-1H-indene

3,4,4-trimethyl-2,4-dihydro-1H-cyclopenta[cd]indene

Fig. 1.15 Inclusion of the identity of the sides of the ring when there is an adjacent ring

1.3.1.2 Heterocyclic Structures

This group represents one of the largest libraries of small molecules in organic chemistry and imaging probes. In general, heterocyclic compounds have a main cyclic chemical structure with one or more different atoms present in the ring. The cyclic ring can be of any size, starting from three carbons, and the heteroatoms can be any elements. Heterocyclic compounds play crucial roles in the biological system and cell metabolism. They are key components in DNA and RNA syntheses, such as purine, pyrimidine, and the protein building process, such as tryptophan, proline, or histidine. They also constitute a large reservoir in antioxidant compounds and vitamins. Furthermore, heterocyclic compounds are the building blocks to diversify a large number of drugs for use as anticancer, anti-inflammation, as well as fluorescence dyes for application in high technology and molecular imaging. Not every heterocyclic compound has an aromatic feature, but if they have, then all the atoms of the ring must have a p atomic orbital for cyclic overlap and a Huckel number of electrons. Given their diversity, naming heterocyclic compounds is usually a highly complex task. A number of information should be factored in the process, including ring size, what elements are present in the ring, and their relative position. Fortunately, a great amount of effort has been dedicated to categorizing (Stowell 1988) a very logical set of rules on how to name the heterocyclic compounds. When the heteroatom displaces one or more carbons of the cyclic system, the newly formed compound will be indicated with the prefixes ending with "a" as shown in Fig. 1.16.

The prefixes for oxygen-, silicon-, nitrogen-, sulfur-, boron-, phosphorus-, sulfur-nitrogen-, and two oxygen-containing heterocyclic rings are ox-, sil-, az-, thi-, bor-, phosph-, thiaz-, and diox-, respectively. Figure 1.17 shows some of the common heterocyclic systems for practicing purposes. Aside from the prefix, another way to name heterocyclics by using a combination of prefixes and suffixes (names in parenthesis). Where the former indicates the replacement of the carbocycle with the

Element	Valence	Prefix	Element	Valence	Prefix
O	II	Oxa	Sn	IV	Stanna
S	II	Thia	Pb	IV	Plumba
Se	II	Selena	B	III	Bora
Te	II	Tellura	Al	III	Alumina
N	III	Aza	Ga	III	Galla
P	III	Phospha	In	III	Inda
As	III	Arsa	Tl	III	Thalla
Sb	III	Stiba	Be	II	Berylla
Bi	III	Bismutha	Mg	II	Magnesa
Si	IV	Sila	Zn	II	Zinca
Ge	IV	Germa	Cd	II	Cadma
			Hg	II	Mercura

Fig. 1.16 Prefixes of heterocyclic systems in descending order of priority. Data obtained from Mcnaught (1976) with permission from Elsevier

new atom. The latter denotes the ring size, and whether saturated or unsaturated (Mcnaught 1976).

To label the numerical positions of a heterocycle, one would start with the non-carbon heteroatom first, then increase the number counterclockwise. However, if the saturated carbon presents in the ring, that carbon should be counted next regardless of the direction (Fig. 1.17). When a substituent is present in the ring, then it will be counted next or the lowest possible numbers after the heteroatom. For compounds with more than one heteroatom, the counting should start with the saturated one. For

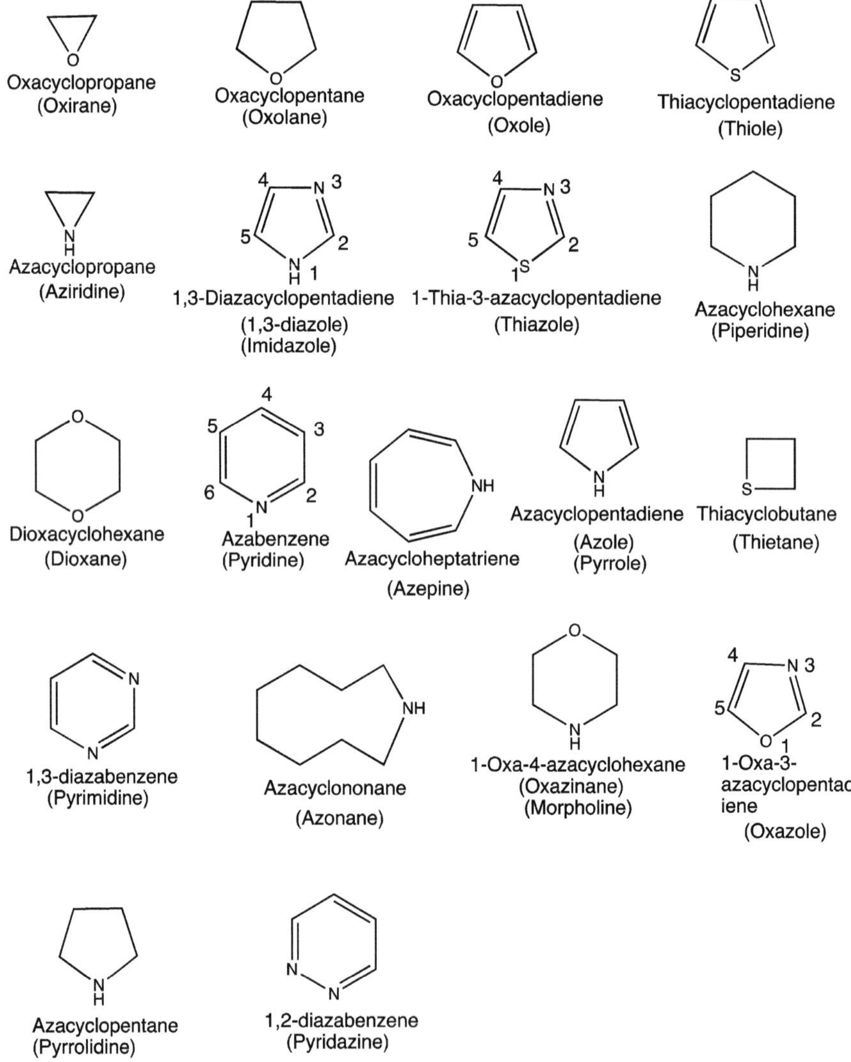

Fig. 1.17 Naming heterocyclic molecules using prefixes or a combination of prefixes and suffixes

heterocycles with more than one type of heteroatoms, the counting would start with the priority order, such as O > S > N as shown in Fig. 1.16.

1.3.2 Types of Chemical Reactions and Mechanisms

Chemical synthesis involves converting starting materials into a different entity using one or more reaction steps. In many cases, the reaction mechanism can be predicted, in which case, the intermediates were identified, and sometimes, they are stable enough for isolation and characterization. Although there are thousands of reaction names and mechanisms for the synthesis of myriads of compounds, in general, there are only four major types of chemical transformations that can explain most, if not all, the reactions for creating imaging probes. These include addition, substitution, elimination, and rearrangement. In the following discussion, these reactions will be discussed in order.

1.3.2.1 Addition Reactions

As the name suggested, this type of reaction involves the conversion of two or more smaller entities into a larger product. An addition reaction usually involves organic molecules with double and triple bonds. The cycloaddition reaction is a good example of this class. Particularly, the Diels–Alder reaction, in which the four π electrons of the diene and two π electrons of the alkene (dienophiles) are involved in the bond formation via mostly *syn* addition. Besides, there are other types of cycloaddition reactions, such as the [3 + 2], [2 + 2], [2 + 8], [4 + 3], and [6 + 4] addition reaction. Figure 1.18 demonstrates a [8 + 2] cycloaddition reaction between a lactone with the in situ-generated vinyl ether from 2,2-dimethoxypropane during the thermolysis reaction. The product of azulene served as a key intermediate for generating dyes, which can be used in optical recording technology and molecular imaging.

Another popular addition reaction involves the electrophilic addition of the hydrogen halide to the alkenes. According to Markovnikov's rule, the addition of halides into unsaturated bonds favors the most substituted carbon. The mechanistic study showed that the addition of the proton at the π electrons at the less substituted carbon atom leads to a more stable carbocation intermediate. More substituted carbocation is more stable than less substituted counterparts because carbocations are electron-deficient moieties; thus, they are stabilized by the electron-releasing alkyl groups, which reduce their electron deficiency (Fig. 1.19). This mechanism can be elaborated the same for carbon radicals. Thus, Markovnikov's reaction is a regioselective reaction.

The addition of alkyl halides, particularly hydrogen bromide or hydrogen chloride to alkenes, is a third-order reaction with a rate equal to k[alkene][HX]^2. It is also worthwhile to note that the stereochemistry of hydrogen halide addition to olefins dominantly favors *anti*-addition when the reaction occurs at room temperature. *Syn*

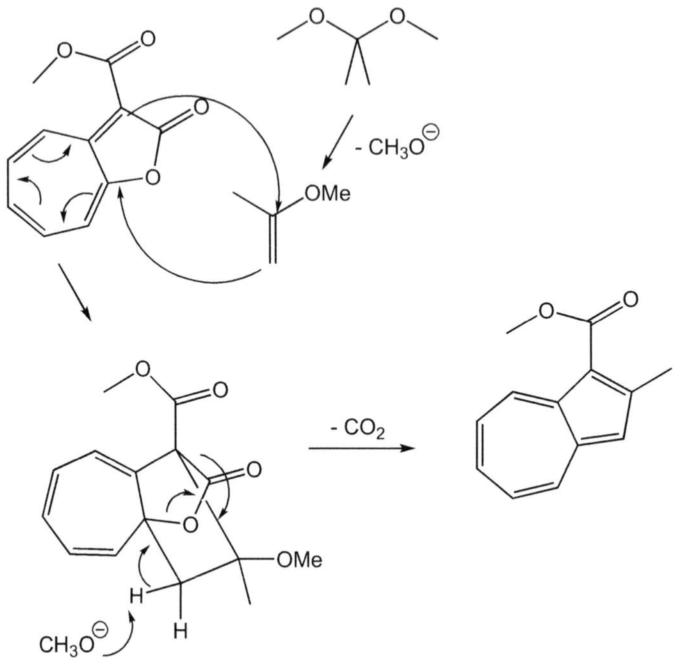

Fig. 1.18 [8 + 2] cycloaddition reaction to make an azulene analog. Data obtained from Nolting et al. (2009)

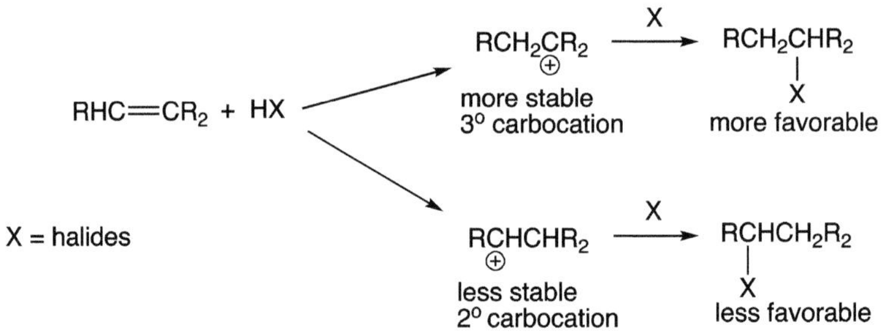

Fig. 1.19 Markovnikov reaction and mechanism

addition, on the other hand, can be achieved when the reaction is carried out at − 78 °C (Becker and Grob 1973). The generated carbocation is very sensitive with nucleophiles, and thus side reaction is unavoidable if the reaction occurs in the presence of nucleophilic solvents. Besides halides, other nucleophiles can also react readily with alkenes such as thiol, amine, and alcohol groups.

Radical reaction on the olefin is an extension of the addition reaction. In the next chapters, some discussions will focus on the application of radical mechanisms to design very mild methods for labeling proteins/peptides for molecular imaging applications, as well as in biochip fabrication technology. Free-radical molecules are electron-deficient species since they possess one or more unpaired electrons. For this reason, the chemistry of radical reactions is very different from charge molecules like carbocations or carbenes. In most cases, the alkyl radicals have lower activation energy; consequently, the radical reactions are rapid and neat.

How to Generate Radicals

Since radicals are unstable, they are generated in situ, and often the generated radical would be trapped by reaction with other intermediates upon formation since it is challenging, if not possible, to isolate these unstable species. We can generate radical species via sigma bond homolysis. When bond cleavage occurs homolytically, the products of which are radicals. They are electron-deficient and unstable to the point that the isolation of such radicals is difficult. However, it is possible to detect radical intermediates using the electron spin resonance (ESR) technique. Basically, the method is similar to NMR, albeit electron spin, instead of nuclear spin, will be involved in the process. An unpaired electron (radicals) has a spin like a proton, with an associated spin state $m_s = \pm 1/2$. When a magnetic field is applied, two non-degenerate spin energy states occur, i.e., the low energy state (more stable) corresponds to the quantum number $m_s = -1/2$, while the high energy state (less stable) with $m_s = +1/2$. The absorption of energy that gives rise to structural information is expressed as:

$$\Delta E = h\upsilon = g_e \beta_B B_0$$

g describes the separation of the two spin states of a free electron, β_B is the Bohr magneton and is equal to 9.2740×10^{-24} JT^{-1}, while B_0 represents the external magnetic field. Another structural feature that can be deduced from the ESR spectrum is the hyperfine interaction, which involves the interaction between the magnetic moment of the electron and nuclear spins. Each nuclear spin I induces a splitting into $[2I + 1]$ the number of lines. For example, a proton (^{1}H) has $I = 1/2$ will split the energy level into a spectroscopic signal of a doublet.

Since only radicals give rise to an ESR spectrum, this technique is quite useful for the detection of short-lived and unstable radicals, as well as their concentration.

Sigma-Bond Homolysis

This is also called homolytic fission, where the bond is broken down by chemical means, using radical generator, heat, or UV light energy (Fig. 1.20). In this reaction mechanism, each fragment contains an equal number of unpaired electrons. To generate a radical readily on carbon atoms, we usually choose strain molecules since the C–C bond on these small-angle molecules is very unstable.

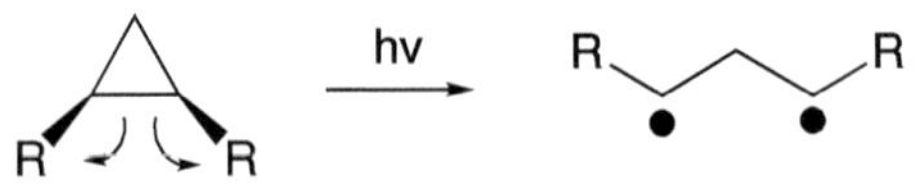

Fig. 1.20 Homolytic
cleavage of a sigma bond

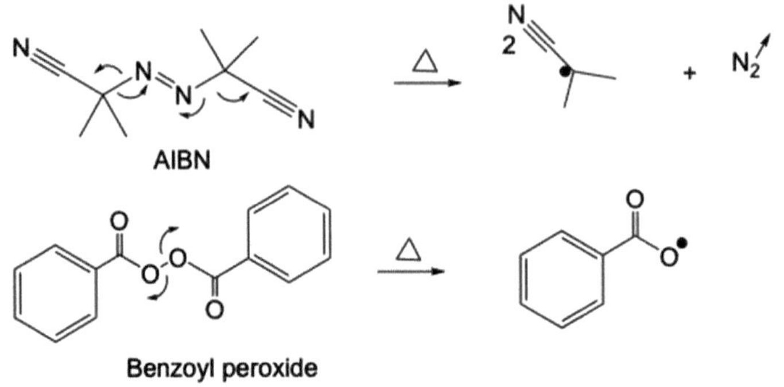

Fig. 1.21 Common radical initiators

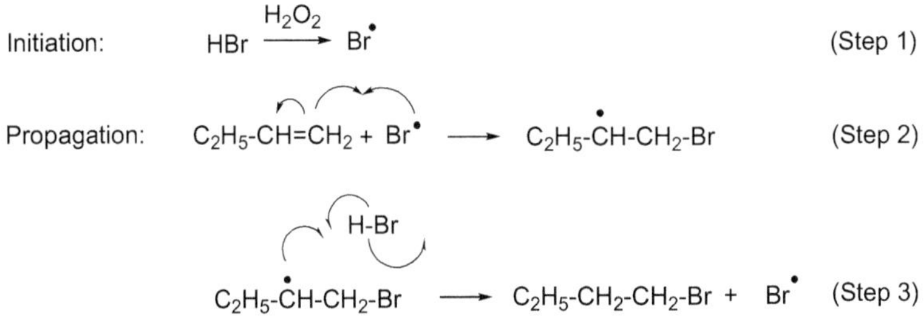

Initiation: $HBr \xrightarrow{H_2O_2} Br^{\bullet}$ (Step 1)

Propagation: $C_2H_5\text{-}CH=CH_2 + Br^{\bullet} \longrightarrow C_2H_5\text{-}\overset{\bullet}{C}H\text{-}CH_2\text{-}Br$ (Step 2)

$C_2H_5\text{-}\overset{\bullet}{C}H\text{-}CH_2\text{-}Br \longrightarrow C_2H_5\text{-}CH_2\text{-}CH_2\text{-}Br + Br^{\bullet}$ (Step 3)

Fig. 1.22 Radical initiated addition to olefin

Or other ways, utilizing chemical reagents to generate free radicals given the availability of the materials, such as hydrogen peroxide, 2,2′-azobis(2-methylpropionate) (AIBME), benzoyl peroxide, or azobisisobutyronitrile (AIBN) (Fig. 1.21).

Now, let us get back to our continuing discussion regarding the addition of radicals to olefin. As shown in Fig. 1.22, in the initiation process, homolysis of hydrogen peroxide resulted in hydroxyl radical, which then abstracts a hydrogen atom from hydrogen bromide to generate a bromine atom (or also called bromine radical). The reactive bromine radical propagates a chain reaction with 2 distinguish steps; first is the addition of the radical to olefin; this addition is regioselective to form a more stable carbon radical as an intermediate. Then, the second step involves carbon radical abstraction of hydrogen from another hydrogen bromide to form an anti-Markovnikov alkyl bromide and a new bromine atom, which will propagate another round of reaction until the radicals are quenched.

This type of addition of halides to an alkene is very useful, and thus, it merits further examination. It is well documented in the literature that the incorporation

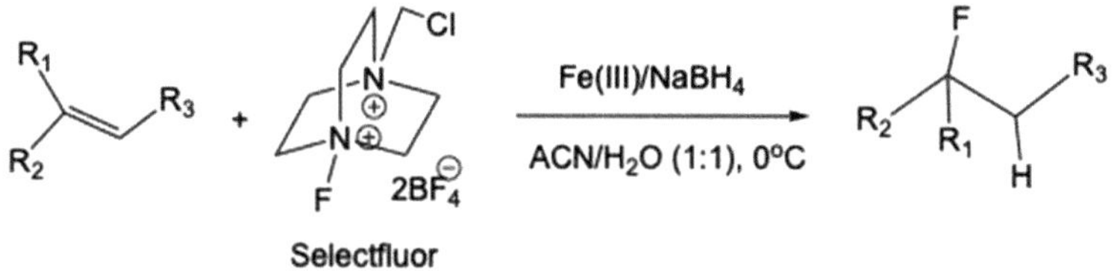

Fig. 1.23 Fe(III)NaBH$_4$-promoted free-radical hydrofluorination of alkenes. Data derived from Barker and Boger (2012) with permission from the American Chemical Society

of halides into active molecules helps selectivity, including enhanced lipophilicity, binding specificity, metabolic stability, and bioavailability (Chatterjee et al. 2016).

Another more robust method using Fe(III)/NaBH$_4$-mediated process for the addition reaction to an unactivated olefins via a free-radical intermediate was reported recently (Barker and Boger 2012). As shown in Fig. 1.23, this generalized method of hydrofluorination (addition of hydrogen fluoride to olefins) of olefins employing Selectfluor as a source of fluorine has been used to develop several important synthons in medicinal chemistry. But most importantly, this chemistry using Fe(III)/NaBH$_4$-mediated reaction for the hydrofluorination of alkenes contributes significantly to the molecular imaging probe development. It paves a new strategy to improve the ^{18}F labeling for positron emission tomography. When it comes to working against time for short half-life isotopes, this reaction is impeccable because it is fast and occurs under very mild reaction conditions using water as a cosolvent, offering the potential for ^{18}F labeling with bioactive and labile protein, peptides, and antibodies. Further, the labeling occurs with unactivated alkenes. Unlike the stringent reaction condition in conventional free-radical hydrofluorination of alkenes, this new reaction could happen in an open-air. Furthermore, regiospecific fluorination can be manipulated since the reaction is amenable to Markovnikov's rule. More information about this work will be discussed in Chap. 3.

Another metal-based catalytic addition of halides to olefins was developed recently using a cobalt catalyst (Shigehisa et al. 2013). This exclusive Markovnikov hydrofluorination reaction has great implications for regioselective fluorination of pharmaceutical compounds, and molecular probes, particularly for PET radioligands. Experimental data suggested the involvement of a radical intermediate in the process (Fig. 1.24). The reaction condition is mild, and the hydrofluorination is tolerant to functional groups, and scalable. Altogether, this new mechanism for fluorination is another yet versatile method for addition to olefins.

Another type of radical addition to olefin is called addition polymerization, a prevalent method for developing nanotechnology-based imaging probes. If the addition is initiated by a radical, the process is called radical polymerization. In this design, the initiator is a radical, which adds to the carbon–carbon double bond to form an intermediate, with the propagating site as a carbon radical. This highly reactive intermediate goes on further to interact with other alkene monomer providing another reactive intermediate, and the process keeps repeating until the radicals are quenched. One example of this type or reaction is shown in Fig. 1.25; the styrene monomer

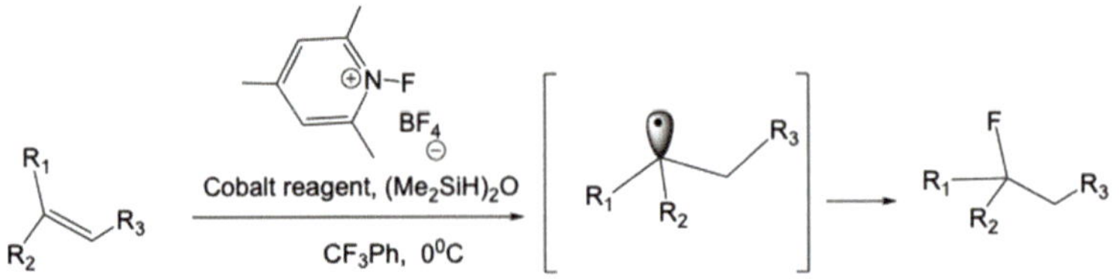

Fig. 1.24 Cobalt-mediated free-radical addition to alkenes. Data obtained from Shigehisa et al. (2013) with permission from the American Chemical Society

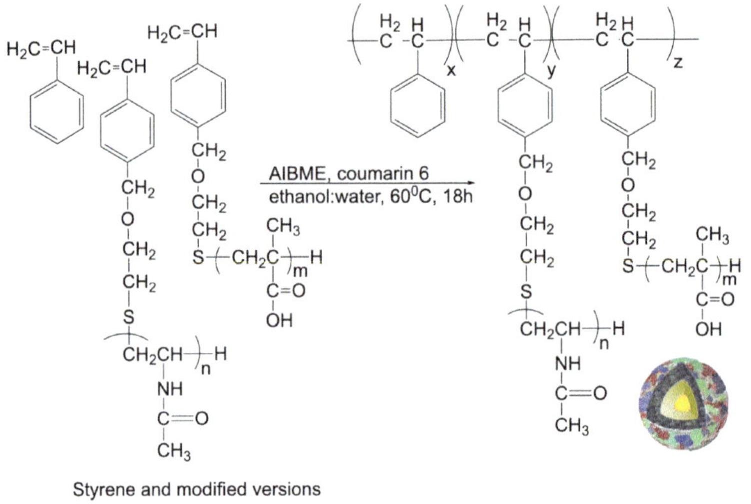

Fig. 1.25 Radical addition to olefins to generate polymers. Data obtained from Barton et al. (2018)

and its modified derivatives went through a copolymerization process using 2,2′-azobisisobutyronitrile (AIBN) or a more friendly radical generator, such as dimethyl 2,2′-azobis (2-methylpropionate) (AIBME) as the radical initiator to generate a multi-modal nanobeacon for imaging applications. In this work, polystyrene and its modified versions were polymerized simultaneously in the presence of a fluorescence dye, such as coumarin 6. During the process of polymerization, coumarin 6 dye was encapsulated inside the nanoparticles providing a fluorescent signal readout. The free carboxylic groups were derivatized on the surface of the nanoparticles to provide a handle for conjugation with biomolecular recognition ligands via the use of coupling reagent, such as dicyclohexylcarbodiimide (DCC) for targeted imaging (Kumagai et al. 2013). This approach has great in vivo applications because not only is the strong fluorescent signal emitted by thousands of dyes encapsulated inside each nanoparticle, but the process protects the dyes from degradation and exposure.

1.3.2.2 Substitution Reactions

A substitution reaction occurs when one chemical group is displaced by the other group. This reaction can be generalized in any chemical transformation where the sigma bond is displaced. In a nucleophilic substitution, the nucleophile attacks the $\delta+$ carbon via S_N1 or S_N2 mechanism. Both reaction types require an effective design of the leaving groups so that they can drive the substitution process.

In an S_N2 reaction, as the name suggests, a nucleophilic substitution involves a bimolecular reaction in which a concerted, single-step mechanism without an intermediate adduct is formed. This is a single transition state reaction, where a simultaneous attack of the nucleophile occurs on the opposite side of the leaving group. The transition state has five groups attached to the carbon atom arranged in a trigonal bipyramidal geometry. As shown in Fig. 1.26, this means that there is a steric occupation of other groups on this carbon determines the rate of the reaction. Thus, we can consider S_N2 reaction as stereospecific (producing only one and stereochemically different product) and stereoselective (producing only one of the possible diastereomers). The larger substituted groups will hinder the approaching nucleophiles. As a result, the reactivity order of an S_N2 reaction ranks as follows: $CH_3- > CH_3CH_2- > (CH_3)_2CH- > (CH_3)_3C-$. In another words, methyl $>1° >2°$ $>3°$. Nevertheless, it is worthwhile remembering that, in an S_N2 reaction, the rate of the reaction is also determined by the choice of the nucleophiles; the more reactive the nucleophiles, the higher the chance to drive the reaction to completion. In this regard, polar aprotic solvents with large dipole moments, such as dimethylformamide (DMF) or dimethyl sulfoxide (DMSO), are ideal for rendering strong nucleophiles, generally strong bases. Since the nucleophiles approach from the backside of the transitioning carbon, an S_N2 reaction always results in an inverted configuration, and this can be observed when the affected carbon is a chiral center (Fig. 1.27). So far, we have discussed the substituted groups, nucleophilic effects, and solvents as key factors to drive S_N2 reaction. How about the leaving group? As a matter of fact, the choice of a good leaving group also improves the rate of the reaction. In contrast to the nucleophiles, which should be strong bases, as mentioned earlier, the weakest base is the best leaving group. Thus, the order of leaving groups is as follows: RSO_3^- (mesylates, tosylates) $> I^- > Br^- > Cl^- > F^- >$ p-nitrobenzoate $> CH_3CO_2^-$ (Carey and Sundberg 2007).

For S_N1 substitution reaction, as the name suggests, this is a substitution nucleophilic unimolecular reaction. In stark contrast to S_N2 reaction, S_N1 reaction does not require a strong nucleophile, albeit a good leaving group is essential. Obviously,

Fig. 1.26 Mechanism of S_N2 reaction

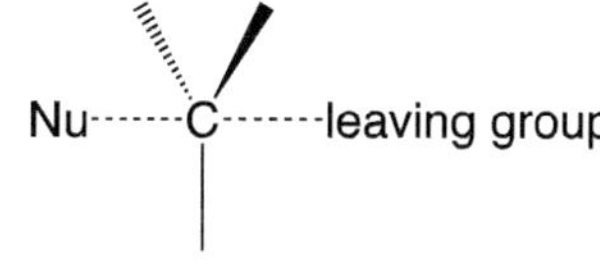

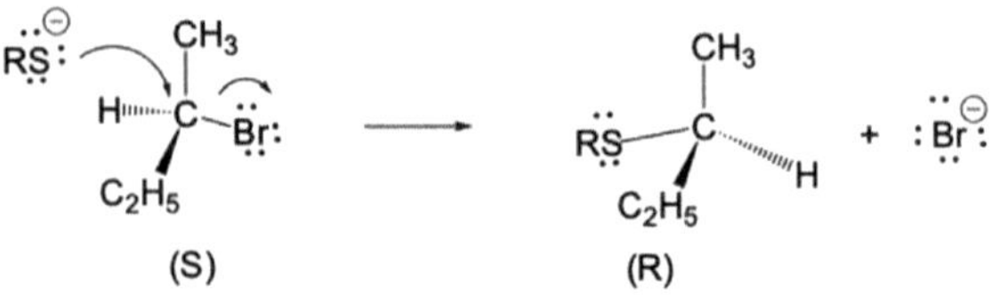

Fig. 1.27 S_N2 reaction mechanism; large nucleophiles attack carbon from backside resulted in inversion of configuration

the mechanism of the S_N1 reaction is unique since it requires two steps in the process. The first step involves the slow formation of the carbocation intermediate, and this is the rate-determining step. The second step involves a rapid reaction between the carbocation and the nucleophile. The stereochemical analysis of the S_N1 reaction demonstrated that the carbocation owns a planar geometry; thus, the nucleophile can attach it on either side of the plane, resulting in a complete racemic product. The electron-donating inductive effect stabilizes the stability of the carbocation intermediate; thus, the more R groups attached to the carbon atom, the greater is the stabilizing effect. Therefore, different from S_N2 reaction, in S_N1 the reactivity order is $(CH_3)_3C- > (CH_3)_2CH- > CH_3CH_2- > CH_3-$.

Electrophilic substitution, S_E1 (substitution electrophilic unimolecular), involves the displacement of hydrogen moiety on the ring by an electrophile, which is usually related to Lewis acids, one of the typical reactions is aromatic substitution. This type of reaction is very different from S_N1 and S_N2 reactions, where the attacking group is a nucleophile or Lewis base and the leaving group is a weaker base. The reaction starts with an electrophilic attack of arene to form the benzenium ion, a very stable allylic carbocation, which is stabilized by the contributing resonance structures (Fig. 1.28).

The rate-determining step of electrophilic aromatic substitution is the formation of the intermediate benzenium ion or also called arenium cation since this is a slow step. The intermediate is stabilized via the resonance structure. In contrast, the elimination of H^+ is much faster in order to convert the less stable cyclohexadienes into a stable aromatic π structure. Overall, this reaction is no more than a hydrogen from the aromatic system substituted by an electrophilic group to preserve the aromaticity. Typically, these reactions are useful for the development of intermediates for imaging probes, including nitration, Friedel–Crafts alkylation/acylation, aromatic sulfonation, and bromination/chlorination (Fig. 1.29). The modification of aromatic rings via electrophilic substitution reactions is indispensable in developing near-infrared dyes, which will be discussed more in Chap. 2. It is worth noting that

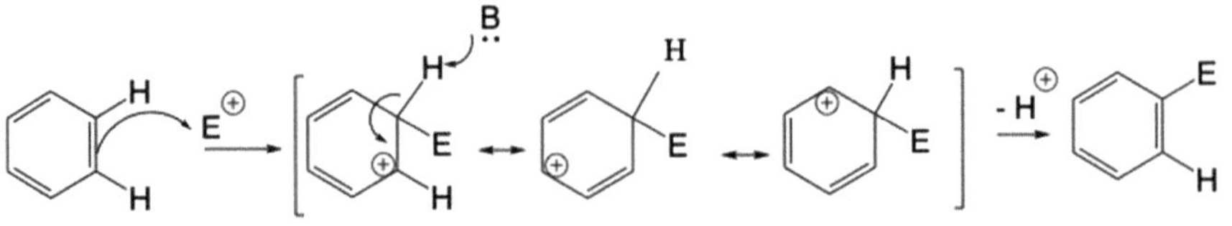

Fig. 1.28 Mechanism of electrophilic aromatic substitution reaction

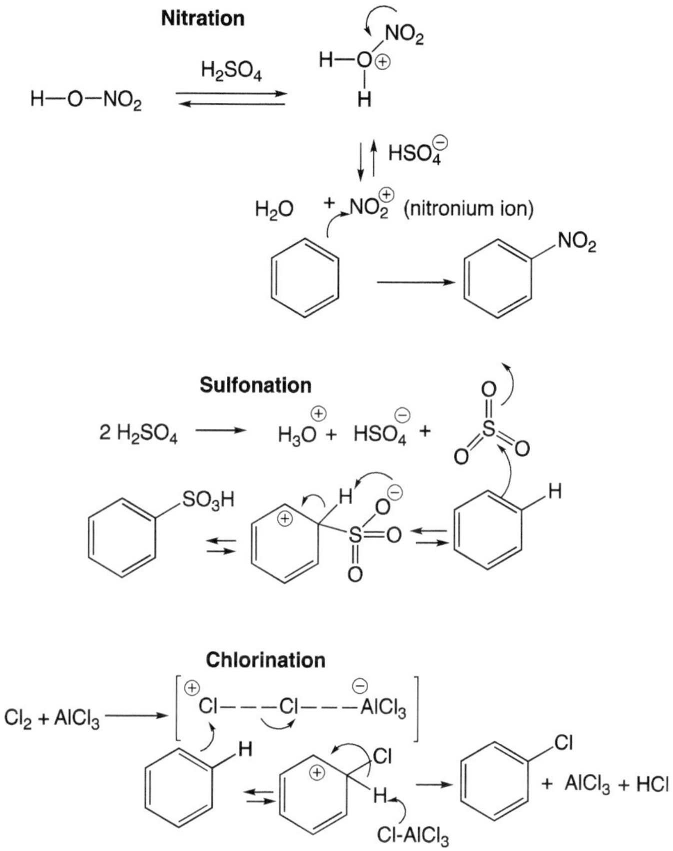

Fig. 1.29 Typical examples of electrophilic aromatic substitution

most aromatic electrophilic substitutions are irreversible processes; however, a few reactions are reversible; for example, sulfonation is one of them.

When an electrophilic substitution reaction is performed on the monosubstituted aromatic ring, the new group will take the position either ortho, meta, or para to the primary group, which can dictate the reaction rate. The reaction rate can be faster or slower than benzene, depending on the substituent groups. If the substituent groups enhance the reaction rate more than benzene, it is called activating group, while deactivating groups retard the reaction rate. A general list of these activating/deactivating groups is shown in Fig. 1.30.

1.3.2.3 Elimination Reactions

This reaction is opposite to addition; elimination occurs when two or more atoms are dissociated from the main molecule. To put it in a context, elimination reaction

ortho/para				meta	
Strong activating	**Moderate activating**	**Weak activating**	**Deactivating**	**Strong deactivating**	**Moderate deactivating**
-OH, -OR, -NH$_2$, -NHR, -NR$_2$	-OCOR -NHCOR R=alkyl groups	-R, -Ar, -CR=CR2	-F, -Cl, -Br, -I, -CH$_2$X, -NO X = halides	-NO$_2$, -CF$_3$, -$^+$NH$_3$, -$^+$NH$_2$R, -$^+$NR$_3$	-CN, -SO$_3$H, -COR R = H, OH, Cl, OR

Fig. 1.30 Classification of substituent groups that affect electrophilic aromatic substitution reaction

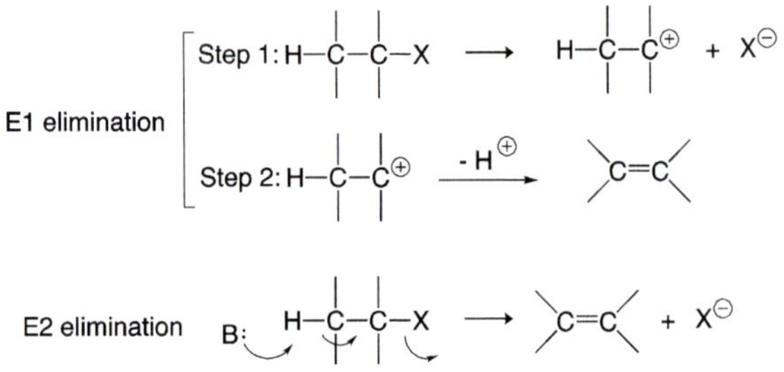

Fig. 1.31 Mechanisms of elimination reactions

E1 elimination

Step 1: $H-\overset{|}{\underset{|}{C}}-\overset{|}{\underset{|}{C}}-X \longrightarrow H-\overset{|}{\underset{|}{C}}-\overset{|}{\underset{|}{C}}^{\oplus} + X^{\ominus}$

Step 2: $H-\overset{|}{\underset{|}{C}}-\overset{|}{\underset{|}{C}}^{\oplus} \xrightarrow{-H^{\oplus}} _{\backslash}C=C_{\diagup}$

E2 elimination $B{:} \quad H-\overset{|}{\underset{|}{C}}-\overset{|}{\underset{|}{C}}-X \longrightarrow _{\backslash}C=C_{\diagup} + X^{\ominus}$

simply means two single bonds in the starting material are converted into a double bond. Elimination reactions occur in many mechanisms. In the context of this book, we will discuss two major types: unimolecular reactions called E1 and bimolecular reactions called E2. The former involves 2 steps, elimination to form carbocation intermediate followed by deprotonation of the carbocation (Fig. 1.31). The latter, in contrast, is a concerted reaction where a base removes a proton neighboring to the leaving group, subsequently forming a double bond. The kinetics of E1 is first-order or unimolecular since the rate depends on the ionization process, while in the E2 reaction, the rate of the reaction depends on the concentrations of the alkyl halide and the base. Therefore, the kinetic is second-order overall or bimolecular.

In practice, there are a few types of elimination reactions. For instance, α-elimination involves the removal of the most acidic proton and a leaving group from the same carbon atom, resulting in a carbene, while β-elimination involves the removal of 2 groups from 2 adjacent atoms. The β-elimination reactions are very popular in molecular probe chemistry, and a number of reaction names are associated with this process called dehalogenation, dehydration, and dehydrohalogenation. In these reactions, since the atoms are removed with a net loss of 2 electrons from each of the vicinal carbons, it is also called 1,2 eliminations.

The nucleophilic addition–elimination reaction is also important for the design of imaging probes, particularly, the addition of primary amino groups to carbonyls of ketone or aldehyde. Depending on the nucleophilic species, if the primary amine reacts with a ketone, it will form imine in a reaction called the Schiff base reaction, while the same reaction of ketones with hydroxylamine, hydrazine, or hydrazine amide will generate oxime, hydrazone, or semicarbazone, respectively. These reactions are reversible since the mechanism for the formation of imines is the reverse of hydrolysis. In order to drive the reaction to completion, the removal of water is essential in this operation. This can be accomplished by using azeotropic distillation or carrying out the reaction in the presence of dehydrating reagents.

1.3.2.4 Rearrangement Reactions

Chemical rearrangement involves scrambling chemical bonds to form new bonds and subsequently creating a structural isomer of the original molecule. This is the most interesting type of reaction since the chemistry involves the formation of a new product with distinct and reclusive chemical structures. Fortunately, there are general rules that help to identify how the bonds are cleaved and formed. One of the approaches involves tracking the electron movement using a reaction mechanism shown in Fig. 1.32.

Rearrangement reactions are unique, albeit difficult to recognize from time to time, because they happen after other types of reactions that we just discussed above,

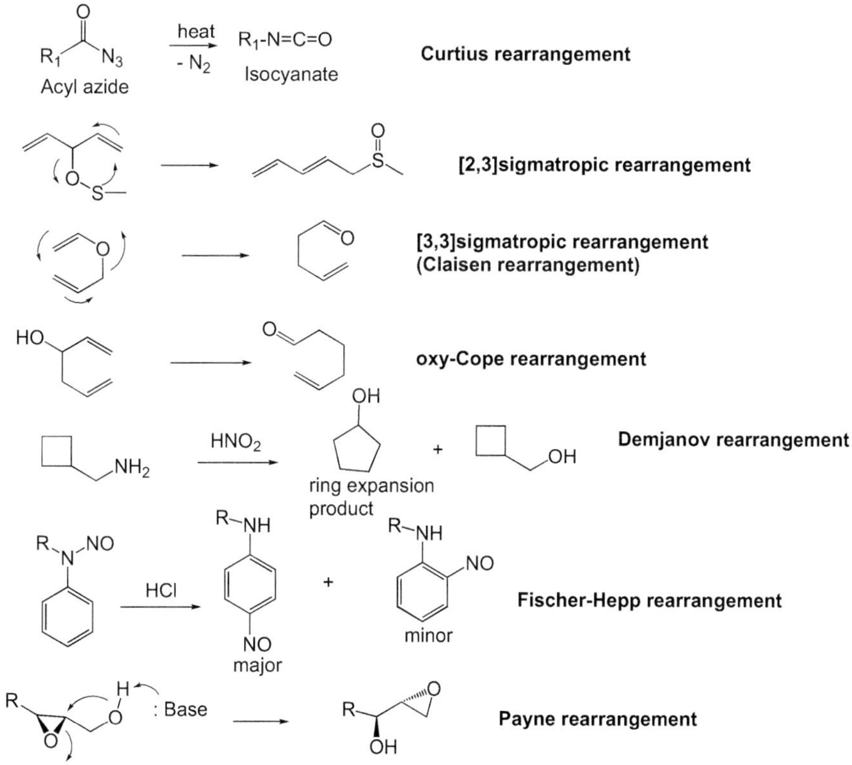

Fig. 1.32 Typical rearrangement reactions

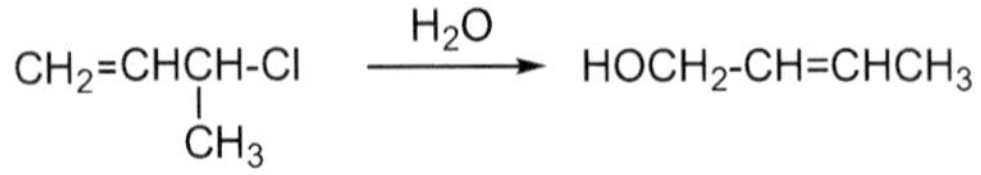

Fig. 1.33 Allylic rearrangement

including substitution, addition, and elimination reactions. For instance, in the substitution reaction shown in Fig. 1.33 (or sometimes called hydrolysis reaction), there is a movement of a double bond, usually called allylic rearrangement. In this reaction, the OH group replaces the Cl leaving group as an S_N1 mechanism, but on a different carbon.

For a more detailed mechanistic explanation, let us examine why a substitution reaction of 2-bromo-1,1-diphenylpropane provides only 1,1-diphenylpropan-1-ol instead of 1,1-diphenylpropan-2-ol. As shown in Fig. 1.34, one would recognize that hydrolysis of the secondary alkyl bromide does not provide 1,1-diphenylpropan-2-ol, but a rearrangement reaction accompanied by a hydride shift occurs instead in order to establish a more stable carbocation intermediate. The formation of the final product of 1,1-diphenylpropan-1-ol suggests that the nucleophile ends up in a different carbon from where the leaving group departs.

Another type of rearrangement reaction usually can be found accompanying an elimination reaction. For example, during the condensation of 2,2-dimethylcyclohexan-1-ol, the R_2CH^+ is generated, but it is not as stable as compared to the R_3C^+ intermediate (Fig. 1.35). To have this happen, the methyl group from C2 position would shift to C1 in a process called methide relocation, where the methyl group is associated simultaneously with both C1 and C2 position, then followed by the loss of a proton to form the condensation product.

Rearrangement accompanying an addition reaction also involves the establishment of a more stable carbocation intermediate. As shown in Fig. 1.36, the addition of the halide component, HBr, to the alkene follows Markovnikov's rule. Thus, the

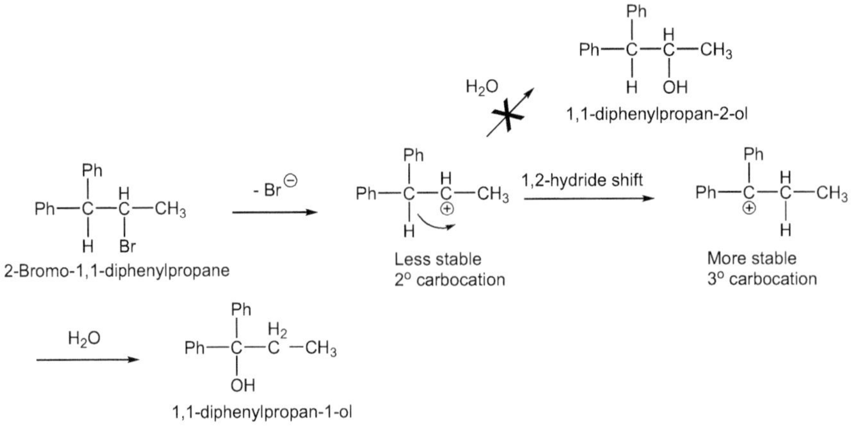

Fig. 1.34 Complex rearrangement reaction

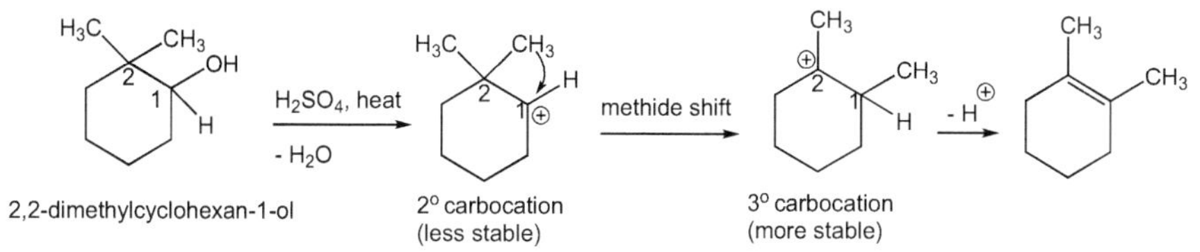

Fig. 1.35 Elimination-rearrangement reaction

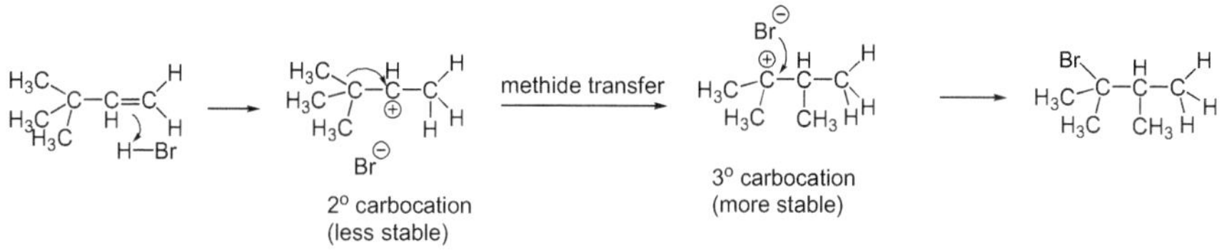

Fig. 1.36 Addition-rearrangement reaction

addition prefers the more highly substituted carbon within the alkene. However, for this particular structure, a methide transfer would render the carbon in the vicinity of the alkene a better and more stable carbocation. As a result, the addition of the halide occurs in this carbon, which is outside of the alkene.

1.4 Useful Reaction Names for the Chemical Development of Molecular Probes

As mentioned earlier, there are only four reaction types, but they cover a myriad of chemical reactions. Since the number of chemical reactions associated with new and complex reaction conditions has increased dramatically over the years, organic chemists chose to communicate by using the reaction names instead of describing detailed reaction conditions each time. Expression of chemical reactions using specific names helps simplify the process and is also clear, concise, and avoids confusion. In the context of this chapter, some common reaction names will be mentioned. However, more detailed reaction names can be found in other books on reaction names.

Baeyer–Villiger oxidation: Conversion of ketone to ester in the presence of peroxyacid, such as mCPBA.

Beckmann rearrangement: Acid-catalyzed conversion of an oxime to an amide, in which an R group trans to the leaving group migrates to nitrogen.

Birch reduction: Involving aromatic rings conversion into unconjugated 1,4-cyclohexadienes in the presence of sodium or lithium in liquid ammonia and an alcohol.

Borch reduction: Reductive amination of aldehydes or ketones by cyanoborohydride to make amines.

Buchwald–Hartwig amination: Conversion of aryl halide into an aryl amine, using a catalytic amount of palladium and a base.

Cannizzaro reaction: A redox reaction in which two equivalent aldehydes react to produce a primary alcohol and a carboxylic acid using a base.

Claisen condensation: To create a β-keto ester starting from two esters moieties in the presence of an alkoxide base in alcohol.

Claisen rearrangement: Conversion of vinyl ether into a γ, δ-unsaturated carbonyl compound in the presence of a Lewis acid.

Cope rearrangement: Conversion of 1,5-diene into another 1,5-diene isomer under thermal condition.

Corey–Kim oxidation: Conversion of an alcohol to an aldehyde or ketone, using N-chlorosuccinimide, dimethylsulfide, and triethylamine.

Curtius rearrangement: Conversion of an acyl azide into an isocyanate.

Dieckmann condensation: Creation of a cyclic β-keto ester from two joined esters in the presence of a base.

Diels–Alder condensation: Reaction between a diene and a dienophile to form a cyclic olefin under thermal conditions.

Eschenmoser–Claisen rearrangement: Formation of γ, δ-unsaturated amide when an allylic alcohol is treated with dimethylacetamide dimethyl acetal under thermal conditions.

Finkelstein reaction: Conversion of an alkyl halide into another alkyl halide in the presence of a metal halide salt.

Fischer indole condensation: Conversion of phenyl hydrazine and ketone to form an indole in the presence of an acid as a catalyst.

Friedel–Crafts alkylation/acylation: Alkylation or acylation of aromatic analogs using alkyl or acyl halides in the presence of Lewis acids.

Gabriel synthesis: Conversion of alkyl halides into primary amines using phthalimide.

Grignard reaction: Conversion of aldehyde and ketone to alcohols, using Grignard reagent, which is an organomagnesium compound.

Heck reaction: Reaction between an organohalide with an alkene to create a substituted alkene, using a catalytic amount of palladium and a base.

Hofmann elimination: Conversion of an amine with a β-hydrogen to an alkene using silver oxide, methyl iodide, and water under thermal conditions.

Hofmann rearrangement: Reduction of a terminal amide into a primary amine using a halogen, base, and water under thermal conditions.

Horner–Wadsworth–Emmons reaction: Generation of an olefin with good selectivity via the reaction of an aldehyde/ketone with a phosphorus ylides.

Huisgen cycloaddition reaction: Involving a 1,3-dipolar cycloaddition to generate a 5-membered heterocycle.

Johnson–Claisen rearrangement: Allylic alcohol and trialkyl orthoacetate are heated under the mild acidic conditions to produce a γ, δ-unsaturated ester.

Jones oxidation: Oxidation of an alcohol into carboxylic acid using in situ-generated chromic acid.

Knoevenagel condensation: Conversion of an aldehyde or ketone and an activated methylene into a substituted olefin using an amine base as a catalyst.

Kolbe–schmitt reaction: Conversion of a phenol into a hydroxy benzoic acid using CO_2 gas and a base.

Kumada cross-coupling: Involving the reaction between an organohalide with a Grignard reagent to provide a coupled product using palladium as a catalyst.

Liebeskind–Srogl reaction: Coupling reaction between the thioether or thioester electrophiles with boronic acid or stannane nucleophiles, catalyzed by metal to form a carbon–carbon bond.

Leuckart reaction: Conversion of ketones or aldehydes to amines by reductive amination using either formamide or ammonium formate as source of nitrogen.

Mannich reaction: Aminomethylation of activated methyl groups by in situ-generated iminium salt. This salt is synthesized briefly from aldehyde with secondary amine salt.

McMurry coupling: Formation of olefins by cross-coupling of ketones; the reaction is mediated by titanium.

Michael addition: Addition of nucleophiles to α, β-unsaturated esters, ketones.

Mitsunobu reaction: Conversion of primary and secondary alcohol into esters or ether employing triphenylphosphine and diethyl azodicarboxylate (DEAD).

Mukaiyama aldol addition: Conversion of an aldehyde and a silyl enol ether into a 1,3 ketol using a Lewis acid as a catalyst.

Negishi cross-coupling: Generating C–C bond as a coupled product using organohalide and organozinc in the presence of palladium as a catalyst.

Pauson–Khand reaction: Reaction between an alkyne and an alkene to make a substituted cyclopentenone in the presence of cobalt as a catalyst.

Pictet–Spengler reaction: β-arylethylamine undergoes condensation with an aldehyde or ketone in strong acidic condition to initiate a ring closure to form a product called β-carboline.

Reformatsky reaction: Zinc-mediated reaction between α-haloester with a ketone to make a β-hydroxyester.

Robinson annulation: Creating a bicyclic system by reaction of hexanones with vinyl ketones in the presence of a base.

Sandmeyer reaction: Copper-catalyzed reaction to convert aryl diazonium salt into aryl halide.

Sharpless epoxidation: Stereocontrolled conversion of an allylic alcohol to an epoxy alcohol using diethyl tartrate, t-butyl hydroperoxide, and a catalytic amount of titanium isopropoxide.

Sonogashira coupling: Coupling of terminal alkynes with aryl or vinyl halides using palladium and copper (I) as catalysts and an organic amine base.

Staudinger reaction: Conversion of organic azide into a primary amine using triphenylphosphine and water.

Stille reaction: Generating a C–C product using organohalide and organostannane starting materials in the presence of palladium as a catalyst.

Strecker amino acid synthesis: Creating an amino acid using ketone/aldehyde with a primary amine in the presence of metal cyanide.

Suzuki coupling: Palladium-catalyzed reaction between organoboronic acid and alkyl halides. The reaction is useful for making biphenyl structures.

Swern oxidation: Conversion of alcohols to ketones or aldehydes using oxalyl chloride and DMSO.

Ullmann reaction: Conversion of 2 equivalents of aryl halide into a biaryl product using copper metal under thermal conditions.

Vilsmeier–Haack reaction: Formylation of aromatic or activated carbons by generating iminium reagent using DMF and phosphorus oxychloride in situ.

Williamson synthesis: Conversion of an alcohol into an ether using a base.

Wittig reaction: Olefin synthesis using phosphorane ylides and ketone or aldehydes.

Wolff–Kishner reduction: Reductions of ketones to alkanes using hydrazine and potassium hydroxide.

Yamaguchi esterification: Conversion of a carboxylic acid into an ester using triethylamine.

Ziegler–Hafner synthesis: A versatile synthesis of azulene via substituted cyclopentadienes.

1.5 Conclusion

Organic chemistry is a large field that is impossible to cover everything. Still, surely a quick and meaningful review of the principles of organic chemistry will be very helpful for comprehension in the next topic, which focuses on the discussion of fluorescence dyes' physical and chemical design.

References

T.J. Barker, D.L. Boger, Fe(III)/NaBH4-mediated free radical hydrofluorination of unactivated alkenes. J. Am. Chem. Soc. **134**, 13588–13591 (2012)

S. Barton, B. Li, M. Siuta, J. Vaibhav, J. Song, C.M. Holt, T. Tomono, M. Ukawa, H. Kumagai, E. Tobita, K. Wilson, S. Sakuma, W. Pham, Specific molecular recognition as a strategy to delineate tumor margin using topically applied fluorescence embedded nanoparticles. Precis Nanomed **1**, 194–207 (2018)

K.B. Becker, C.A. Grob, Stereoselective cis and trans addition of hydrogen chloride to olefins. Synthesis **1973**, 789–790 (1973)

A. Bhinge, P. Chakrabarti, K. Uthanumallian, K. Bajaj, K. Chakraborty, R. Varadarajan, Accurate detection of protein:ligand binding sites using molecular dynamics simulations. Structure **12**, 1989–1999 (2004)

G.N. Cantor, Newton on optics: the optical papers of isaac newton. Science **224**, 724–725 (1984)

F.A. Carey, R.J. Sundberg, *Advanced Organic Chemistry: Part A: Structure and Mechanisms*, 3rd edn. (2007)

T. Chatterjee, N. Iqbal, Y. You, E.J. Cho, Controlled fluoroalkylation reactions by visible-light photoredox catalysis. Acc. Chem. Res. **49**, 2284–2294 (2016)

A.M. Davis, S.J. Teague, Hydrogen bonding, hydrophobic interactions, and failure of the rigid receptor hypothesis. Angew. Chem. Int. Ed. Engl. **38**, 736–749 (1999)

J.J. Deadman, S. Elgendy, C.A. Goodwin, D. Green, J.A. Baban, G. Patel, E. Skordalakes, N. Chino, G. Claeson, V.V. Kakkar et al., Characterization of a class of peptide boronates with neutral P1 side chains as highly selective inhibitors of thrombin. J. Med. Chem. **38**, 1511–1522 (1995)

K.C. Kemp, T.L. Brown, H.E. LeMay, B.E. Bursten, *Chemistry-The Central Science*, 9th edn. (Prentice Hall, New Jersey, 2003)

H. Kumagai, W. Pham, M. Kataoka, K. Hiwatari, J. McBride, K.J. Wilson, H. Tachikawa, R. Kimura, K. Nakamura, E.H. Liu, J.C. Gore, S. Sakuma, Multifunctional nanobeacon for imaging Thomsen-Friedenreich antigen-associated colorectal cancer. Int. J. Cancer **132**, 2107–2117 (2013)

R.G. Kurumbail, A.M. Stevens, J.K. Gierse, J.J. McDonald, R.A. Stegeman, J.Y. Pak, D. Gildehaus, J.M. Miyashiro, T.D. Penning, K. Seibert, P.C. Isakson, W.C. Stallings, Structural basis for selective inhibition of cyclooxygenase-2 by anti-inflammatory agents. Nature **384**, 644–648 (1996)

A.D. Mcnaught, The nomenclature of heterocycles, **20**, 175–319 (1976)

S. Mizukami, S. Watanabe, Y. Hori, K. Kikuchi, Covalent protein labeling based on noncatalytic beta-lactamase and a designed FRET substrate. J. Am. Chem. Soc. **131**, 5016–5017 (2009)

D.D. Nolting, M. Nickels, M.N. Tantawy, J.P. Xie, T.E. Peterson, B.C. Crews, L. Marnett, J.C. Gore, W. Pham, Convergent synthesis and evaluation of 18F-labeled azulenic COX2 probes for cancer imaging. Front Oncol **2**, 1–8 (2013)

D.D. Nolting, M. Nickels, R. Price, J.C. Gore, W. Pham, Synthesis of bicyclo[5.3.0]azulene derivatives. Nat. Protoc. **4**, 1113–1117 (2009)

O.C. Olson, J.A. Joyce, Cysteine cathepsin proteases: regulators of cancer progression and therapeutic response. Nat. Rev. Cancer **15**, 712–729 (2015)

S.K. Panigrahi, Strong and weak hydrogen bonds in protein-ligand complexes of kinases: a comparative study. Amino Acids **34**, 617–633 (2008)

A. Patronov, E. Salamanova, I. Dimitrov, D.R. Flower, I. Doytchinova, Histidine hydrogen bonding in MHC at pH 5 and pH 7 modeled by molecular docking and molecular dynamics simulations. Curr. Comput. Aided. Drug. Des. **10**, 41–49 (2014)

R.G. Pearson, Hard and soft acids and bases-the evolution of a chemical concept. Coord. Chem. Rev. **100**, 403–425 (1990)

I.M. Serafimova, M.A. Pufall, S. Krishnan, K. Duda, M.S. Cohen, R.L. Maglathlin, J.M. McFarland, R.M. Miller, M. Frodin, J. Taunton, Reversible targeting of noncatalytic cysteines with chemically tuned electrophiles. Nat. Chem. Biol. **8**, 471–476 (2012)

S. Shaik, The Lewis legacy: the chemical bond–a territory and heartland of chemistry. J Comput Chem **28**, 51–61 (2007)

S.Y. Sheu, D.Y. Yang, H.L. Selzle, E.W. Schlag, Energetics of hydrogen bonds in peptides. Proc. Natl. Acad. Sci. USA **100**, 12683–12687 (2003)

H. Shigehisa, E. Nishi, M. Fujisawa, K. Hiroya, Cobalt-catalyzed hydrofluorination of unactivated olefins: a radical approach of fluorine transfer. Org. Lett. **15**, 5158–5161 (2013)

R.B. Silverman, *The Organic Chemistry of Drug Design and Drug Action.* AP Academic Press (1992)

J.C. Stowell, Intermediate Organic Chemistry (1988).

L. Stryer, *Biochemistry*, 4th edn. 912 (1995)

D. Tsvirkun, Y. Ben-Nun, E. Merquiol, I. Zlotver, K. Meir, T. Weiss-Sadan, I. Matok, R. Popovtzer, G. Blum, CT Imaging of enzymatic activity in cancer using covalent probes reveal a size-dependent pattern. J. Am. Chem. Soc. **140**, 12010–12020 (2018)

O. Vasiljeva, D.R. Hostetter, S.J. Moore, M.B. Winter, The multifaceted roles of tumor-associated proteases and harnessing their activity for prodrug activation. Biol. Chem. (2019)

R.C. Wade, K.J. Clark, P.J. Goodford, Further development of hydrogen bond functions for use in determining energetically favorable binding sites on molecules of known structure. 1. Ligand probe groups with the ability to form two hydrogen bonds. J. Med. Chem. 36, 140–147 (1993)

J.C. Way, Covalent modification as a strategy to block protein-protein interactions with small-molecule drugs. Curr. Opin. Chem. Biol. **4**, 40–46 (2000)

K.W. Whitten, K.D. Gailey, R.E. Davis, *General Chemistry*, 3rd ed. (1988)

Chapter 2
Principles for the Design of Fluorescent Dyes

2.1 Introduction

The chemical development and literature of fluorescent dyes are not only colossal but are also intriguing. It is not the dyes themselves but precisely their colors, which became a subject of research with significant economic implications. Humans are always fascinated with natural colors since the beginning of civilization. There is an old saying that there is no bad color and no bad flowers. In the old days, dyes were collected from natural sources, such as colored plants and flowers. There is convincing evidence suggesting that humans have already mastered the technique of extracting dyes from plants, and dying methods dated back more than 4000 years ago. Dyed jewelry, fabrics, cosmetics, and clothing were found in ancient Egyptian tombs. Since then, dyes have been used in all walks of life. As wisdom gained throughout development, aside from plants and flowers, dyes were extracted from vegetables, legumes, lichens, and insects.

Nearly two centuries ago, a breakthrough in chemistry changed the whole business and scientific development of the dye landscape. It started in 1856 when British chemist William Perkin reported the first synthesis of dyes from coal tar (Titford 2007). This work jumped to start the industrial revolution, dyes, and the chemical industry in general. The race between countries began as the implications of synthetic dyes in the culture and society, along with their profound economic benefits, were soon realized. Overall, Perkin's discovery contributed to the expansion of many other dye-dependent industries, such as textiles, paper, leather, ink, food, and more. The lucrative benefits generated from the dye industry threatened the fading natural dyes in the world markets; during that time, it was tantamount to what "clean" energy has done to coal nowadays.

As predicted, up to date, the majority of dyes used in the markets in our times come from synthetic sources. Not too long after Perkin's development of methods to synthesize dyes, in 1884, Danish microbiologist Hans Christian Gram discovered crystal violet (Fig. 2.1), a versatile dye used in textile, paper, and publishing industries, can distinguish stain bacteria (Coico 2005). This robust biological assay

W. Pham, *Principles of Molecular Probe Design and Applications*,
https://doi.org/10.1007/978-981-19-5739-0_2

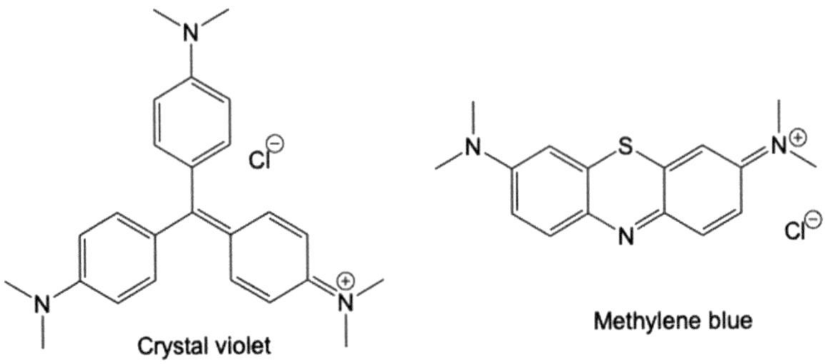

Crystal violet

Methylene blue

Fig. 2.1 The early dyes used in biomedical research

still plays a dominant role in the development of antibiotics and bacteriology in our modern days. Gram-positive bacteria are those that can retain the dye after staining and appear purple. While gram-negative bacteria do not retain the dye, thus the stain can be washed away and appear pink. And all of these observed phenomena have to do something with the bacterial cell wall. Regardless of possessing a strong cell wall, gram-positive bacteria are more receptive to cell wall-targeted antibiotics compared to gram-negative counterparts due to the absence of a cell wall. Another discovery is made by German Nobel laureate Paul Ehrlich, who showed that methylene blue (Fig. 2.1) could stain living nerve cells, not the peripheral tissues. This observation prompted him to propose a theory regarding the specific delivery of chemicals for therapy, which is now called chemotherapy (Hunter 1995). However, dyes were not so popular in biomedical research until the discovery of fluorescence property of the dyes by Sir Frederick William Herschel in 1845, followed by a detailed and systematic explanation of this phenomenon by George Stokes in 1852 (Stokes 1852). Then, the emergence of fluorescence microscopy at the beginning of the twentieth century developed by Carl Zeiss and Carl Reichert gave a considerable impetus to the shift of using dyes for the visualization of biological samples.

It is noteworthy to mention here, in this course, that organic dyes covered in this lecture differ from pigments. Dyes are organic molecules, and they get colors through the electronic absorption of light via the unsaturated carbon chains. While the pigments are inorganic composites, they are primarily metal-based substances.

Many different fluorescent dyes are used in biomedical imaging and other advanced industries, including optical recording technology. In the context of premedical and medical applications of this course, this chapter will cover the chemical design and synthesis of only the most practical, useful, and popular dyes, such as those belonging to the family of rhodamine and cyanine dyes and the burgeoning modifications of these dyes that might have profound implications in molecular imaging. Each type has its unique characteristics, including the pros and cons. Due to the chemical constraints, the idea of designing a dye that meets all of the desired features for in vivo study is unachievable. There is always a trade-off between extended wavelength and utilities. For example, a long polymethine chain reduces the stability of the dye due to increased rotation around the unsaturated system, resulting in reduced quantum yields. These challenges coupled with new opportunities for those who started to enter into this field for making a breakthrough.

In the past two decades, the number of optical probes using these dyes, primarily for preclinical imaging work, increased significantly due to the emergence of innovative chemistry. Aside from the development of fluorescent dyes, the availability of small animal imaging devices that enable the detection of fluorescent signals with improved sensitivity, resolution, and tissue penetration, altogether made this optical imaging technique a burgeoning field. The growing interest in optical imaging is not inadvertent; its simplicity, swiftness, cost-effectiveness, safety, and non-association with radiation energy insinuate powerful future clinical implications. The probe design has unlimited features, including small organic molecules, peptides, DNA, RNA, aptamers, nanoparticles, antibodies, and nanobodies. For in vitro studies, cell-based or ex vivo imaging, the use of visible dyes would suffice, while near-infrared (NIR) versions dominate in vivo applications since long-wavelengths penetrate tissue deeper. Imaging at the NIR range, approximately from 700 to 800 nm, can navigate the tissue matrix at a depth of a few centimeters (Zhang et al. 2012).

As being said, the development of an optical probe is probably the most complicated process among the types. The process involves the bioconjugation of targeted ligands with designated fluorescent dyes. It seems easy as it sounds, but the nature of unstable fluorescent dyes, particularly the NIR versions, due to their extended conjugated backbone necessitates optimal labeling reaction conditions, in which temperature, pH, solvents, and reaction times, must be taken into careful consideration. Experimental data indicate that the successful development of optical probes relies on the chemical design of fluorescent dyes. These molecules have extended conjugated carbon chains embedded in their chemical structures so that they can absorb light energy and emit fluorescent signals at longer wavelengths. According to the molecular orbital theory, this phenomenon occurs due to the electron transition from the highest occupied molecular orbital (HOMO) to the lowest unoccupied molecular orbital (LUMO) (Nolting et al. 2011). The smaller the energy gaps between the HOMO and LUMO, the more absorption possesses the bathochromic shift (extending to a longer wavelength). This can be achieved experimentally by increasing the conjugated carbon chain, albeit at the expense of stability and poor quantum yield. Another challenge exacerbates these issues that is to synthesize such small organic molecules with the expectation of desired excitation/emission wavelengths, functional groups for bioconjugations, and enhanced water solubility for biological or in vivo studies. In this chapter, we will focus our discussion on the design and synthesis methods of fluorescent dyes with these criteria in mind.

Before delving into the chemical development of fluorescent dyes, a mechanistic explanation of how dyes emit fluorescence after obtaining excitation energy warrants some discussion. According to the molecular photochemistry theory, upon the absorption of a photon, in a matter of a femtosecond, the excited electrons will leave the initially occupied ground state energy level S_0 for the higher energy level, previously unoccupied orbital, called the excited state S_1, in discrete amounts called quanta (Turro 1991). The energy of a quantum is defined as $E = h\nu = hc/\lambda$, where

h is Planck's constant (6.63×10^{-34} Js), c is the speed of light in a vacuum (3.0×10^8 m/s), and v and λ are the frequency in Hertz and wavelength of the excited photons, respectively. From the excited state, the electrons will start to cascade back to the ground state S_0 via two mechanisms, called the radiation and radiationless mechanism. In the radiation mechanism, if the excited electrons relax to the ground state from the S_1 energy level emitting light (singlet–singlet emission-$S_1 \rightarrow S_0 + hv$), it is called fluorescence. It is worth noting from Planck's equation that the radiation energy is inversely proportional to the wavelength. This means that for the given excited electrons transitioning from the high to lower energy level, such as fluorescence, the wavelength lambda maximum of the emission is always larger than that of the excitation wavelength. The distance of lambda maximal between the excitation and emission spectra is called Stokes shift. Usually, for imaging dyes, this can range from 30 to 150 nm.

Meanwhile, if the excited electrons relax to the ground state from the T_1 energy level (triplet–singlet emission-$T_1 \rightarrow S_0 + hv$), the process is called phosphorescence. In the radiationless mechanism, the transition from $S_1 \rightarrow S_0$ generating heat instead of light is called internal conversion. Meanwhile, the intersystem crossing is characterized as the transition from $S_1 \rightarrow T_1$ or $T_1 \rightarrow S_0$ and generating heat.

The excited electrons in the higher energy levels have different electronic states depending on the spin orientation. In the excited singlet state (S_1), the electrons are paired, while in the triplet state (T_1), the electrons have a parallel orientation (unpaired). Based on the multiplicity equation learning from the previous chapter, $2s + 1$, where s is the total spin angular momentum. So, for the S_1, the expression would be $2(+1/2 + -1/2) + 1 = 1$, thus making it a singlet state. For the T_1, the calculation would be $2(+1/2 + +1/2) + 1 = 3$, which makes it a triplet state as expected. Experimentally, the paired electrons remain in a single state if exposed to an external magnetic field, called the singlet state (S_1). In contrast, the unpaired electrons split into three quantized states, called triplet state (T_1).

2.2 Xanthene Backbone Dyes

This is the largest family of fluorescent dyes; they have been long used as coloring materials in the food, cosmetics, and textile industries (Kato et al. 2012) before they started to play essential roles in biomedical research and imaging technology. The chemistry for the synthesis of xanthene-based dyes is robust. Besides being a robust and straightforward chemistry, these dyes are known for their unmatched stability, along with high molar extinction coefficient and quantum yield. These dyes contain a xanthene backbone where the amino or hydroxyl groups are located meta to the oxygen. There are several fluorescent dyes in this group covering a wide range of wavelengths for a countless number of applications, from blue to green, yellow-to-red, far-red, and near-infrared regions. Rhodamine and fluorescein dyes dominate in this group.

2.2.1 Rhodamine Dyes

As mentioned, the rhodamine family of dyes comprised several commercial fluorescent dyes used in biomedical imaging, covering an entire wide range of photon spectra from ultraviolet to visible red and NIR. Probably, the Alexa Fluor series exemplifies this notion, including the Alexa Fluor series, such as Alexa 350, 488, 514, 532, 546, 568, 594, 610, and 647. Rhodamine dyes possess high molar extinction coefficients (ε), with some versions that can reach approximately 300,000 M^{-1} cm^{-1}. Rhodamine dyes are also well recognized with a great quantum efficiency ($\phi \sim 0.4$–1.0) suitable for all kinds of available optical imaging devices and filters in fluorescence microscopy among the fluorescent dyes. Because of its versatility, the modification to optimize the chemical structures for multi-application using rhodamine backbone (Fig. 2.2) has been a subject of intense study in the past decades. Changing the substituents on positions R_1, R_2, R_3, R_4, R_6, or the counterion X^- will influence the dye's photophysical properties, including absorption, emission, fluorescence lifetime (τ), and fluorescence quantum yield (ϕ) (Beija et al. 2009). For example, alkylation of rhodamine 110, whose absorption/emission (abs/em) at 501/525 nm with tetramethyl groups as seen in tetramethylrhodamine elicits a bathochromic shift with abs/em equivalent to 552/575 nm, albeit with a significantly reduced quantum yield.

Fine-tuning dyes for molecular imaging is a daunting task. Particularly, the operation lingers with an inadvertently diminished fluorescent signal. This is the reason why the fluorescent property of rhodamine dyes varies significantly between compounds, which is dictated by the non-radiative deactivation caused by the internal conversion process. This process involves activated and/or non-activated mechanisms. The activated process governs by the twisted intramolecular charge transfer (TICT), that occurs by fast intramolecular electron transfer from the nitrogen (donor) to the xanthene ring (acceptor) resulted in intramolecular donor–acceptor twisting around the single bond. Following intramolecular twisting, the TICT state returns

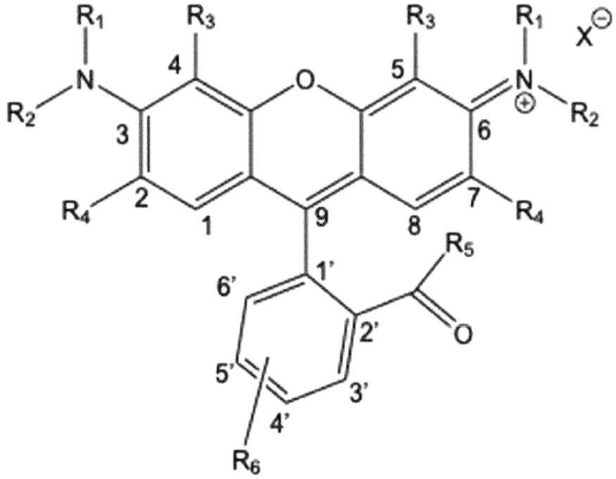

Fig. 2.2 General structure of rhodamine dyes

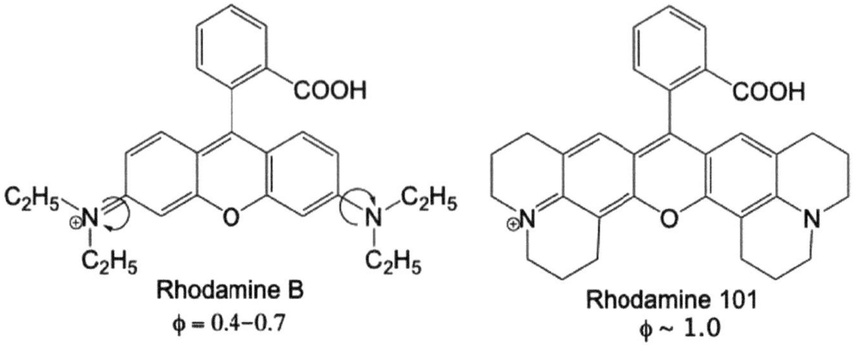

Fig. 2.3 Rigidizing the electron-donor/acceptor moieties to improve photostability

to the ground state by non-radiative relaxation (Sasaki et al. 2016), while the non-activated process involves energy dissipation by C–H and N–H stretching modes coupled with high-frequency vibration modes of the solvents (Beija et al. 2009). Understanding this phenomenon enables the design of rhodamine dyes with more stable structures and thus improved quantum yield. For instance, if the terminal nitrogen moieties on rhodamine dyes are rigidized by association them within the ring structures, such as the ones shown in Rhodamine 101 (Fig. 2.3), the dye will not only improve the photostability, which can survive high temperature but the quantum yield is enhanced remarkably as well (Karstens and Kobs 1980).

When labeling rhodamine dyes for biological study, both the labeling conditions and buffer media where the probe will be tested need careful design since these dyes are sensitive to the microenvironment generated by solvent effects, including micropolarity and microviscosity. The color of the dye might change, or the abs/em spectra might shift; that does not mean the dyes decompose or deteriorate by any means, but rather these dyes can exist in different isomeric forms as a function of solvent composition and solvent system (Chang and Cheung 1992). For instance, in an acidic condition, the dye exhibits a cationic character, in which the carboxylic group is protonated. However, in basic conditions, the carboxylic group is deprotonated, rendering the dye in the zwitterionic form (Fig. 2.4). Although both the cationic and zwitterionic forms share the same chromophore, the negative charge on the carboxylic acid has an inductive effect on the central carbon atom of xanthene chromophore, leading to a hypsochromic shift (shifting to shorter wavelength) of both absorption and fluorescence maxima and a slight reduction of the extinction coefficient (Beija et al. 2009). In contrast, it is the negative charge of the zwitterion structure increases the π-electron density near the remote amino groups leading to a slightly larger quantum yield ($\phi = 0.49$) and lifetime ($\tau = 2.48$ ns) compared to cationic isoform (0.4/2.21 ns) (Chang and Cheung 1992). Zwitterionic rhodamine can also form the lactone in a reversible manner at equilibrium. Particularly, this happens at elevated temperatures or in the presence of less polar organic solvents. As shown in Fig. 2.4, the formation of lactone analogs resulted in disrupting the unsaturated

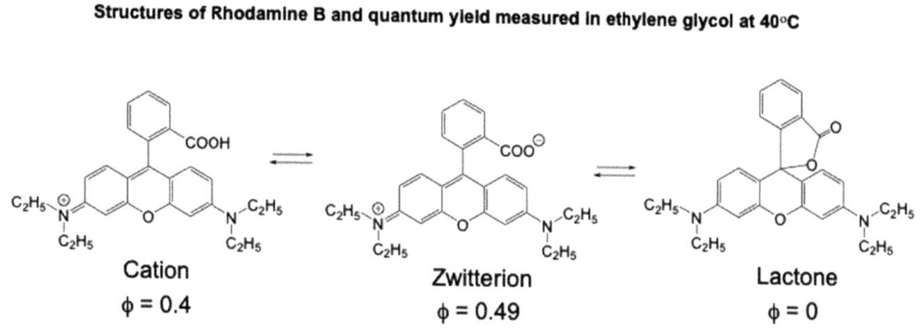

Structures of Rhodamine B and quantum yield measured in ethylene glycol at 40°C

Cation
$\phi = 0.4$

Zwitterion
$\phi = 0.49$

Lactone
$\phi = 0$

Fig. 2.4 The pH-dependent isomeric transformations. Data obtained from Chang and Cheung (1992) with permission from the American Chemical Society

carbon network; thus, lactone products have no color, nor do they possess absorbance or fluorescent signal.

Most of the rhodamine dyes available in the market so far possess symmetric structures. At the least, the synthesis of symmetric dyes is simple compared to asymmetric counterparts. Aside from additional steps of synthesis, asymmetric operations will end up with more side products; thus, an extensive effort for product purification is inevitable. In general, symmetric rhodamine dyes can be synthesized via electrophilic aromatic substitution with repeated Friedel–Crafts mechanisms. As shown in Fig. 2.5, the phthalic anhydride is protonated under acidic conditions to generate an acyl cation, which reacts with the substituted phenol to create a substituted benzophenone similar to Friedel–Crafts acylation. Under strong acidic conditions with elevated heating conditions, the benzophenone will be protonated as a cationic intermediate, which fosters the substituted phenol to react in a Friedel–Crafts alkylation type of reaction. Then, the cationic moiety induces ring closure to form the main xanthene backbone, followed by condensation and lactonization to form the desired dye.

This reaction mechanism serves not only for the synthesis of rhodamine dyes but also works for the development of fluorescein. The same reaction scheme is applicable, albeit the nitrogen moieties in rhodamine dyes will be replaced by the hydroxyl groups. In that case, the substituted version of resorcinol should be used.

Designing the fluorescent dyes for biomedical innovation requires an understanding of the mechanisms, which are amenable to the general rules for improving physical properties. Aside from that, activation of the dyes with desired functional groups for bioconjugation is also an important task. It is apparent from the reaction mechanism that the substituted phenols are critical in optimizing the dyes with improved photostability, quantum yield, and tuning the excitation and emission wavelengths. All will focus on modifications on positions 3 and 6 of the xanthene ring. Meanwhile, the functional group-laden phthalic anhydride in the synthesis will confer a handle for activation. So far, many activated rhodamine dyes have been achieved for labeling a spectrum of biological materials using this strategy, including the succinimide esters, maleimides, and isothiocyanates (NCS) (Fig. 2.6). More later, other new versions of the activated dyes also emerged, such as azide or alkyne moieties,

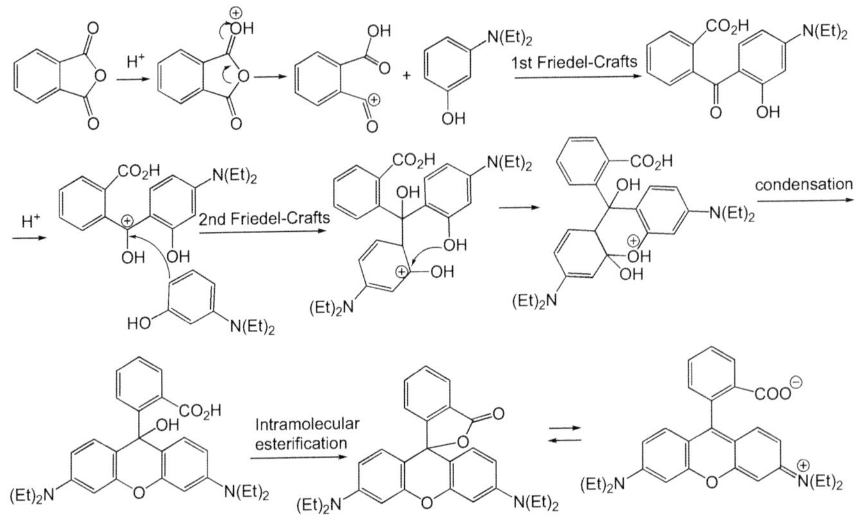

Fig. 2.5 Mechanism-based synthesis of rhodamine dyes. Derived from McCullagh and Daggett (2007)

which are derivatized with the extended linkers or polymers for aqueous conjugation via copper-catalyzed Click chemistry. Rhodamine dyes with azide terminals can also react with strained cyclooctyne in a copper-free Click reaction to form a stable triazole in very mild reaction conditions. Some rhodamine dyes were also activated with biotin for affinity labeling using the tetramer streptavidin approach. The biotin-streptavidin complex is the strongest non-covalent binding in nature. It is resistant to extreme temperatures or pH and unwavering to denaturants or detergents with a Kd of approximately 10^{-14} mol/L.

In another approach, direct conjugation to the rhodamine dye ester (2' position) could be achieved in a mechanism in which the primary (alkyl/aryl) amine could form a reversible reaction at position 9 on the xanthene ring of non-alkylated or mono-N-alkylated rhodamine ester. The reaction will not stop there; instead, the amino group will continue to initiate the intramolecular cyclization to form a spirolactam ring, which will subsequently undergo ring opening under the acidic condition to afford the desired conjugated product (Adamczyk and Grote 2000).

Rhodamine dyes can be synthesized using other intermediates or reaction mechanisms, but from a practical point of view, a short and straightforward reaction is crucial for scaling up the products, cutting costs, and facilitating in vivo applications. In that regard, using a cheap, isomerically pure, and commercially available fluorescein dye as a starting material to make isomerically pure rhodamine dyes attests as one of the most creative and robust ways to generate the most needed dyes, the holy grail for biomedical and high-technological research. Another incentive for generating more robust chemistry is the challenge of modifying the rhodamine backbone for labeling purposes, as well as to improve the optical property, particularly at the

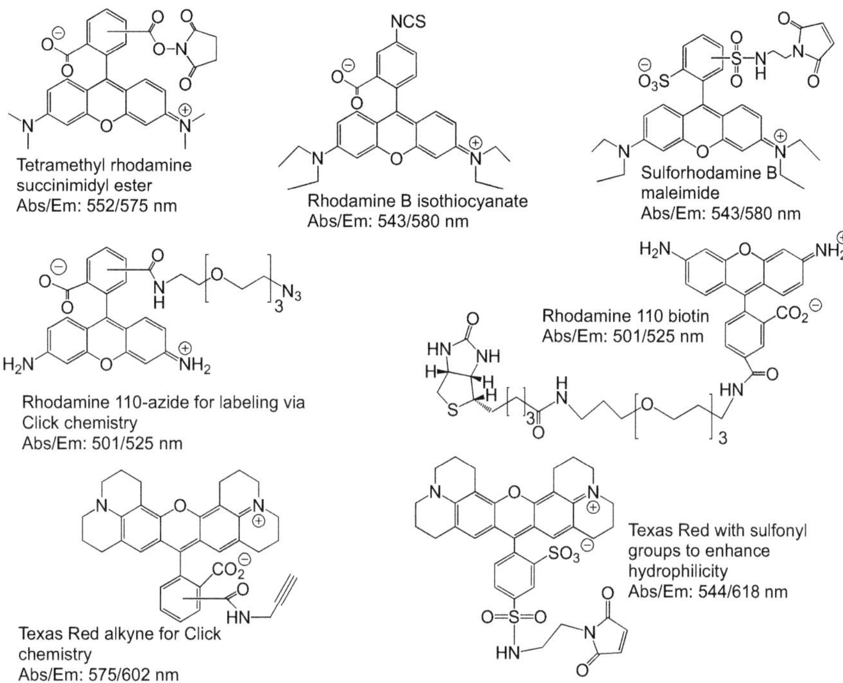

Fig. 2.6 Activated rhodamine dyes for labeling applications

rhodamine amines. The delocalization of the nitrogen's free electron pair into the aromatic ring renders amines as weak nucleophiles. This new approach is based on the Buchwald–Hartwig reaction, in which palladium catalyzes the C–N cross-coupling mechanism (Grimm and Lavis 2011). Buchwald–Hartwig conditions, including Pd(OAc)$_2$, BINAP, and Cs$_2$CO$_3$ in toluene at 100 °C were effective to couple fluorescein ditriflate with different versions of amines, including primary and secondary amines of aliphatic and cyclic analogs to create N-alkyl and N-aryl rhodamine dyes (Fig. 2.7). The reaction conditions using Pd(OAc)$_2$/BINAP and Cs$_2$CO$_3$ are excellent for cyclic amines. However, for secondary acyclic and primary aliphatic amines, the use of Pd$_2$dba$_3$ with active biaryl ligand XPhos provided a better yield. In contrast, nitrogen-containing heteroaromatics are better with organophosphorus compounds, like Xantphos.

It is found that using amides or carbamates (Boc group), or protected amine analogs as substrates in C–N cross-coupling with fluorescein ditriflates is more favorable than free amines since the rhodamine products, now lock in the lactone forms (Fig. 2.7), which facilitates purification and manipulation than free rhodamines. After the cross-coupling reaction, the acidic protected rhodamine, such as the Boc groups and *tert*-butyl esters, could be removed in TFA to afford free rhodamines in good yield (>90%).

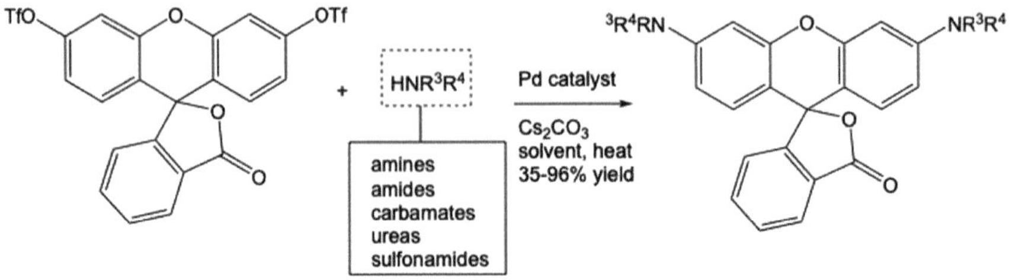

Fig. 2.7 Synthesis of rhodamines from fluorescein ditriflates using Pd-catalyzed C–N cross-coupling with amines, amides, carbamates, and other nitrogen nucleophiles. Data obtained from Grimm and Lavis (2011) with permission from the American Chemical Society

As mentioned earlier, it has been known that alkylation of rhodamine dyes resulted in bathochromic shifts, but this is at the expense of quantum yield unless terminal amines are rigidized in ring structures. One of the alternatives to circumvent this limitation is developing a hybrid rhodamine approach, in which silicon-substituted xanthene dyes were developed (Fig. 2.8). Basically, the replacement of the central xanthene oxygen by alkylsilane moiety would tune the dye with observed bathochromic shifts for both absorbance and emission (Grimm et al. 2017). Whether the strength of the Si–C bond contributes to structural stability and consequential bathochromic shift remains to be seen, but it is known that Si–C bonds are as strong as, or sometimes stronger than, corresponding C–C counterparts. Given silicon atoms are more electropositive than carbons, a nucleophilic substitution at silicon is more effective than carbon. A typical reaction to make Si–C using chlorosilane with in situ-generated organolithium to generate diphenyl silane. Then, regioselective bromination can be achieved using N-bromosuccinimide (NBS) to provide the key intermediates bis(5-amino-2-bromophenyl)silanes. The dibromides were treated with t-BuLi followed by slow addition of phthalic anhydride to provide the Si-rhodamine, albeit the yield is low. It is observed that probably the lithium abduct is unstable, which leads to poor yield. If the intermediate is treated with $MgBr_2 \cdot OEt_2$ in an in situ metalation process, the yield improves significantly. However, it is worth noting that the direct formation of the aryl magnesium species from the dibromide using Grignard reagents, such as i-PrMgCl·LiCl and PrBu$_2$MgLi resulted in sluggish reaction or very poor yields. Another notable improvement in this design is incorporating rhodamine nitrogen in azetidine rings, thus improving the quantum yield and fluorogenicity compared to their tetramethyl rhodamine congeners (Birke et al. 2021).

So far, we discuss different methods for synthesizing rhodamine dyes to improve the biophotonic property, as well as different modes of activation for labeling purposes. However, it is crucial to mention that by nature, rhodamine dyes are very hydrophobic. They have the proclivity to aggregate in buffers. One of the approaches to overcome this issue focuses on the carboxyphenyl ring. It has been shown that the inclusion of polar groups, such as the sulfonate moieties, can render the dyes in an aqueous environment easily, suitable for labeling biologics using common buffers.

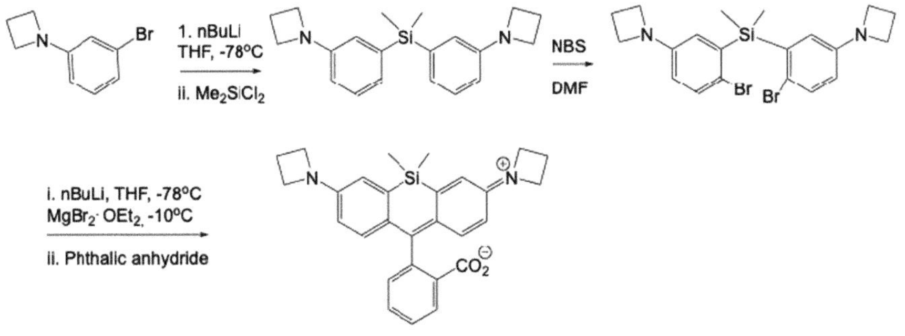

Fig. 2.8 Dual improvement of rhodamine dye with silicon-substituted xanthene core scaffold and rigidizing the nitrogen-based donor/acceptor termini. Data obtained from Grimm et al. (2017) with permission from the American Chemical Society

Texas Red shown in Fig. 2.6 has two sulfonate groups; thus, it offers great biological applications. Recently, other versions of sulfonated rhodamine dyes have been developed using the Buchwald–Hartwig reaction as described in the past (Grimm and Lavis 2011). The synthesis started with incorporating the azetidine-3-carboxylate via the bistriflate, to serve as a handle for conjugation with taurine (Fig. 2.9). Finally, the dye was activated with an amine- or thiol-reactive linker. All conjugation steps in this work were carried out using water-soluble coupling reagent TSTU (2-succinimido-1,1,3,3-tetramethyluronium tetrafluoroborate) in polar aprotic solvent in the presence of a strong organic base (DIEA), particularly in light of polarity of the dyes.

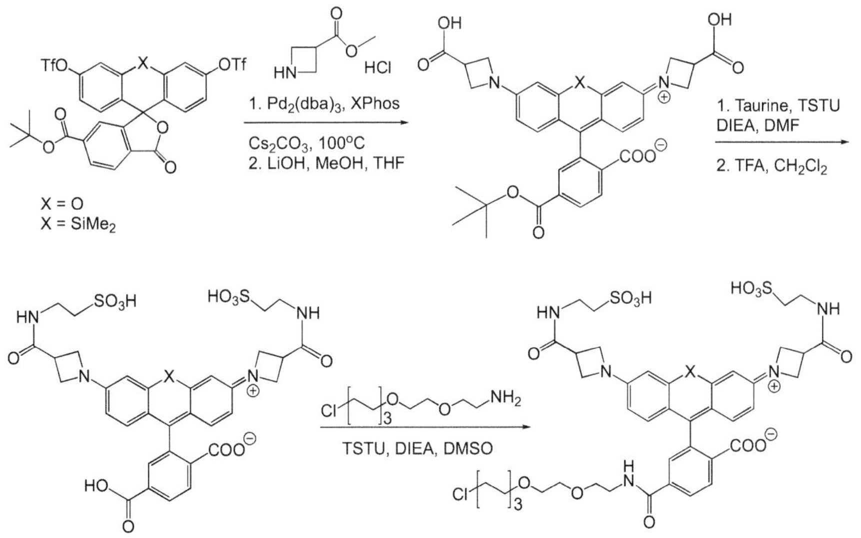

Fig. 2.9 Synthesis of water-soluble sulfonated rhodamine dyes. Data obtained from Birke et al. (2022) with permission from the author and the Royal Society of Chemistry

Besides the development of these di-sulfonated rhodamines, a handful of commercial and activated rhodamine-based Alexa dyes also have two sulfonates on each dye. These include Alexa Fluor®488, Alexa Fluor®532, Alexa Fluor®546, Alexa Fluor®568, Alexa Fluor®594, and Alexa Fluor®610.

2.2.2 Fluorescein Dyes

Fluorescein was first synthesized by Bayer in 1871 (Ziarani et al. 2018). The difference between rhodamine and fluorescein is merely the displacement of amino moieties of the former with the hydroxyl counterparts; nevertheless, this constitutes unique characteristics for fluorescein dyes. As mentioned earlier, the synthesis of fluorescein is identical to those used for rhodamine dyes. Simple fluorescein can be obtained by treating resorcinol with phthalic anhydride in the presence of sulfuric acid or a Lewis acid at elevated temperature (>300 °C). Therefore, fluorescein is also called resorcinolphthalein. Since the hydroxyl groups are weak nucleophiles compared to the amine counterparts of rhodamine dyes, thus modification at the hydroxyl groups via alkylation to improve the dye optical property is less prolific than rhodamine dyes. However, the hydroxyl group is an excellent asset for creating labile esters, which makes fluorescein one of the most favorite intermediates for creating "smart" on/off activated probes for the detection of metals and proteases. In a similar approach to develop hybrid xanthene dyes, oxygen moiety is replaced by silicon to create Si-fluorescein (Fig. 2.10). Surprisingly, Si-fluorescein dyes exert approximately 100-nm bathochromic shifts in both absorption maximum and fluorescence maximum, even better than Si-rhodamine dyes (Grimm et al. 2017).

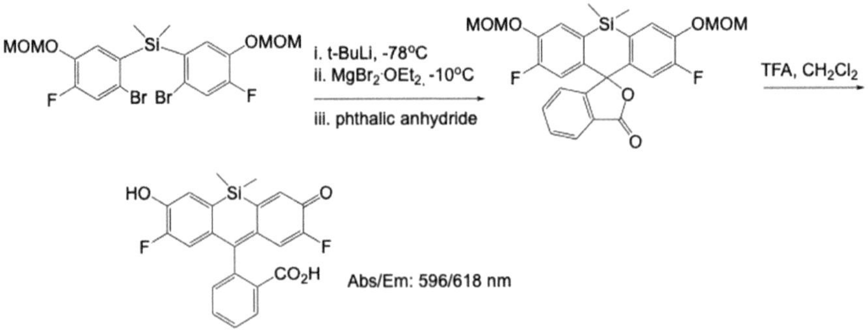

Fig. 2.10 Synthesis of silicon-based fluorescein dyes. Data obtained from Grimm et al. (2017) with permission from the American Chemical Society

2.3 Cyanine Dyes

Greville Williams first synthesized cyanine dyes in 1856 using quaternary quinolinium salts with 1-iodopentane in an alkaline condition (Hamer 1950). The unintentionally discovered dye (Fig. 2.11) got a very intense dark blue color, called "cyanine," which means dark blue in Greek.

This is another family of fluorescent dyes that have robust applications in biomedical imaging. Cyanine dyes can be readily fine-tuned to the near-infrared spectrum, with which imaging penetrating deep tissue is possible. Thus, this family of dyes dominates preclinical imaging, including some limited clinical operations. With the advent of molecular imaging, never before has so much effort been put forth to create biologically compatible cyanine dyes. The anticipation for in vivo study necessitates new and creative synthetic methods to design small organic dyes fitting essentially multiple criteria, such as to improve water solubility, quantum yield, stability, and most importantly near-infrared (NIR) capabilities (Nolting et al. 2012). The dyes are characterized as the heterogenous or homogenous nitrogen-containing heterocyclic systems, including but not limited to indoles, quinolines, isoquinolines, benzothiazoles, and benzoxazoles. These rings are linked together by an unsaturated carbon chain, resulting in a quaternary amine in one ring and the tertiary amine on the other, and they serve alternatively as a donor and an acceptor. The electron propagation between the amino groups along the unsaturated linkage or polymethine bridge alternating π electron density along the bridge is the hallmark of cyanine dyes. As shown in Fig. 2.12, the wave function of the dyes has equal contributions from two cationic resonance structures (Nolting et al. 2011). This dynamic push–pull mechanism defines the absorbance, emission, and photostability of cyanine dyes. As mentioned earlier, one of the approaches to fine-tune the dyes for NIR range is to reduce the energy gaps ΔE between the π (HOMO) to π^* (LUMO) excitation of the conjugated system. Further, reducing bond-length alternation also helps the bathochromic shift of the wavelength. And cyanine's polymethine bridge offers all of these unique characteristics. Different from polyenes (unsaturated carbon chains) that have a significant bond-length alternation, the bonds cannot be represented as single and double bonds in equivalent resonance structure, thus lowering the energy of the HOMO (Autschbach 2007). Meanwhile, the association of the donor–acceptor moieties in polymethine bridge (unsaturated carbon chains with the presence of the electron sinks, like nitrogen, at the ends) decreases the bandgap by increasing the HOMO and decreasing the LUMO energy and does not have bond-length alternation, ideal for tuning the dyes in the NIR range with the shortest possible conjugated

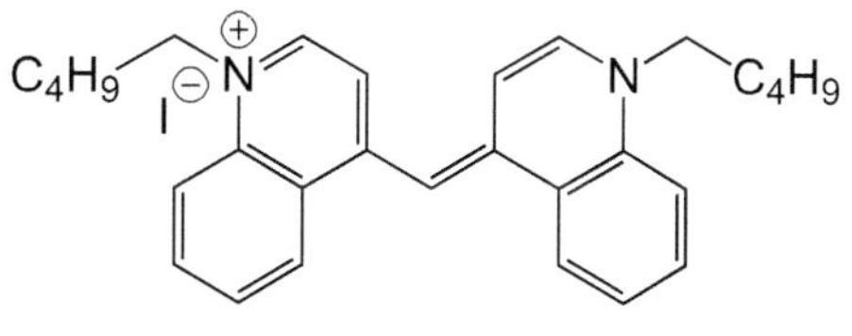

Fig. 2.11 First cyanine dye called Quinoline Blue

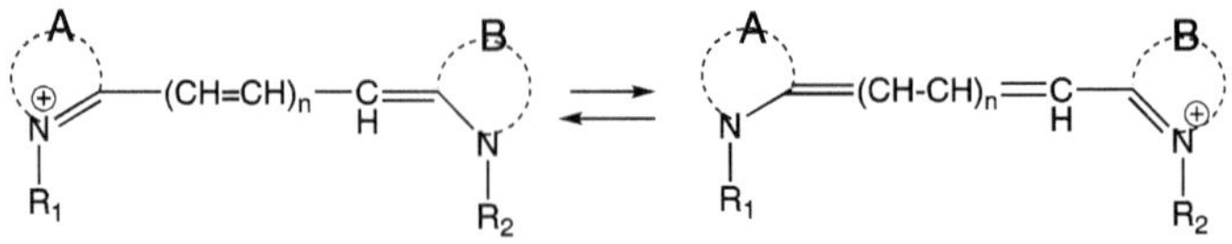

Fig. 2.12 Resonance structure of cyanine dye isoforms

carbon length. Further, what makes cyanine dyes unique compared to other dyes is the robust chemistry that allows for multiple chemical modifications in such a tiny molecule from the heterocyclic rings to the polymethine bridge.

The synthesis of cyanine dyes can be accomplished through different types of condensation depending on the use of the starting materials. For example, if the starting materials are alcohols and aldehydes, then aldol condensation would rule. Let us discuss the synthesis of a simple cyanine dye, such as hemicyanine structures. This type of dye has versatile applications in textile industry. They also work as photographic sensitizers in photocells (Stathatos et al. 2001). Most relevant to this book, hemicyanine dyes serve as fluorescent probes for imaging cell membrane potentials and intracellular pH changes (Gaspar et al. 2006). One of the early synthesis of hemicyanine dyes employed the condensation of Fischer's base aldehyde with anilines in the presence of acetic acid (Gaspar et al. 2006). In this reaction, the right proportion of acetic acid must be carefully calibrated to enhance the electrophilicity of aldehyde carbon while not too acidic that will decrease the nucleophilicity of the free amine of aniline (Fig. 2.13). The reaction rate decreases with increasing concentrations of acetic acid (Gaspar et al. 2006). Further, the substitutions on the aniline can be exploited to improve the reaction rates as far as the high concentration of free amine is needed to drive the condensation reaction. Theoretically, anilines substituted with electron-donor groups would increase the condensation rate because of greater generation of the nucleophilic character of anilines. However, this hypothesis is contradictory to reality, specifically when the acetic acid content in the solvent increased from 30 to 60%. At high acidic conditions, the diminished concentration of free amine overshadows the benefic effect of the electron-donor substituents on the reaction rate (Gaspar et al. 2006).

In another approach, Knoevenagel condensation reaction has been reported for the design of hemicyanine dyes using the quaternary salts of picoline or 2-methylbenzoxazole with benzaldehyde in the presence of piperidine (Fig. 2.14). The mechanism suggests that piperidine acts as a strong base, abstracting the acidic proton on the methyl group of the starting materials to form the methenyldihydropyridine. This reactive intermediate readily reacts with the aldehyde to provide the cationic hemicyanine dyes (Jedrzejewska et al. 2003).

The next group of cyanine dyes in the hierarchy in terms of emission wavelength is trimethine cyanine dyes, or simply called Cy3, because of the presence of three carbons in the polymethine bridge. Trimethine cyanine dyes can be synthesized by the condensation of a reactive quaternary ammonium salt with an orthoester, such as triethyl orthoformate in the presence of sodium tetrafluoroborate (Fig. 2.15).

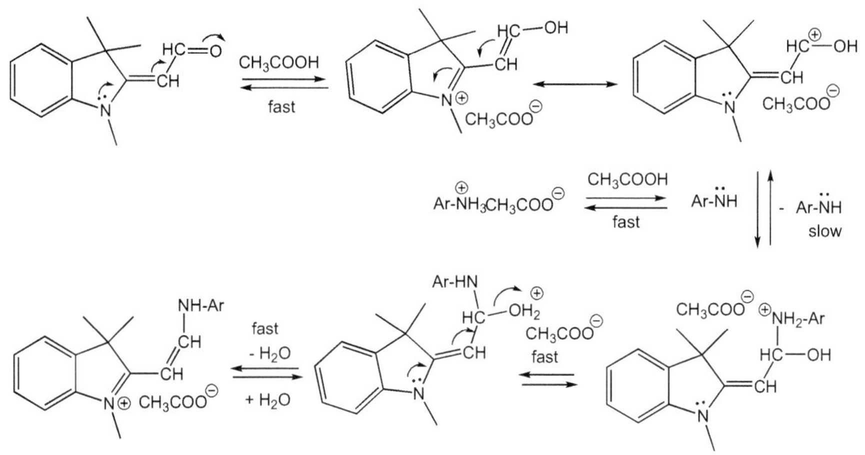

Fig. 2.13 Mechanistic explanation for the formation of hemicyanine dye by the condensation of Fischer's base aldehyde with an aniline. Data obtained from Gaspar et al. (2006) with permission from Elsevier

This approach is also employed for the synthesis of a number of different versions of Cy3 using heterocyclic intermediates, such as benzothiazoles, N-alkylquinoline, N-alkylpyridine, and alkylbenzoxazoles. This type of reaction usually provides a good yield (~70%). With the emission λ_{max} of approximately 600 nm, Cy3 dyes are useful for in vitro assays, such as cell sorting analysis, high throughput screening, and fluorescence microscopy due to their high extinction coefficient. Aside from limited tissue penetration of visible light, imaging signal interference is another drawback given high background signals caused by intrinsic biological materials such as hemoglobin and others can emit light in the range of 400–600 nm. It is for these reasons; there is a need to develop more extended emission dyes somewhere in the near-infrared region (700–900 nm) for in vivo work.

One of the most generalized and approachable methods for the synthesis of extended polymethine cyanine dyes is via the condensation between the activated methyl group on the indole or similar analogs (Fischer's base) with malonaldehyde dianilide hydrochloride salt in the presence of a catalytic amount of sodium acetate (Fig. 2.16).

This mechanism is more robust than aldol condensation since the carbon of the iminium is a more reactive electrophile than a carbonyl moiety of ketone or aldehyde. Basically, in a slightly basic condition, an alkaline, such as sodium acetate, will abstract a proton on the methyl group of the indole molecule, rendering the methyl group as an anion (enamine), a strong nucleophile which will react with malonaldehyde dianilide hydrochloride, in the presence of a catalyst, such as sodium acetate. This anion is very stable because the free pair of electrons can be delocalized inside the indole ring via the resonance structures (Fig. 2.14). This approach has been used to develop Cy5 and Cy7 dyes using a symmetric or asymmetric synthetic approach.

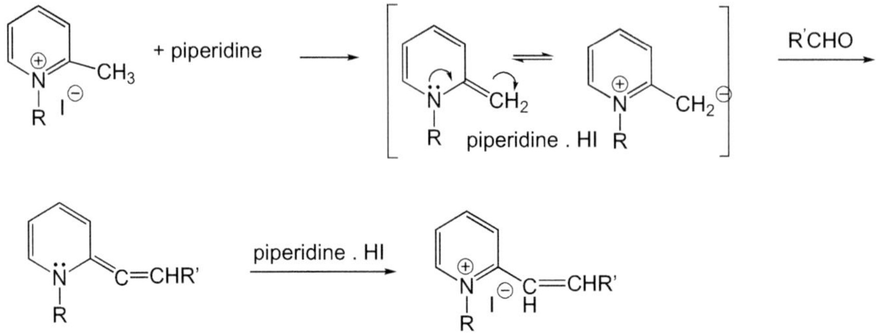

Fig. 2.14 Knoevenagel condensation for the synthesis of hemicyanine dyes. Data obtained from Jedrzejewska et al. (2003) with permission from Elsevier

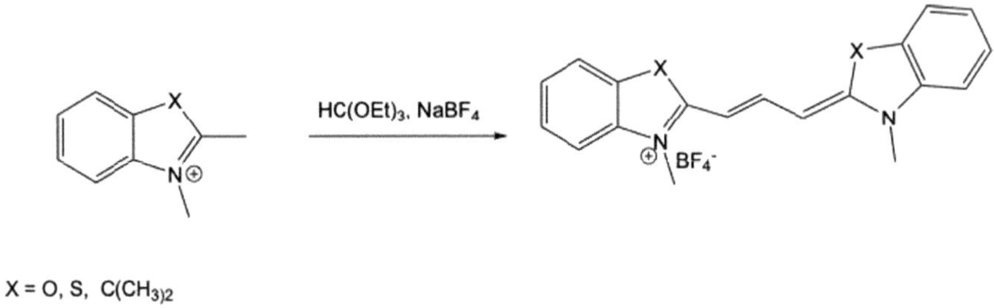

Fig. 2.15 Synthesis of Cy3 dyes. Data obtained from Tietze and Eicher (1988)

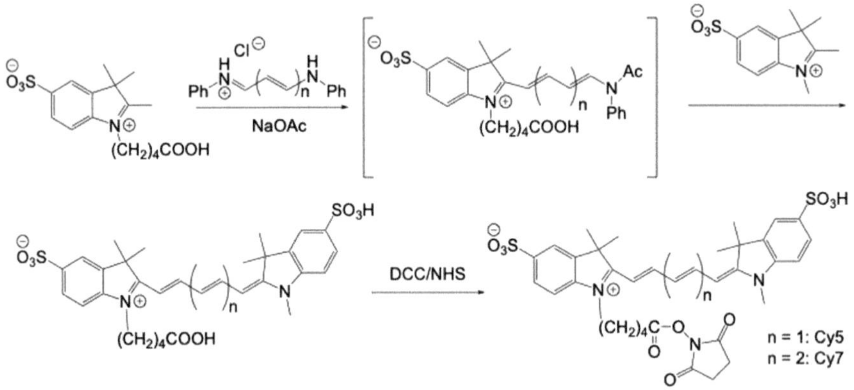

Fig. 2.16 Condensation reaction between Fischer's base and malonaldehyde dianilide hydrochloride for generating NIR cyanine dyes. Data obtained from Mujumdar et al. (1993) with permission from the American Chemical Society

Every method has advantages and disadvantages, and it is not exceptional in this case. Although the reaction is very robust, the limitation of this chemistry is the challenge of purification, particularly the presence of by-product anilines and sodium acetate as a catalyst proved to be the main obstacle for obtaining neat and pure products.

To overcome this issue, a new synthetic pathway was derived without the use of a catalyst (Narayanan and Patonay 1995). The reaction was carried out using a quaternary salt of a heterocyclic base containing an activated methyl group and an unsaturated bisaldehyde in a mixture of 1-butanol and benzene, and the condensation was driven to completion by prolonged refluxing followed by removal of water as an azeotrope by a Dean-Stark condenser, in an equilibrium step (Fig. 2.17). Since no by-products or catalysts were involved in the process, purification of the dye product can be achieved by simply washing the dye with diethyl ether or using silica gel chromatography. One of the initial thoughts attempted to explain why the reactions happen without a catalyst pointing out that the chlorine in the cyclic pentamethine derivative may make the aldehyde protons more acidic that drive the reaction to completion. However, the experimental results disproved this hypothesis as the reaction proceeded successfully when the chlorine is displaced by hydrogen. The slow pace of equilibrium reaction enables the synthesis of asymmetric dyes. Stepwised condensation of the first N-alkyl-substituted quaternary salts with bisaldehyde in refluxing butanol/benzene with the removal of water for 2 h followed by addition of the second quaternary salt resulted in asymmetric dye with good yield. The undesired symmetric dyes were apparently formed as well, but the product can be purified and isolated from the unwanted dyes with good chemical yield using flash chromatography.

Aside from the approaches mentioned above for the synthesis of near-infrared cyanine dyes, the modification of the polymethine chain with an electronic effect can influence the absorbance of the dye. Thus, it provides a capability to fine-tune the wavelength for a particular application, whether the intension is for a bathochromic or hypsochromic shift. For instance, introducing the electronegative nitro group in the meso position of the pentamethine causes a hypsochromic shift of 50–150 nm (Reichardt 1968). This happens due to the disturbance of the

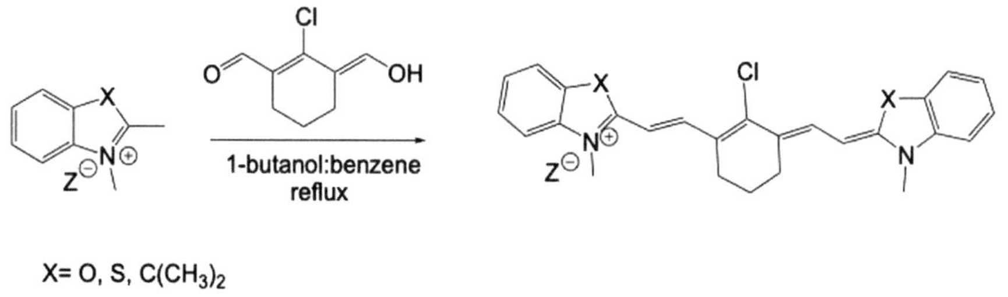

Fig. 2.17 Synthesis of NIR cyanine dyes without a catalyst. Data obtained from Narayanan and Patonay (1995) with permission from the American Chemical Society

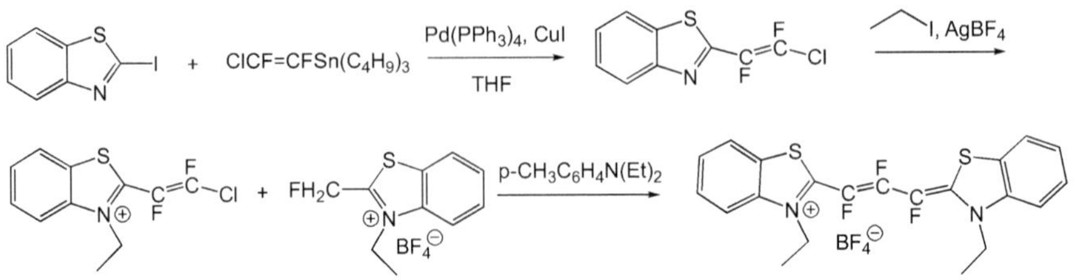

Fig. 2.18 Synthesis of a Cy3 dye with a perfluorinated polymethine chain. Data obtained from Yagupolskii et al. (2008) with permission from Elsevier

electronic propagation along the methine chain by the electron-withdrawing group, resulting in the negative inductive and mesomeric effect that promotes hypsochromic anomaly of the dye. Moderate or weak electron-withdrawing groups will cause a smaller hypsochromic shift than those generated by the nitro group. These serve as the options for the custom design of dyes. However, it is worthwhile to note that this rule is inapplicable to halogens. Besides they serve as a weak electron-withdrawing moieties, halogens with non-bonding electrons can also donate the electrons to the π-bond of the polymethine chain, resulting in a bathochromical shift. One of the examples of this phenomenon is documented in the synthesis of a cyanine dye containing a perfluorinated polymethine chain (Fig. 2.18). The development of the intermediate 2-(2-chlorodifluorovinyl)benzothiazole and related analogs is critical for the synthesis of these peculiar dyes. They can be obtained via Stille coupling reaction, which occurred between 2-iodobenzothiazole and (2-chlorodifluorovinyl)tributyltin. Next, N-ethyl-2-fluoromethylbenzothiazolium tetrafluoroborate and 2-(2-chlorodifluorovinyl)benzothiazole were mixed with p-diethylaminotoluene in the nitromethane (CH_3NO_2) for 10 min, and the product was precipitated in ether before purification.

The polymethine chain is also an ideal place to derivatize a spacer for bioconjugation. Particularly, there is a need to equip the dyes with handles for bioconjugation toward the development of probes for biomedical imaging. The halide moiety (F, Cl, Br) can readily be substituted with a nucleophile without a catalyst in aprotic solvents. In another development, the perturbation of the electronic effect on the polymethine bridge resulted in cyanine dyes with unusually large Stokes shift, which we will discuss more later.

When it comes to using the dyes for biological assay or in vivo imaging application, the solubility of the dyes in aqueous conditions remains the key criteria in the operation. Obviously, for a given excitation wavelength, cyanine dyes are less hydrophobic compared to rhodamine and other dyes, which rely mainly on vast aromatic ring scaffolds. Nevertheless, indoles, benzothiazoles, and oxazoles or other heterocyclic systems used in cyanine dyes alone cannot effectively bring the dyes into aqueous solution; consequently, aggregation and optical property reduction occur.

In most biomolecule-labeling studies, a significant portion of polar organic solvents (5–20%, such as acetonitrile, ethanol, DMF, or DMSO) was used along with aqueous diluents. This causes a spectrum of issues, including cellular toxicity or toxicology concerns, depending on whether the study involves cells or living subjects. One of the approaches to developing water-soluble cyanine dyes focuses on using commercially available heterocyclic starting materials with polar head groups, such as amine, alcohol, or sulfonate groups.

The underlying mechanism to enhance water solubility is to form hydrogen bonding. Although all these groups can form hydrogen bonding, only the sulfonate derivative has up to three oxygen atoms, thus making it the most polar group among the water-soluble enhancing moieties used in dye chemistry. The strong electronegativity of the oxygen atom on sulfonate can form strong dipole interaction with the hydrogen of water. Disulfonated and tetrasulfonated cyanine dyes were reported in response to these needs (Mujumdar et al. 1996). Notably, the tetrasulfonated cyanine dyes became among most favorite dyes for in vivo imaging applications.

Although cyanine dyes are small analogs, modification of the dye with two sulfonate groups is still insufficient to prevent aggregation in an aqueous solution. For example, it has been demonstrated that indocyanine green (ICG) dye, which has two sulfonate groups, regardless of that aggregation of ICG in the buffer is notorious, resulting in the quenching of ICG fluorescence. Further, the evidence of ICG aggregation is confirmed by the hypsochromic shift of absorbance lambda max from 780 nm to 700–720 nm, thus limiting its translation to in vivo work unless a further modification is necessary (Villaraza et al. 2010).

The tetrasulfonated near-infrared cyanine dyes were synthesized starting with the commercially available disodium salt of 6-amino-1,3-naphthalenedisulfonic acid (Fig. 2.19) (Mujumdar et al. 1996). However, due to the poor solubility of this starting material in acidic conditions, such as concentrated hydrochloric acid, the sodium-free material was generated after passing the disodium salt through a cation-exchange resin. The sodium-free precursor was treated with sodium nitrite in the presence of hydrochloric acid to convert the aromatic amine into hydrazine, which serves as a precursor for making an indole. After alkylating the amino group of the indole ring with 6-bromohexanoic acid, the key intermediate N-(carboxypentynyl)-1,1,3-trimethylbenzindolenimium 6,8-disulfonate went through a reaction with 1,3,3-trimethoxypropene to form the desired water-soluble cyanine dye.

Aside from designing fluorescent dyes with an intrinsic water-soluble property, some innovative works focus on fostering the effective delivery of hydrophobic dyes in the biological milieu. These dyes can be shielded from an aqueous environment to prevent micro-aggregation by simply encapsulating inside macromolecules. One of the initial reports showed the effectiveness of PEGylation to repurpose fluorescent dyes for biological applications. For instance, by conjugating the amine-reactive indocyanine green (ICG) dye with the unproportionally large PEG polymer with an average molecular weight of 3400 g/mol, ICG is rendered more water-soluble, and thus, the whole complex can be used for labeling with antibodies. The absorption spectra of the labeled PEGylated conjugates suggest the dye is more stable due to increased hydrophilicity and reduced aggregation (Villaraza et al. 2010). Another innovative approach exploits the versatility of nanotechnology to overcome

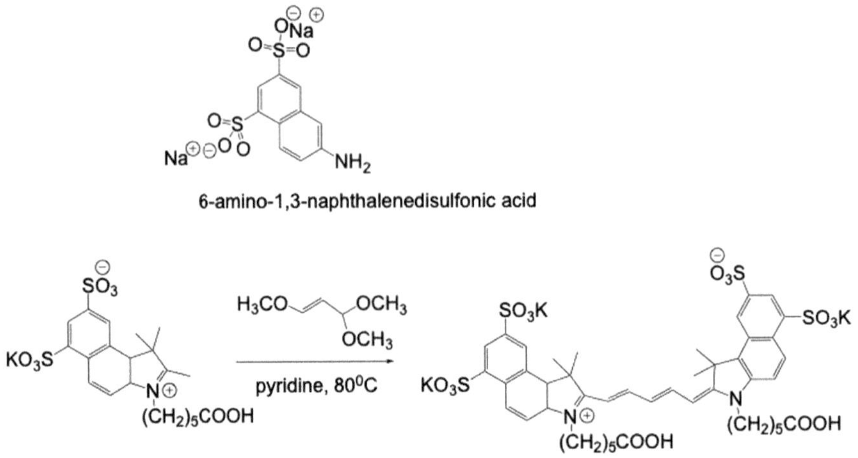

6-amino-1,3-naphthalenedisulfonic acid

H3CO   OCH3
OCH3
pyridine, 80°C
SO3
KO3S
N
(CH2)5COOH
SO3K
KO3S
N
(CH2)5COOH
O3S
SO3K
N
(CH2)5COOH

Fig. 2.19 Synthesis of water-soluble near-infrared cyanine dyes. Data obtained from Mujumdar et al. (1996) with permission from the American Chemical Society

the hydrophobicity of cyanine dyes for in vivo study. The work described the design of an ICG version flanked between the biodegradable dendrimer scaffold via covalent encapsulation. The probe reports an insight into the fate of nanoparticles in vivo via non-invasive whole-body fluorescence lifetime. As the dendritic shell biodegrades, the dye becomes exposed, enabling monitoring of fluorescence lifetime changes (Almutairi et al. 2008).

Fine-tuning the dyes to the near-infrared range with acceptable quantum efficiency for in vivo work does not just stop there with the extension of the polymethine bridge or increasing the hydrophilicity. More attention should also be focused on the choice of the counterions. And this might be the case for other types of dyes as well. The incorporation of appropriate counterions has fundamentally provided the dye with better photostability and absorption characteristics. For instance, in non-polar solvents, like dichloromethane, small and hard anion, such as bromide, is able to polarize the polymethine chain. Meanwhile, a softer anion does not polarize the conjugated chain on the same type of dye (Bouit et al. 2010). In stark contrast, in a polar solvent, the bulky soft anion of TRISPHAT or tetrakis(pentafluorophenyl)borate ($B(C_6F_5)_4^-$) maintains the best and ideal configuration of the polymethine bridge with minimal bond-length alternation. This suggests that anion/cation interactions may potentially control the electronic structures of the polymethine chain. Particularly, this counterion mechanism should be included in any design related to the encapsulation of dyes inside nanoparticles. The advantages of loading dyes into nanoparticles include improving tissue distribution and stability. The complex also emits an exceptional fluorescent signal. However, overloading can lead to aggregation and subsequent quenching. It has been demonstrated experimentally that $B(C_6F_5)_4^-$ pairing with rhodamine B octadecyl ester enables the dye encapsulation with minimal aggregation-caused quenching, resulting in 5–100-fold brighter than the corresponding quantum dots (Andreiuk et al. 2019).

2.3.1 Merocyanine Dyes

So far, we have discussed the development of cyanine dyes with an odd number of polymethine carbons to make Cy3, Cy5, and Cy7 dyes. How about cyanine dyes with an even number of methine carbons? Although less prevalent in biomedical imaging, it is possible to synthesize cyanine dyes with even numbers of methine carbons. For instance, the merocyanine dyes belong to this category (It is noteworthy that there are merocyanine dyes with an odd number of carbons on the polymethines as well). Several di-, tetra-, and hexamethinemerocyanines have been developed by the condensation of barbituric acids with heterocyclic analogs, such as indoles or benzothiazoles. In contrast to cationic cyanines, merocyanines are neutral dyes, and thus, they tend to have less degree of aggregation. Further, cyanine dyes are insensitive to solvent polarity, while the fluorescence property of merocyanines is environment-dependent; particularly, temperature, solvent polarity, and viscosity can form complexes or associates with dye molecules (Kulinich and Ishchenko 2009). Thus, they have been used as probes for chemical analysis but have less applications in molecular imaging due to inferior fluorescence quantum yields compared to other cyanine cousins. Nevertheless, they have been studied widely as photosensitizers for their implications in photodynamic therapy, probably due to their known long-lived triplet excited states.

Merocyanine dyes can be synthesized via a condensation reaction between an activated methylene group of a heterocyclic compound that serves as a nucleophile with a polymethine fragment containing the carbonyl groups or its synthetic equivalents serve as an electrophile (Kulinich and Ishchenko 2009). The synthesis of spiropyran exemplifies the general method for making merocyanine dyes. This dye has been the subject of widespread study in the past two decades due to its unique photochemical property. Spiropyran can be considered as zero methine cyanine or dimethine cyanine. When the dye locks into spiro structure, two end groups link directly; there is no methine carbon in between, color or fluorescence signal. However, in the open form, where the end groups are flanked by a dimethine, as with the merocyanine, the color is restored as well as the quantum yield. Because this dynamic conversion is attributed to controlled inputs, thus spiropyran can potentially be used in information processing or by acting as a logical on/off switch on nanodevices for high-technology applications. Using available starting materials, spiropyran was synthesized in a one-step reaction (Fig. 2.20).

Meanwhile, the synthesis of merocyanine with dimethine structure can be prepared in a one-step reaction involving the generation of a formylating reagent in situ using DMF in acetic anhydride (Fig. 2.21) (Wurthner 1999). The reaction

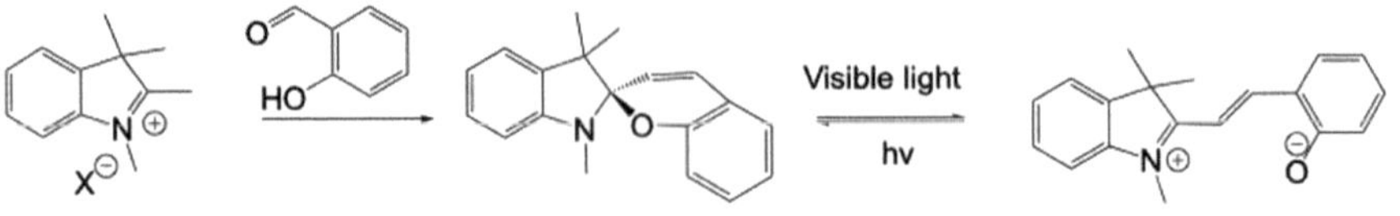

Fig. 2.20 Synthesis of spiropyran dye. Data obtained from Kulinich and Ishchenko (2009) with permission from Russian Chemical Reviews

proceeds through reactive enamines of the CH-acidic heterocycles, such as barbituric acid analogs, and the subsequent condensation with the electron-rich heterocycles or methylene bases, like the Fischer bases of the indole rings. This one-step versatile reaction can be used to generate a large repertoire of monomethine (if using CH-acidic heterocycles) or dimethine (Fischer indoles) merocyanine dyes. It is noteworthy that the reaction provided a very high product yield and purity. In most cases, the products can be purified by recrystallization in a mixture of water and alcohol.

This type of reaction must be carried out under strictly anhydrous conditions. In the presence of water, hydrolysis will occur, in which one of the ketones of the barbituric acid will be converted into alcohol. Alternatively, in some severe cases, deformylation took place, returning to the starting material. One another potential problem that may arise in this reaction condition is the acetylation of methylene base.

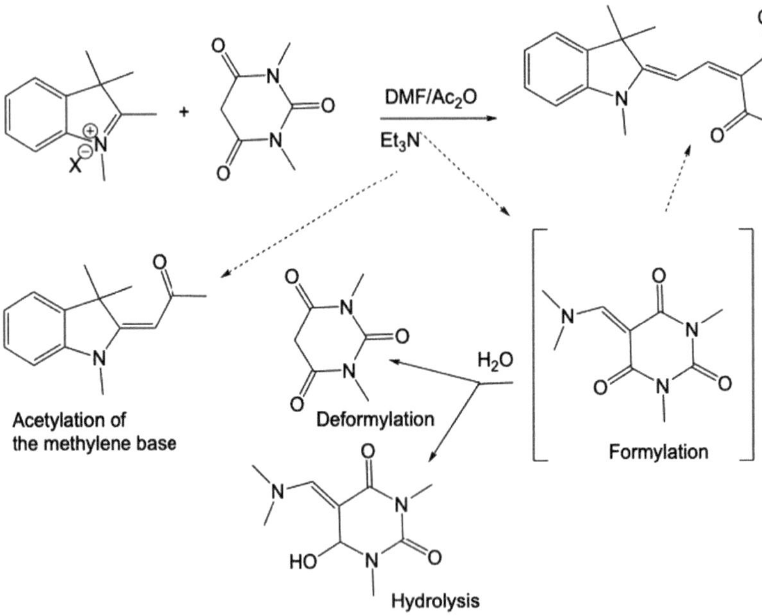

Fig. 2.21 One-pot synthesis of dimethine merocyanine dye via in situ-generated formylated intermediate and potential side products. Data adapted from Wurthner (1999)

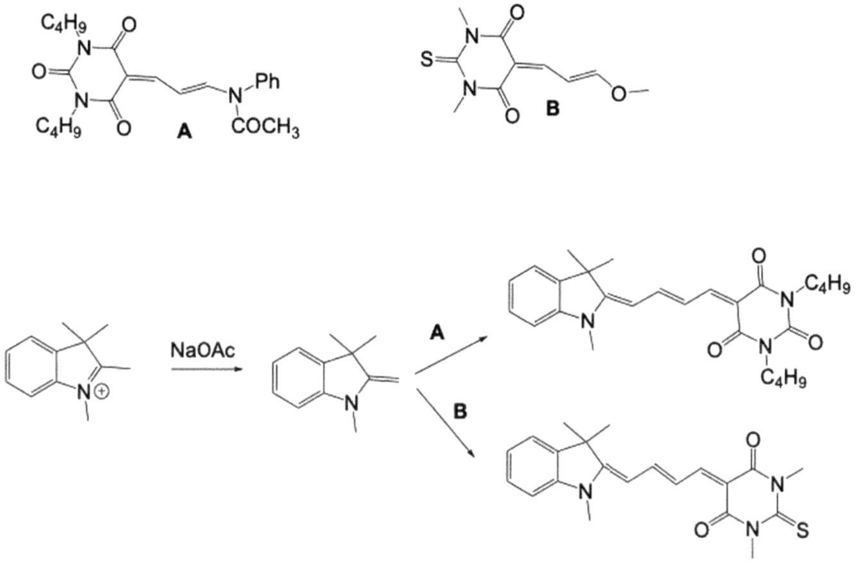

Fig. 2.22 Synthesis of longer polymethine intermediates for the development of NIR merocyanine dyes. Data obtained from Toutchkine et al. (2007) with permission from the American Chemical Society

For these reasons, this chemistry has not been very successful for the synthesis of tri- and tetramethine merocyanine dyes.

Like every common dye chemistry, shorter merocyanine dyes are easier to develop than longer versions. Asides from the unavailability of the starting materials, the synthesis of long polymethine merocyanines usually ended up with several side reactions. The bottleneck is the lack of robust vinylene chains of different lengths to be attached to the heterocyclic end structures. To overcome this issue, a number of innovative approaches focused on the chemistry generating barbituric acid-based vinylenes as synthons (Fig. 2.22) using N,N'-diphenyl formamidine or malonalde-hyde dianilide of other types of orthoesters for near-infrared merocyanine dyes (Ernst et al. 1989; Toutchkine et al. 2007). A longer methine bridge can be developed using longer vinylene structures, such as glutaconaldehyde anilides and similar analogs.

2.3.2 Strategies to Fine-Tune the Near-Infrared Capabilities of Cyanine Dyes

We discussed earlier the role of polymethine carbons of cyanine dyes in the extension of the absorbance wavelength, partly due to reduced bond alternation. It is apparent that the biomedical imaging field would benefit greatly if the dyes' optical parameter could be manipulated in the near-infrared region. In vivo imaging using near-infrared channels will improve tissue penetration and help reduce the interference caused by tissue autofluorescence, which mostly happens in the visible spectrum. In general, a few positions on cyanine dyes fit as candidates for modification. The chemical manipulations can focus on either expanding the backbone with additional aromatic rings on both ends of the dyes or extending the polymethine carbon chains. While the former approach sounds attractive given the availability of a vast amount of commercially available fused aromatic rings, supposedly can contribute to diversifying the cyanine dyes repertoire. However, one major issue associated with the large aromatic structures is the enhanced hydrophobicity, thus making in vivo application more problematic. Overall, the approach to extend the near-infrared feature for cyanine dyes via an extension of hydrophobic rings seems to be impractical. Aside from the hydrophobic issue, it had been shown that for every additional aromatic ring attached to the cyanine dye, the absorbance would experience a bathochromic wavelength shift of merely 20 nm (Fig. 2.23). In stark contrast, for every extra methine carbon increased in the vinylic system, the emission will benefit 100-nm bathochromic shift (Pham et al. 2004). As the length of the polymethine chain increases, this leads to dramatic changes in the electron distribution and molecular geometry when the width of the charge wave becomes comparable to the length of the π-system. Then, both charge-isomer forms are responsible for the change in the spectral characteristics of near-infrared dyes (Ko et al. 2005). Using this robust strategy, a significant number of near-infrared dyes were achieved, along with the incorporation of sulfonate groups to enhance the water solubility, thus preventing dye aggregation under labeling conditions. This is a very important design for cyanine dyes, which have a high degree of structural aggregation caused by the intrinsic property of cyanines with elongated chromophore structures with π-electron density along the polymethine chain (Bricks et al. 2017).

With the further extension of the polymethine bridge to a limit where such dyes can be achieved, the longest wavelength of a cyanine dye was reported at 1600 nm (Tolmachev et al. 1998). As far as the near-infrared imaging window (700–900 nm) is a concern, and if it has been taken care of, there is no need for developing a longer polymethine chain than Cy7 dyes. The reasons are (i) beyond the near-infrared window, signal interference is inevitable. Water absorption is notorious from 1000 nm and above; (ii) longer polymethine chain induces more rotation around the polymethine bonds, thus reducing stability and quantum yield. Further, the absorbance band intensity of cyanine dyes with vinylene groups > 5 drastically reduces along with broadening spectral bands as the result of the change in the electronic structure in the ground and excited states occurring in long cyanine dyes compared to shorter counterparts (Kachkovski et al. 2005). So far, the longest polymethine chain used in biomedical imaging has extended from 5 to 7 methine carbons. This includes the FDA-approved ICG dye, called Cy7 dye, for multiple clinical applications, including determining tumor margins, hepatic function and liver blood flow, myocardial perfusion, retinal perfusion and imaging biliary ducts, lymph nodes tracking or visualization of cervical and uterine tumors, and tissue perfusion in several surgical procedures, including arthritic diseases (Muller et al. 2013).

Fig. 2.23 The advantage of increasing the methine carbons versus aromatic rings to extend the dye in the near-infrared window. Data obtained from Nolting et al. (2012)

Ring system	Bridge Length (n)	λ_{abs}	λ_{em}
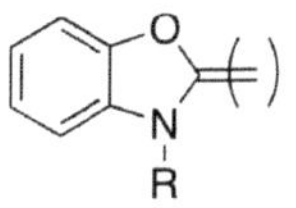	1	490	515
	2	585	620
	3	680	740
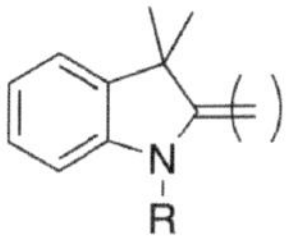	1	550	590
	2	650	690
	3	750	790
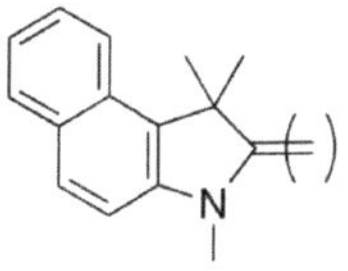	1	570	610
	2	680	720
	3	795	835
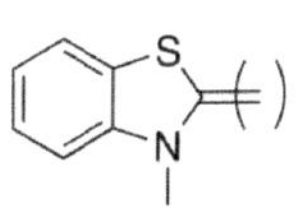	1	560	590
	2	660	690
	3	765	810
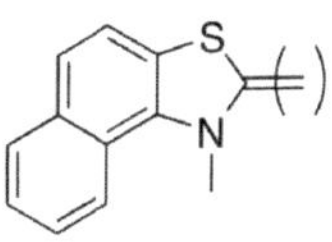	1	600	630
	2	700	740
	3	815	855

In order to explain the spectral broadening phenomenon of cyanine dyes with long polymethine chains, it is crucial to examine the solitonic wave of the polymethine bridge of which the electrons traverse alternatively with registered amplitude $\Delta q = q_\mu - q_{\mu+1}$, where μ is the number of the π-center (Kachkovski et al. 2005). As seen in Fig. 2.24, that describes the charge alternation Δq in the ground state of the electrons of the polymethine bridge of cyanine dyes; the electronic wave remains symmetrical when the number of the vinylene groups of the polymethine bridge is up to 4. However, longer dyes experience an unsymmetrical solitonic shape. In the case of a hypsochromic shift of long cyanine dyes, the routine solitonic electron waves distributing alternatively between the odd and even positions along the polymethine bridge, there is apparently a transition with the intramolecular transfer of the charge from one side of the dye molecule to the other. Such a transition, called the intramolecular charged transfer transitions, is associated with the widest and low-intensity spectral bands (Rettig 1994).

Spatial rotation is another problem associated with long cyanine dyes. To reduce rotation, particularly in Cy7 dyes, a cyclic polymethine system was introduced in the design. The addition of one, two, or three cyclic polymethines in the vinylog chain prevents the dye molecule from conformational transformations, which results in

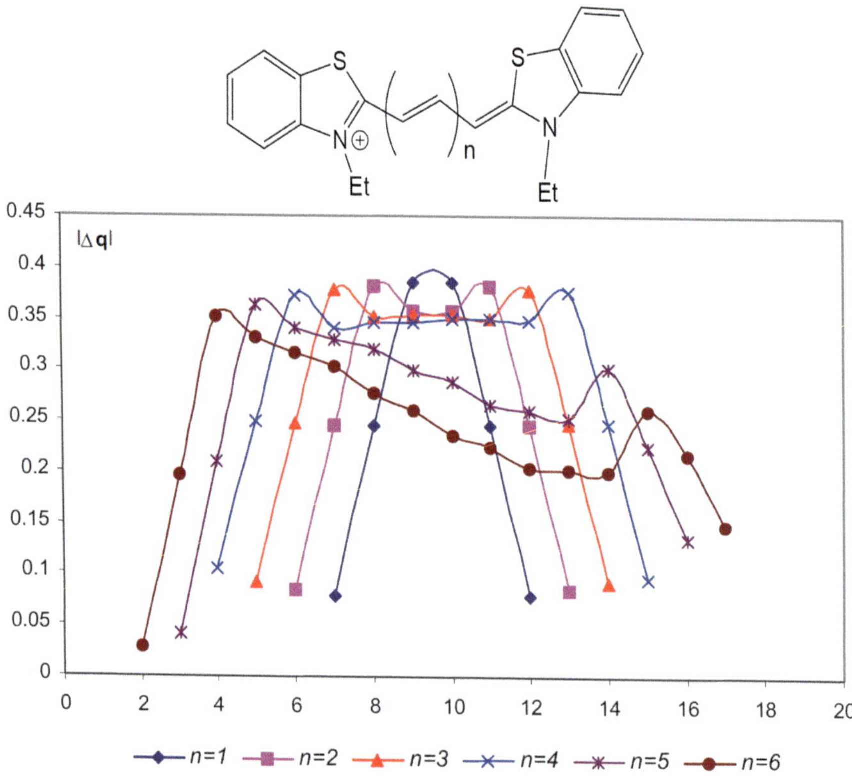

Fig. 2.24 Charge alteration in the ground state vinylogous series. Data obtained from Kachkovski et al. (2005) with permission from Elsevier

increased fluorescence quantum yield and registered bathochromic shift of spectra (Tolmachev 1998). Further, the rigidity of cyclic polymethine carbons reduces degradation in a severe environment. It has been demonstrated that Cy7 dye with a linear polymethine bridge cannot survive strong acidic conditions in solid-phase peptide chemistry. However, after modification of the dye with cyclic groups, the dyes survive and thus foster direct labeling of NIR dye in solid-phase synthesis to improve the production and scaling up (Pham et al. 2005). Although it is not applicable to all cases, as mentioned earlier, in general, the introduction of cyclic methine into the poly chain of cyanine dyes has the potential to increase the fluorescence quantum yield (Kulinich and Ishchenko 2009).

2.3.3 Development of Near-Infrared Cyanine Dyes with a Large Stokes Shift

Cyanine dyes possess very narrow absorption and emission bands due to their unique in forming J-aggregates over the broad spectral range, from visible to NIR region (Bricks et al. 2017). This type of aggregation strongly enhances the molar absorbance ε (expressed in M^{-1} cm^{-1}) compared to isolated monomers (Eisfeld and Briggs 2007). Some long polymethine-laden near-infrared cyanine dyes have the molar extinction coefficient equivalent to several million. Consequently, the Stokes shift of cyanine dyes is usually small, with merely 30 nm maximum; they are one of the narrowest spectral bands among reported fluorescence dyes. This has advantages and disadvantages. The use of narrow Stokes shift dyes with distinct excitation/emission parameters enables imaging multiple targets simultaneously. With the appropriate deployment of well cutoff filters, multiple excitations and emissions corresponding to different dyes can be collected exclusively representing different targets. The drawback is that not every filter set can provide a clear cutoff profile, and thus, signal bleaching is inevitable. In that case, it is intuitive to fine-tune the dyes with a larger Stokes shift. For example, as mentioned earlier, when the perturbation of the π-electron distribution along the polymethine vinylogs resulted in the transition of the intramolecular transfer of the charge from one side of the dye molecule to the other, leading to the significant widening of the long-wavelength spectral bands and, simultaneously, to considerable decreasing of the absorption band intensity (Kachkovski et al. 2005).

In a typical experiment, the nucleophilic substitution (S_{NR1}) at the central vinylogous halide carbon ($C(sp^2)$-X) by an amine group can occur at an elevated temperature (Fig. 2.25). Thus, this transition is usually characterized by the hypsochromic shift of the broader absorption spectrum, while the extension of the emission spectrum maintains pretty much in the near-infrared window (Pham et al. 2008). This unique phenomenon offered by cyanine dyes, exhibiting a super large Stokes shift, approximately from 150–180 nm, has a potential for multi-channel imaging applications.

It is worthwhile to mention that the shift of absorbance to a shorter wavelength is amenable with a complementary color, and thus, it is very convenient to monitor the progress of the reaction. Particularly, when the reaction completes, there is a conversion of color from green to blue, which corresponds to a hypsochromic shift of the dye from the initial wavelength of 780 nm to 600 nm.

2.4 Non-fluorescent Near-Infrared Dyes

A small group of dyes falls into the so-called dark fluorescence quencher category since they have the absorbance in the visible or near-infrared windows, albeit the

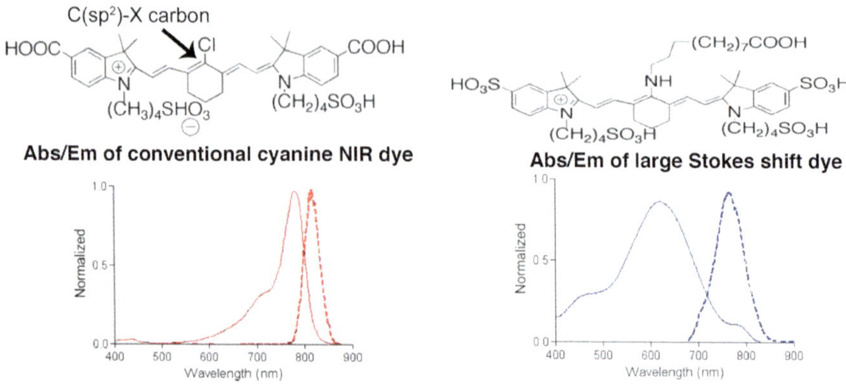

Fig. 2.25 Development of large Stokes shift cyanine dye. Data obtained from Pham et al. (2008)

excited photons cascade down from the S_1 excited levels to the ground state generating heat, not light. These dyes have extensive applications in bioanalytical analysis and recently in molecular imaging as a key component in the molecular reporter construct. The most recognition of quencher dyes is for sensing protease activities and in nucleic acid work. They are ideal for assembly into activated FRET (fluorescence resonance energy transfer) probes, serving as an electron sink. In fact, highly sensitive and quenchable fluorochromes have become essential for acquiring genomic and proteomic data from biological samples both for in vitro and in vivo applications. When tethering a fluorescently compatible dye and a quencher onto a biological-based backbone, the excited photons of the fluorescent dye transfer the energy to the quencher, resulting in a signal quenching effect. When the backbone is degraded by a protease, liberating the donor and the acceptor dyes beyond the FRET distance, the fluorescence signal will be detected. The most commonly used quenchers to pair with other dyes in FRET probe are hydrophobic 2,4-dinitrophenyl (DNP) and azobenzene dye, 4-(4'-dimethylaminobenzeneazo)benzoic acid (DABCYL), which have a broad absorption spectrum between 360 and 560 nm.

2.4.1 Azo Dyes

Azo dyes constitute a large family of hundreds of dyes, including monoazo dyes, diazo dyes, polyazo dyes, thioazole-azo dyes, and stilbene-azo dyes. These dyes have vast and traditional applications in the fabric dying industry and recently in biomedical research ranging from pH indicators to subcellular, cellular, and tissue staining. The majority of azo ($-N = N-$) compounds are characterized spectroscopically by a low-energy (n, π^*) state, which is manifested as a weak absorption band at long wavelengths. Further, azo dyes have very weak to no fluorescence in an aqueous environment because the (n, π^*) states of these compounds exhibit unusual coupling

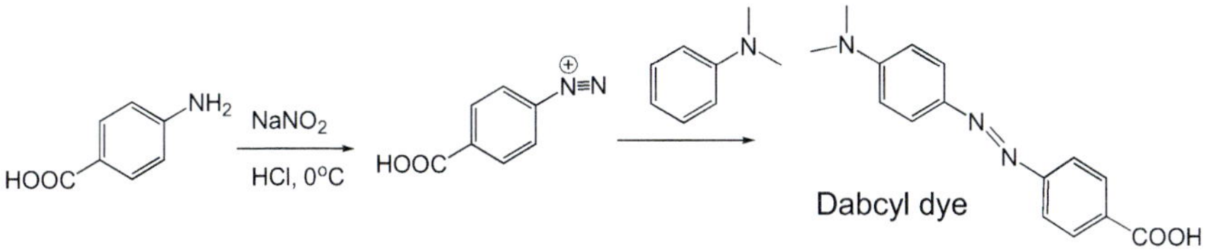

Fig. 2.26 Synthesis of Dabcyl dye. Data obtained from Kempf et al. (2017) with permission from the American Chemical Society

properties and absence of a vibrational structure in the $n \rightarrow \pi^*$ band (Rau 1973). The fluorescence of azo dyes can be restored when cooling the samples to the temperature of liquid nitrogen or embedded in rigid solvents (Barrow 1964). However, this is out of the scope of this discussion. Up to date, many different ways have been developed for the synthesis of azo compounds. The most general method includes diazo coupling reaction, which starts with the in situ generation of the arenediazonium moiety (ArN_2^+) (Fig. 2.26). Diazonium analog is rather a weak electrophile, which can undergo diazo coupling with only highly activated aromatic analogs, such as phenols (Ar–OH), anilines (Ar–NH$_2$), and N-alkylanilines (Ar–NHR or Ar–NR$_2$) via electrophilic aromatic substitution. The generated dyes can exist as cis- and trans-isomers, albeit the trans-isomers are usually more stable.

Azo compounds are sensitive in different pH conditions. Therefore, a large number of azo dyes serve as pH indicators. For instance, at pH higher than 6, the compound is dominant in the solution with yellow color (Fig. 2.27). However, as the pH becomes lower than 3, the protonated isomer with resonance structures will emerge with a red solution. As shown in the figure, under acidic conditions, the nitrogen of the azo group is protonated, the structure of which has a resonance isomer. As a result, the energy difference between the HOMO and LUMO of the protonated form is smaller than that of the neutral form, because the chemical structures of the resonance isomers have lower LUMO energy than the HOMO. This resulted in the change of the color in the protonated form.

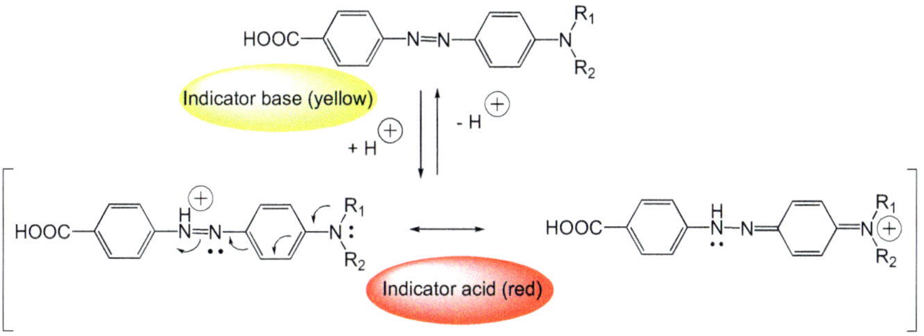

Fig. 2.27 Dynamically sensitive azo dyes in different pH conditions. Data adapted from Kahlert et al. (2016)

The chemistry for making Azo dyes is not only simple; the diazo coupling reaction can be performed with substituted aromatic rings, enabling the design of activated groups for labeling purposes. As shown in Fig. 2.28, the availability of the carboxylic group on Dabcyl dye is activated as a succinimide ester. While, in another derivative called black hole quencher (BHQ1), the alkylated hydroxyl linker on substituted amine serves as a site for creating an ester or phosphoester.

2.4.2 Azulene Dyes

Another dark hole quenchers with no reported fluorescence belongs to the azulene family (Fig. 2.29). It might be untrue to say that azulene does not have fluorescence. However, the right term would be that they have anomalous fluorescence. So far, all of our discussion about fluorescence dyes used in biomedical imaging is restricted to the $S_1 \rightarrow S_0 + h\nu$ fluorescence. In the case of azulene, it is well documented that excited photons are populated in the S_2 excited state, followed by $S_2 \rightarrow S_0$ fluorescence. This phenomenon is attributed to the large S_2-S_1 energy gap, which slows down the normal very rapid rate of $S_2 \rightarrow S_1$ internal conversion by decreasing the Franck–Condon factor for radiationless transitions (Turro 1991). The $S_1 \rightarrow S_0$ fluorescence of azulene is very negligible with quantum yield $< 10^{-4}$ and can be measured and detected under special conditions (Gillispie and Lim 1978).

From a chemistry point of view, azulene is also a unique molecule. As mentioned in Chap. 1, azulene is a bicyclo [5.3.0] system, consisting of 10 π electrons, the same found in naphthalene, which is amenable to the Huckel rule ($4n + 2$, with $n = 2$), thus making it qualified as an aromatic compound. As said, the π-electron cloud in the azulene molecule is unsymmetrical. The quantum–mechanical analysis suggested that the electronic density of azulene possessed a dipolar structure (Brown 1948). This reflects the tendency of the electronic distribution more on the 5-member ring rendering the cyclopentadienyl anion and the 7-member ring of tropylium cation (Fig. 2.29). Although such a separation of charges occurs only to a slight extent, it has a very remarkable effect on the physical and chemical properties of azulene (Volpin 1960). For instance, the cyclopentadiene of azulene readily undergoes electrophilic aromatic substitution, including Friedel–Crafts reaction, Mannich aminomethylations, condensation, and Vilsmeier formylations. On the other hand, the tropylium cation is suitable for use in nucleophilic reactions and is a subject of interest representing a class of non-benzenoid aromatic compounds (Nolting et al. 2009).

Several mechanisms have been derived for the synthesis of azulene given the high demand for this compound in many areas of high-technological research, such as optical recording materials. Azulene can be synthesized by dehydrogenation of bicyclo[5.3.0] decane or from various bicyclo[5.3.0]decenes in the presence of palladium catalysts (Volpin 1960) (Fig. 2.30). These reactions offer low yield, under harsh conditions, along with difficulty in acquiring the starting materials. Ziegler–Hafner methods emerged after that to improve the yield. The reaction involved annelation

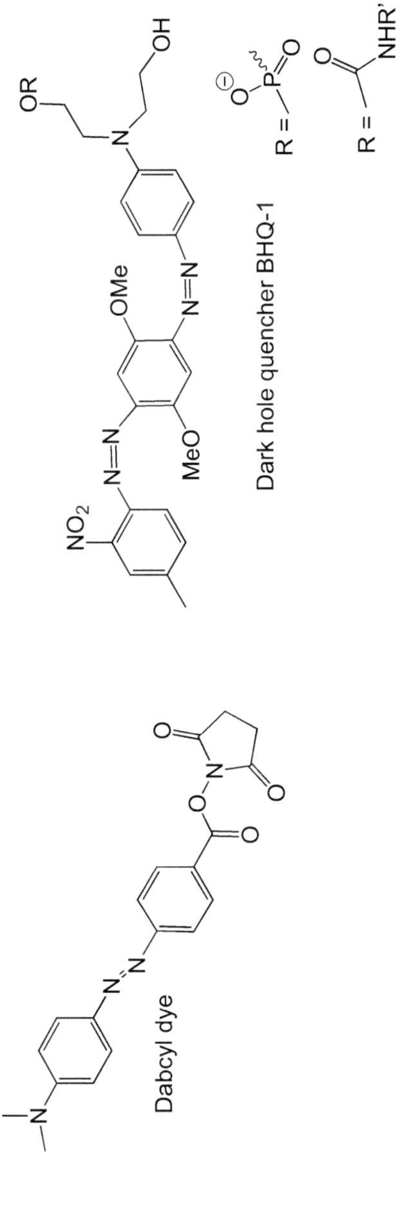

Fig. 2.28 Different modes of activation of azo dyes

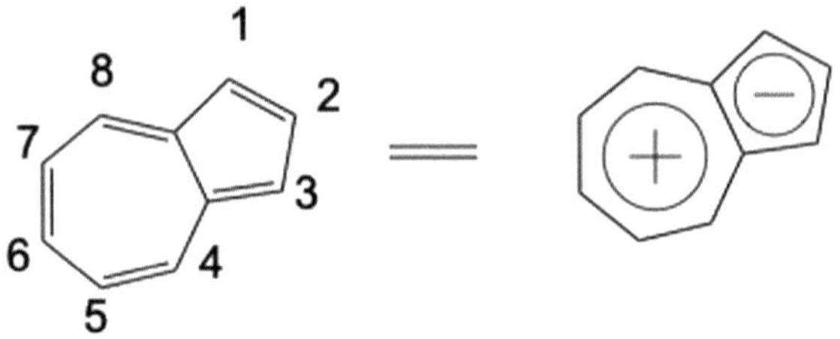

Fig. 2.29 Azulene molecule and its electronic distribution

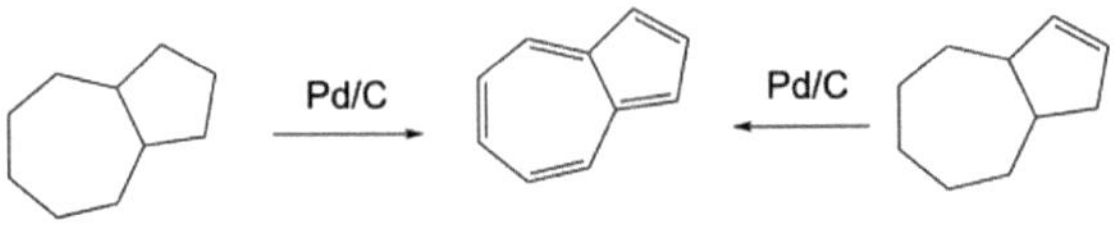

Fig. 2.30 Synthesis of azulene via hydrogenation of saturated precursors. Data obtained from Volpin (1960) with permission from Russian Chemical Reviews

of a 7-membered ring on a 5-membered ring via electrocyclic ring closure with elimination using vinylogous aminopentafulvenes (Jutz 1978). Another method to synthesize azulenes involves cycloadditions of aminopentafulvenes with activated 1,3-dienes through a [4+2] or [6+4] cycloaddition reaction (Sato et al. 1974). A more effective and high-yield synthesis of azulene has been reported using the Mannich reaction between cyclopentadiene and pentamethinecyanine (Hafner and Meinhardt 1984). Another short and effective approach for making azulene requires no dehydrogenation step (Scott et al. 1980). As shown in Fig. 2.31, treating key intermediate diazoketone with a catalytic amount of cuprous chloride resulted in a quick loss of nitrogen to provide bicyclic trienone in a decent yield. This formation was initiated by the intramolecular carbene addition to the 1 and 2 positions of a benzene ring to form the unstable norcaradiene, which transforms into a bicyclic trienone, identified in the reaction mixture. However, this intermediate isomerized the other bicyclic trienone, which forms azulene after dehydration (Fig. 2.31).

To generate more versatile azulene analogs for biomedical imaging with functional groups for conjugation, another reaction was developed in which a [8+2] cycloaddition between a lactone of methyl 2-oxo-2H-cyclohepta[b]furan-3-carboxylate with an in situ-generated vinyl ether (Nolting et al. 2009) (Fig. 2.32).

Most, if not all of the quenchers emit the signal in the visible range. This generates an opportunity for azulene to show its versatility due to possessing a dual ring system, which offers distinct chemical properties. The chemical modifications between the 5- and 7-membered rings can be achieved orthogonally, without interference, and thus, no protection/deprotection steps are necessary. For example, with the π-electronic distribution as of an aromatic system, thus derivatized polymethine carbons flanking between azulene is possible in order to extend the absorbance deep into the infrared range. The 7-membered ring could be modified to afford a handle for bioconjugation.

As the name suggested, azulene is a deep blue color compound, sensitive to the electronic effect. And thus, the progress of the azulene reaction can be monitored

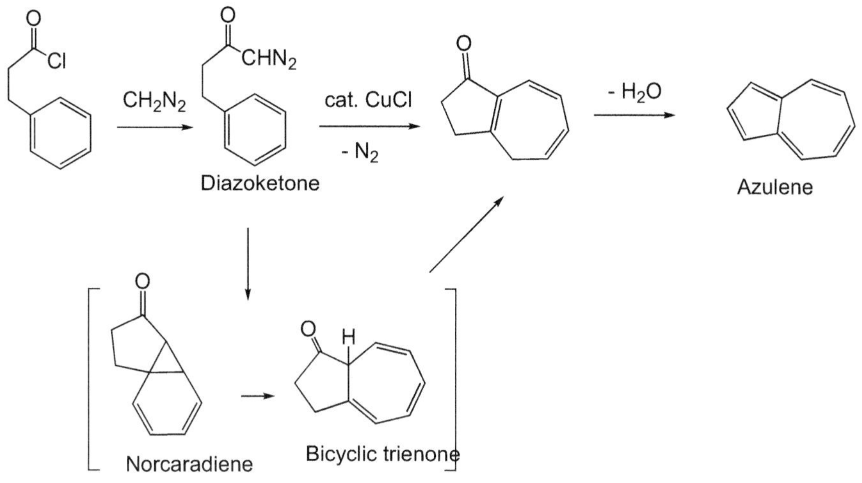

Fig. 2.31 Synthesis of azulene via intramolecular carbene addition mechanism. Data obtained from Scott et al. (1980) with permission from the American Chemical Society

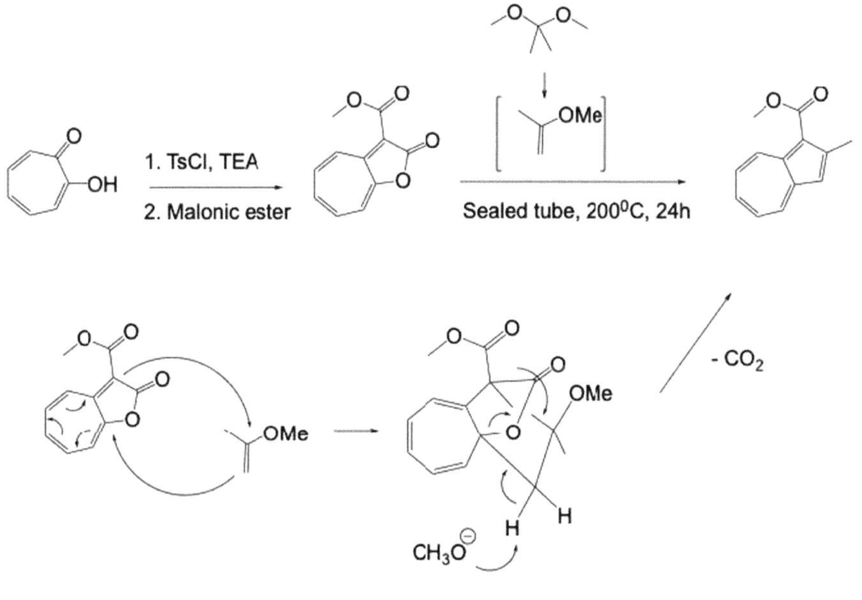

Fig. 2.32 Synthesis of azulene with functional group for bioconjugation using cycloaddition reaction. Data obtained from Nolting et al. (2009).

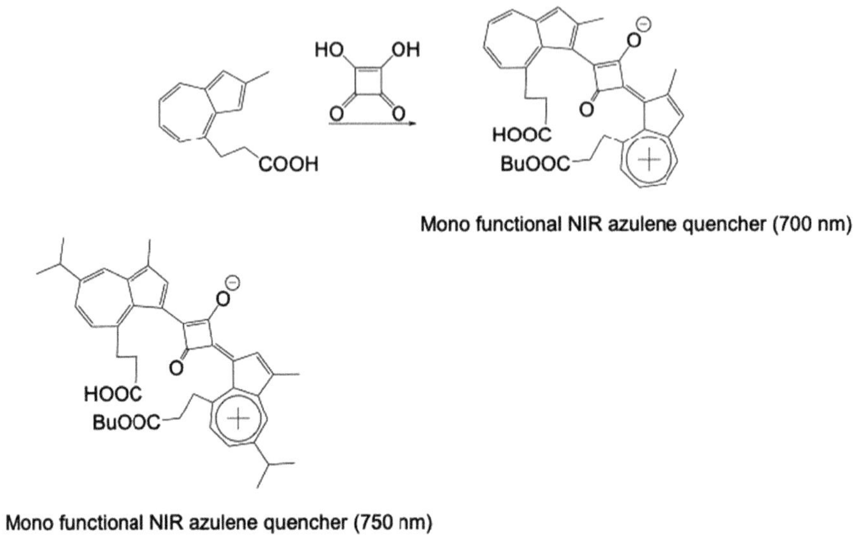

Fig. 2.33 NIR Azulene-based squaraine dyes. Data obtained from Pham et al. (2002)

easily as the color transition from one to another. So far, a whole spectrum of color related to different visible azulene analogs has been reported. Due to its quenching effect, azulene is proposed as an ideal acceptor in FRET work. In order to couple azulene with near-infrared donor dyes in FRET probes, a number of NIR azulene analogs have been developed. As shown in Fig. 2.33, a one-step reaction involving condensation of azulene with squaric acid will provide a NIR azulene dye (700 nm) (Pham et al. 2003). Further, work reported that by using the same procedure but using guaiazulene, the resulting dye had a bathochromic shift of 50 nm compared to azulene. It is likely that the electronic donating isopropyl group on the guaiazulene ring contributes to the redshift (Pham et al. 2002).

2.5 BODIPY Dyes

BODIPY (boron-dipyrromethene) dyes are another popular family of dyes used in biomedical imaging. Since the first discovery reported in 1968, thousands of BODIPY derivatives have been synthesized for various applications (Boens et al. 2012). BODIPYs are very robust dyes that have wide applications from materials chemistry to PDT (photodynamic therapy) and molecular imaging. This family of dyes has a strong light absorption, small Stokes shift, and they emit with high quantum yield along with good photostability. Further, low toxicity profiles make them ideal for in vivo applications. The main backbone of BODIPY was first reported by Treibs and Kreuzer by accident (Treibs and Kreuzer 1968). Their initial goal was to make the acylated pyrroles by treating pyrrole with acetic anhydride under the presence of

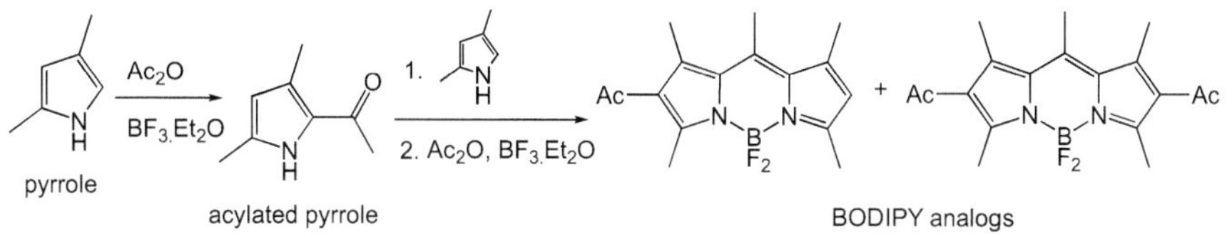

Fig. 2.34 The first synthesis of BODIPY by accident. Data adapted from Treibs and Kreuzer (1968)

boron trifluoride-a Lewis acid as a catalyst. Surprisingly, the product they observed turned out to be a very strong fluorescent molecule, now called BODIPY. So, it turned out that the acylated pyrrole went on reacting with pyrrole to form dipyrrin through a condensation reaction. Then, intermediate dipyrrin was stabilized by complexation with boron difluoride (Fig. 2.34).

The current synthesis of BODIPY dyes (Loudet and Burgess 2007; Ulrich et al. 2008) is simplified in a one-pot acid-catalyzed condensation reaction between pyrrole and a substituted aldehyde or ketones at room temperature, to afford a dipyrromethane (Fig. 2.35). This intermediate is very unstable due to sensitivity with air, light, and acid. Then, dehydrogenation using 2,3-dichloro-5,6-dicyano-1,4-benzoquinone (DDQ) or p-chloranil is performed to convert dipyrromethane to dipyrromethene. It is followed by the removal of a proton from the pyrrole by an organic base, such as trimethylamine (TMA) in preparation to complex with boron trifluoride etherate when it is added to a stirring reaction. The formation of the product is very high yield, and it can be visualized and monitored by the change in reaction color. Depending on the use of substituent on the pyrroles, BODIPY dyes usually have a spectrum of different colors ranging from green to pink, orange, and magenta.

For biological application, BODIPY dyes were functionalized with different activated groups for conjugation. Fortunately, the intrinsic nature of the convergent synthesis approach enables the facile attachment of modified pyrroles via condensation. Further, it is also possible to activate the dyes using *meso* substituents. Molecular Probes first developed creative chemistry for the development of succinimidyl BODIPY for biology work (Fig. 2.36). The approach involved the acid-catalyzed condensation of an ester-laden pyrrole with another pyrrole aldehyde to form a dipyrromethene, from which complexation with borofluoride etherate to form the dye. Saponification to remove the ester, followed activation of the free carboxylic acid, using DCC and NHS to provide the amino activated BODIPY dye.

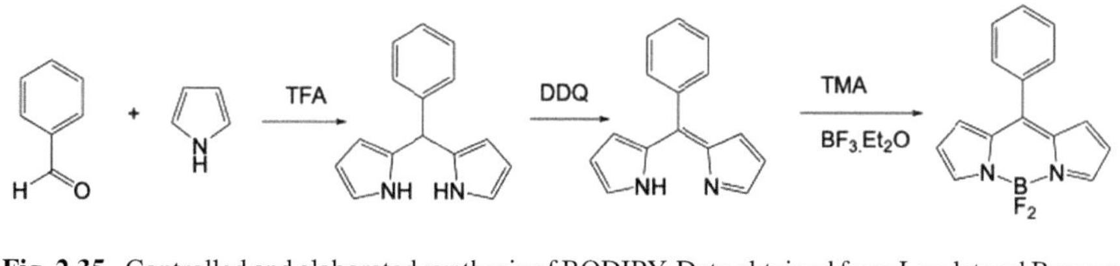

Fig. 2.35 Controlled and elaborated synthesis of BODIPY. Data obtained from Loudet and Burgess (2007) with permission from the American Chemical Society

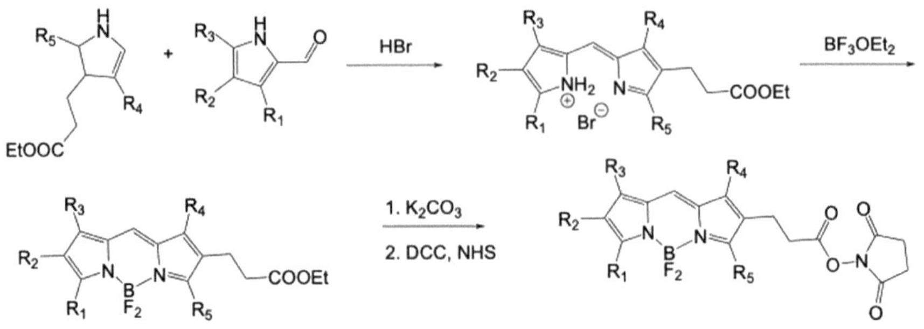

Fig. 2.36 Development of amine-reactive BODIPY. Data obtained from Rezende and Silva (2013) with permission from the author

The more creative synthesis of amino reactive BODIPY dyes using different groups covering a spectrum of alternatives to overcome any potential labeling drawbacks (Fig. 2.37). For example, replacing succinimide ester with isothiocyanate enables conjugating the dyes to peptides or other materials, which usually required harsh conditions, in which succinimide ester would be hydrolyzed.

Other major BODIPY dyes were developed for thiol-reactive biomaterials. The advantages of labeling protein, peptides, or biological materials via maleimide include the following: (i) The reaction proceeds in very mild condition, like the one in the physiological environment (pH7.4), and thus suitable for unstable proteins that may denature at elevated pH conditions; (ii) this reaction requires a strong nucleophile like sulfhydryl group present in cysteine, while it is inert in the presence of amino or hydroxyl group; no protection of these groups is needed while labeling. Exploiting the unique differences between succinimide and maleimide groups for labeling work, particularly the orthogonal coupling strategy generated from this approach, enables complicated manipulation of proteins, peptides, and other reactions. As a matter of fact, the first maleimide BODIPY dye was also designed using this strategy (Haugland et al. 1993), in which the modified version of BODIPY with a short amino-terminated linker tethered asymmetrically was let to react with a bifunctional succinimidyl-maleimidal group to afford the maleimidal BODIPY. While there are many other creative ways to design activated BODIPY dyes with different functional groups, including alkyne, azide, vinyl group, etc., for bioconjugations reported elsewhere, the current trend in this area focuses mostly on how to make very stable dyes and those that can extend to the near-infrared window. It sounds simple as it is, but the reality is that to fine-tune the dyes with redshift, it is necessary to extend the conjugated system; for example, by replacing the alkyl substituents on each pyrrole nucleus with an aryl group, the dye experienced a remarkable bathochromic shift of over 100 nm, albeit the dye suffers low quantum yield θ_F. It is apparent now, to develop stable BODIPY, one would have to identify where is the source that makes the dye vulnerable to degradation or poor fluorescence quantum yield. One of the main mechanisms that reduce fluorescence quantum yield and stability might relate to the non-radiative decay through rotational relaxation. So, similar to the case

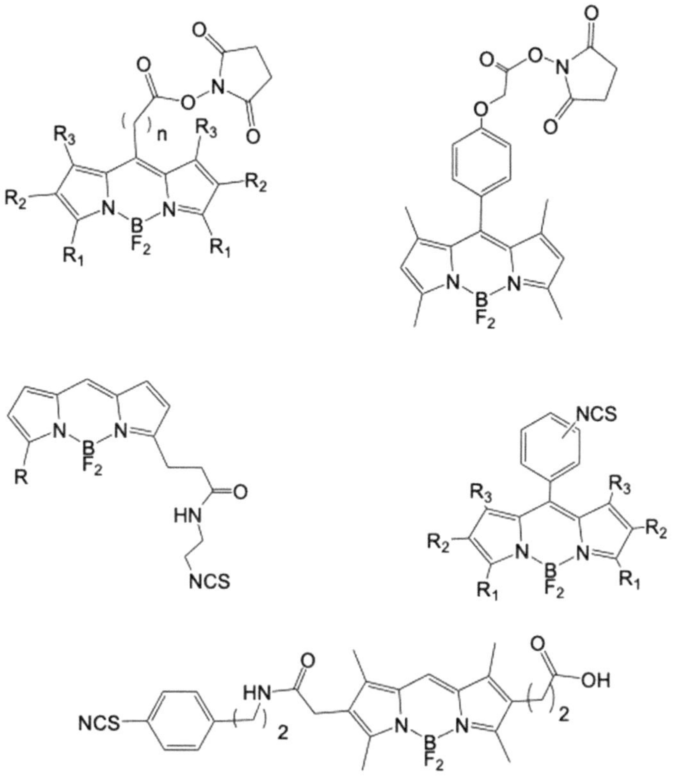

Fig. 2.37 Activated BODIPY dyes for amino-terminal ligands

of cyanine and rhodamine dyes, stable NIR BODIPY dyes can be achieved with the incorporation of chemical structures that facilitate the overall conformationally constrained dyes.

2.6 Activated Dyes

Another innovative chemistry fostering the translation of dyes for biomedical imaging is the activation of these molecules with a variety of functional groups to enable the conjugation of the fluorescent dyes with biological materials via wet-lab/conjugation chemistry (Fig. 2.38).

The uniqueness of this chemistry centers around the art of designing activated groups stable enough for isolation, purification shelving long term for versatile application. These activated dyes have been reported for labeling antibodies, small organic ligands, peptides, avidin, DNA, lipids, polymers, and other materials.

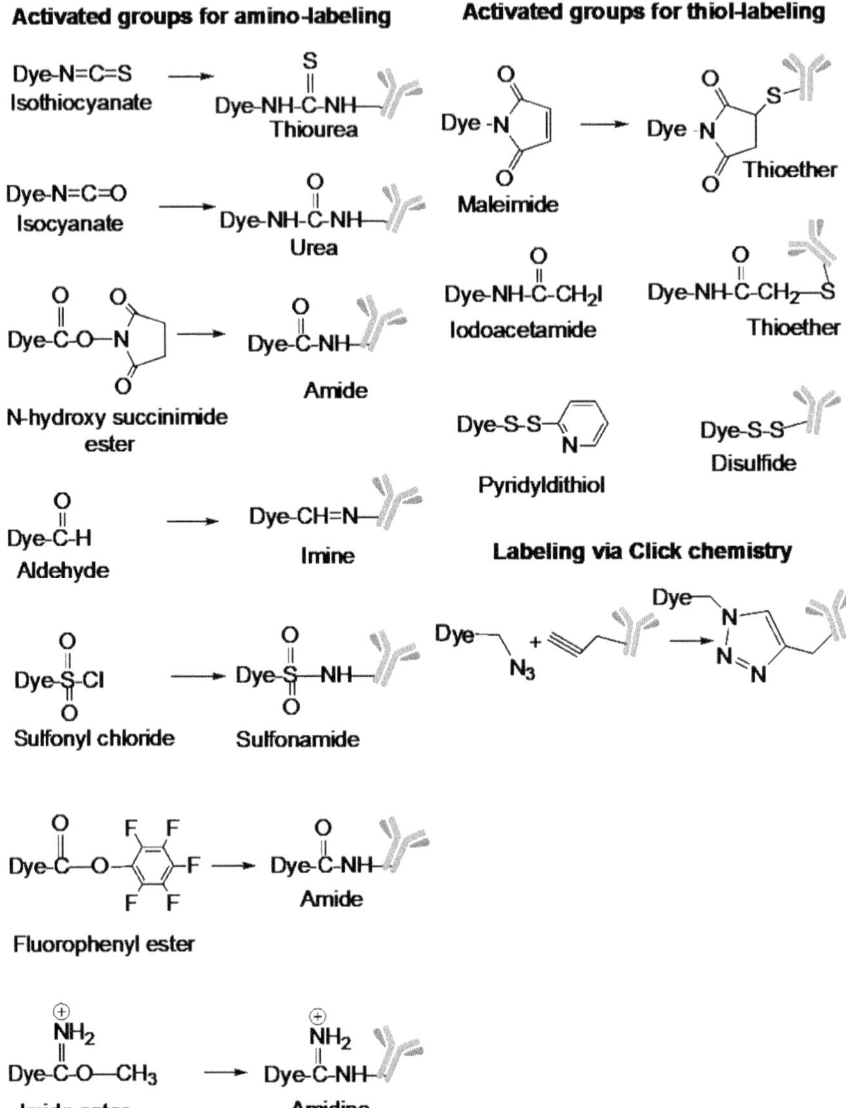

Fig. 2.38 Activated dyes for labeling biological materials

2.7 Conclusion

Although the chemistry for the development of dyes has been developed for over 200 years, the more advanced versions started to make their ways to in vivo and some in clinical applications recently. Thus, there are still a lot of rooms for new discovery. The dynamics of fluorescence technology underlies many innovative and activatable probes that will be discussed in Chap. 6. Now, we move our discussion to the development of radiopharmaceutical probes.

References

M. Adamczyk, J. Grote, Efficient synthesis of rhodamine conjugates through the 2'-position. Bioorg. Med. Chem. Lett. **10**, 1539–1541 (2000)

A. Almutairi, W.J. Akers, M.Y. Berezin, S. Achilefu, J.M. Frechet, Monitoring the biodegradation of dendritic near-infrared nanoprobes by in vivo fluorescence imaging. Mol. Pharm. **5**, 1103–1110 (2008)

B. Andreiuk, A. Reisch, E. Bernhardt, A.S. Klymchenko, Fighting aggregation-caused quenching and leakage of dyes in fluorescent polymer nanoparticles: universal role of counterion. Chem. Asian J. **14**, 836–846 (2019)

J. Autschbach, Why the particle-in-a-box model works well for cyanine dyes but not for conjugated polyenes. J. Chem. Educ. **84**, 1840–1845 (2007)

G.M. Barrow, *The Structure of Molecules: An Introduction to Molecular Spectroscopy* (W. A. Benjamin, Inc., New York, 1964)

M. Beija, C.A. Afonso, J.M. Martinho, Synthesis and applications of Rhodamine derivatives as fluorescent probes. Chem. Soc. Rev. **38**, 2410–2433 (2009)

R. Birke, J. Ast, D.A. Roosen, B. Mathes, K. Roßmann, C. Huhn, B. Jones, M. Lehmann, V. Haucke, D.J. Hodson, J. Broichhagen, Sulfonated rhodamines as impermeable labelling substrates for cell surface protein visualization. bioRxiv, 2021.2003.2016.435698 (2021)

R. Birke, J. Ast, D.A. Roosen, J. Lee, K. Rossmann, C. Huhn, B. Mathes, M. Lisurek, D. Bushiri, H. Sun, B. Jones, M. Lehmann, J. Levitz, V. Haucke, D.J. Hodson, J. Broichhagen, Sulfonated red and far-red rhodamines to visualize SNAP- and Halo-tagged cell surface proteins. Org. Biomol Chem. (2022)

N. Boens, V. Leen, W. Dehaen, Fluorescent indicators based on BODIPY. Chem. Soc. Rev. **41**, 1130 (2012)

P.A. Bouit, C. Aronica, L. Toupet, B. Le Guennic, C. Andraud, O. Maury, Continuous symmetry breaking induced by ion pairing effect in heptamethine cyanine dyes: beyond the cyanine limit. J. Am. Chem. Soc. **132**, 4328–4335 (2010)

J.L. Bricks, Y.L. Slominskii, I.D. Panas, A.P. Demchenko, Fluorescent J-aggregates of cyanine dyes: basic research and applications review. Methods Appl Fluoresc **6**, 012001 (2017)

R.D. Brown, Trans. Faraday Soc. **44**, 984 (1948)

T.L. Chang, H.C. Cheung, Solvent effects on the photoisomerization rates of the zwitterionic and the cationic forms of rhodamine B in protic solvents. J. Phys. Chem. **96**, 4874–4878 (1992)

R. Coico, Gram staining. Curr. Protoc. Microbiol. (2005). **Appendix 3**, Appendix 3C

A. Eisfeld, J. Briggs, The shape of the J-band of pseudoisocyanine. Chem. Phys. Lett. **446**, 354–358 (2007)

L.A. Ernst, R.K. Gupta, R.B. Mujumdar, A.S. Waggoner, Cyanine dye labeling reagents for sulfhydryl groups. Cytometry **10**, 3–10 (1989)

C.L. Gaspar, I. Baldea, I. Panea, Kinetics of the formation of hemicyanine dyes by the condensation of Fischer's base aldehyde with anilines. Dyes Pigm. **69**, 45–53 (2006)

G.D. Gillispie, E.C. Lim, Vibrational analysis of the $S_2 \rightarrow S_1$ fluorescence of azulene in a naphthalene mixed crystal at 4.2 K. J. Chem. Phys. **68**, 4578–4586 (1978)

J. Grimm, L.D. Lavis, Synthesis of rhodamines from fluoresceins using Pd-catalyzed C-N cross-coupling. Org. Lett. **13**, 6354–6357 (2011)

J.B. Grimm, T.A. Brown, A.N. Tkachuk, L.D. Lavis, General synthetic method for Si-fluoresceins and Si-rhodamines. ACS Cent. Sci. **3**, 975–985 (2017)

K. Hafner, K.P. Meinhardt, Organic Synth. **7**, 15 (1984)

F.M.Q. Hamer, Rev. Chem. Soc. **4**, 327–355 (1950)

R.P. Haugland, H.C. Kang, US Patent 5,248,782 (1993)

W.N. Hunter, Rational drug design: a multidisciplinary approach. Mol. Med. Today **1**(31), 34 (1995)

B. Jedrzejewska, J. Kabatc, M. Pietrzak, J. Paczkowski, Hemicyanine dyes: synthesis, structure and photophysical properties. Dyes Pigm. **58**, 47–58 (2003)

J.C. Jutz, Top. Curr. Chem. **73**, 125 (1978)

O.D. Kachkovski, O.I. Tolmachov, Y.L. Slominskii, M.O. Kudinova, N.O. Derevyanko, O.O. Zhukova, Electronic properties of polymethine systems 7: soliton symmetry breaking and spectral features of dyes with a long polymethine chain. Dyes Pigm. **64**, 207–216 (2005)

H. Kahlert, G. Meyer, A. Albrecht, Colour maps of acid-base titrations with colour indicators: how to choose the appropriate indicator and how to estimate the systematic titratin errors. ChemTexts **2**, 7 (2016)

T. Karstens, K. Kobs, Rhodamine B and rhodamine 101 as reference substances for fluorescence quantum yield measurements. J. Phys. Chem. **84**, 1871–1872 (1980)

H. Kato, K. Komagoe, Y. Nakanishi, T. Inoue, T. Katsu, Xanthene dyes induce membrane permeabilization of bacteria and erythrocytes by photoinactivation. Photochem. Photobiol. **88**, 423–431 (2012)

O. Kempf, K. Kempf, R. Schobert, E. Bombarda, Hydrodabcyl: a superior hydrophilic alternative to the dark fluorescence quencher dabcyl. Anal. Chem. **89**, 11893–11897 (2017)

D. KO, O.I. Tolmachov, Y.L. Slominskii, M.O. Kudinova, N.O. Derevyanko, O.O. Zhukova, Electronic properties of polymethine systems 7: soliton symmetry breaking and spectral features of dyes with a long polymethine chain. Dyes Pigm. **64**, 207–216 (2005)

A.V. Kulinich, A.A. Ishchenko, Merocyanine dyes: synthesis, structure, properties and applications. Russ. Chem. Rev. **78**, 141 (2009)

A. Loudet, K. Burgess, Bodipy dyes and their derivatives: syntheses and spectroscopic properties. Chem. Rev. **107**, 4891–4932 (2007)

J.V. McCullagh, K.A. Daggett, Synthesis of triarylmethane and xanthene dyes using electrophilic aromatic substitution reactions. J. Chem. Educ. **84**, 1799–1802 (2007)

R.B. Mujumdar, L.A. Ernst, S.R. Mujumdar, C.J. Lewis, A.S. Waggoner, Cyanine dye labeling reagents: sulfoindocyanine succinimidyl esters. Bioconjug. Chem. **4**, 105–111 (1993)

S.R. Mujumdar, R.B. Mujumdar, C.M. Grant, A.S. Waggoner, Cyanine-labeling reagents: sulfobenzindocyanine succinimidyl esters. Bioconjug. Chem. **7**, 356–362 (1996)

J. Muller, A. Wunder, K. Licha, Optical imaging. Rec. Results Cancer Res. **187**, 221–246 (2013)

N. Narayanan, G. Patonay, A New method for the synthesis of heptamethine cyanine dyes: synthesis of new near-infrared fluorescent labels. J. Org. Chem. **60**, 2391–2395 (1995)

D.D. Nolting, M. Nickels, R. Price, J.C. Gore, W. Pham, Synthesis of bicyclo[5.3.0]azulene derivatives. Nat. Protoc. **4**, 1113–1117 (2009)

D.D. Nolting, J.C. Gore, W. Pham, NEAR-INFRARED DYES: probe development and applications in optical molecular imaging. Curr. Org. Synth. **8**, 521–534 (2011)

D.D. Nolting, M.L. Nickels, N. Guo, W. Pham, Molecular imaging probe development: a chemistry perspective. Am. J. Nucl. Med. Mol. Imaging **2**, 273–306 (2012)

W. Pham, R. Weissleder, C.-H. Tung, An azulene dimer as a near-infrared quencher. Angew. Chem. Int. Ed. **41**, 3659–3662 (2002)

W. Pham, R. Weissleder, C.-H. Tung, A practical approach for the preparation of monofunctional azulenyl squaraine dye. Tetrahedron Lett. **44**, 3975–3978 (2003)

W. Pham, Y. Choi, R. Weissleder, C.H. Tung, Developing a peptide-based near-infrared molecular probe for protease sensing. Bioconjug. Chem. **15**, 1403–1407 (2004)

W. Pham, Z. Medarova, A. Moore, Synthesis and application of a water-soluble near-infrared dye for cancer detection using optical imaging. Bioconjug. Chem. **16**, 735–740 (2005)

W. Pham, L. Cassell, A. Gillman, D. Koktysh, J.C. Gore, A near-infrared dye for multichannel imaging. Chem. Commun. (Camb) 1895–1897 (2008)

H. Rau, Spectroscopic properties of organic azo compounds. Angew. Chem. Int. Ed. **12**, 224–235 (1973)

C. Reichardt, Justus Liebigs Ann. Chem. **715**, 74 (1968)

W. Rettig, Top. Curr. Chem. **169**, 253–299 (1994)

L.C.D. de Rezende, F. da Silva Emery, A review of the synthetic strategies for the development of BODIPY dyes for conjugation with proteins. Orbital Electron. J. Chem. **5** (2013)

S. Sasaki, G.P.C. Drummen, G. Konishi, Recent advances in twisted intramolecular charge transfer (TICT) fluorescence and related phenomena in materials chemistry. J. Mater. Chem. **4**, 2731 (2016)

M. Sato, S. Ebine, J. Tsunetsugu, Cycloaddition of 6-aminofulvenes with coumalic esters, a novel azulene formation. Tetrahedron Lett. **32**, 2769–2774 (1974)

L.T. Scott, M.A. Minton, M.A. Kirms, A short new azulene synthesis. J. Am. Chem. Soc. **120**, 6311–6314 (1980)

E. Stathatos, P. Lianos, A. Laschewsky, Ouari O, P. C (2001) Synthesis of a hemicyanine dye bearing two carboxylic groups and its use as a photosensitizer in dye-sensitized photoelectrochemical cells. Chem. Mater. **13**, 3888–3892 (2001)

G.G. Stokes, On the change of refractibility of light. Philos. Trans. R Soc. London **142**, 463–562 (1852)

L.F. Tietze, T. Eicher, *Reactions and Syntheses* (University Science Books, Mill Valley, California, 1988)

M. Titford, Comparison of historic Grubler dyes with modern counterparts using thin layer chromatography. Biotech. Histochem. **82**, 227–234 (2007)

A.I. Tolmachev, New cyanine dyes absorbing in the NIR region, in *Near-Infrared Dyes for High Technology Applications* (1998), pp. 385–415

A.I. Tolmachev, Y.L. Slominskii, A.A. Ishchenko, *Near-Infrared Dyes for High Technology Applications* (Kluwer Academic Publisher, 1998), pp. 384–515

A. Toutchkine, D.V. Nguyen, K.M. Hahn, Simple one-pot preparation of water-soluble, cysteine-reactive cyanine and merocyanine dyes for biological imaging. Bioconjug. Chem. **18**, 1344–1348 (2007)

A. Treibs, F. Kreuzer, Justus Liebigs Ann. Chem. **718**, 208–223 (1968)

N.J. Turro, *Modern Molecular Photochemistry* (1991)

G. Ulrich, R. Ziesssel, A. Harriman, The chemistry of fluorescent Bodipy dyes: versatility unsurpassed. Angew. Chem. Int. Ed. **47**, 1184–1201 (2008)

A.J. Villaraza, D.E. Milenic, M.W. Brechbiel, Improved speciation characteristics of PEGylated indocyanine green-labeled Panitumumab: revisiting the solution and spectroscopic properties of a near-infrared emitting anti-HER1 antibody for optical imaging of cancer. Bioconjug. Chem. **21**, 2305–2312 (2010)

M.E. Volpin, Non-benzenoid aromatic compounds and the concept of aromaticity. Russ. Chem. Rev. **29**, 129–160 (1960)

F. Wurthner, DMF in acetic anhydride: a useful reagent for multiple-component syntheses of merocyanine dyes. Synthesis **12**, 2103–2113 (1999)

L.M. Yagupolskii, N.V. Kondratenko, O.I. Chernega, A.N. Chernega, S.A. Buth, Y.L. Yagupolskii, A novel synthetic route to thiacyanine dyes containing a perfluorinated polymethine chain. Dyes Pigm. **79**, 242–246 (2008)

X. Zhang, S. Bloch, W. Akers, S. Achilefu, Near-infrared molecular probes for in vivo imaging. Curr. Protoc. Cytom. (2012). **Chapter 12**, Unit12 27

G.M. Ziarani, R. Moradi, N. Lashgari, H.G. Kruger, Chapter 10—Fluorescein dyes, in *Metal-free Synthetic Organic Dyes* (2018), pp. 165–170

Chapter 3
Principles for the Design of PET Probes

3.1 Introduction

Positron emission tomography (PET) is an analytical imaging technology that uses compounds labeled with positron-emitting radioisotopes as molecular probes to image and measure biochemical processes of mammalian biology in vivo (Phelps 2002). Most, if not all, PET imaging probes are developed by labeling known drugs or active biological materials with positron-emitting radioisotopes, sometimes sharing the same element as the drugs. In this regard, developing PET probes have the advantage compared to those from other imaging methods because PET probes can be developed with chemical structures identical to the targeted drug, given the availability of several elemental radioisotopes that can be inserted directly into the parent drug's chemical structures. These popular elemental positron emitters include oxygen (^{14}O, ^{15}O), nitrogen (^{13}N), and carbon (^{11}C). There is no hydrogen positron emitter; therefore, [^{18}F]fluorine is used in place of hydrogen given their proximal sizes (Phelps 2002). Asides from these major radioisotopes, other positron emitters such as ^{64}Cu, ^{62}Zn, ^{38}K, ^{76}Br, ^{75}Br, ^{82}Rb, ^{55}Co ^{52}Mn ^{124}I, ^{122}I, ^{30}P, ^{52}Fe, ^{90}Y, ^{89}Zr, and ^{68}Ga can be used for labeling and imaging as well. As said, in reality, the short half-lives of most positron emitters and the challenging chemistry are the major issues that limited their use; in the context of this lecture, which focuses on the practical translation of the knowledge to lab work, only major positron emitters will be discussed in detail.

Unlike fluorescent dyes mentioned in Chap. 2, the discussion focuses on only the chemical development of the dyes with extensive optimization of the optical properties for imaging. Developing optical imaging probes is a different story, which will be discussed in later chapters. For instance, the bioconjugation process, which also involves the use of the biolinkers, spacers, and so on, must be included because this step necessitates the realization of the probes. In contrast, the discussion of PET chemistry and the design of the probes is inseparable. In other words, the activated fluorescent dyes can be universally labeled any ligand, while labeling the positron emitters to the ligands must go through a custom-designed specifically for that ligand. In many cases, not all, the ability to maintain the very structure of the ligand after

W. Pham, *Principles of Molecular Probe Design and Applications*,
https://doi.org/10.1007/978-981-19-5739-0_3

labeling with the positron emitters makes this chemistry a powerful approach since no further assessment of the K_d, IC_{50}, or pharmacokinetic, biodistritbution is necessary. Aside from deep tissue penetration, the unaltered labeling ligands, altogether facilitate imaging in the brain. In fact, PET imaging is a holy grail in neuroimaging, in both preclinical and clinical operations. From a chemistry perspective, labeling short half-life radioisotopes pose multiple challenges and opportunities. Particularly, developing PET probes requires to optimize robust chemistry to meet that timeframe. In the earlier time, chemists used push/pull catheter lines with 2- or 3-way switches to transfer radioisotopes during the labeling experiment. This tedious work is replaced by automatic labeling modules, nowadays. Yet, there are still many more opportunities for further improvement. One of the future technologies in PET chemistry would focus on developing labeling kits to minimize workforce and cut cost. Another advantage of using short half-life radioisotopes for medical imaging is the quick decay after providing information, thus minimizing exposure to patients and clinical staff, and these half-lives fit well within the clinical procedure window.

First, what is a positron? And how is it generated? A cyclotron is required for the production of new radionuclides. Depending on the beam types, different charged particles are produced, including protons, H^- (negative hydrogen ions) or heavier particles than protons. When the charged particles react with the target nuclei of stable isotopes, a nuclear reaction occurs (interaction between the protons and the target atoms), a neutron will be emitted, generating a new radionuclide. This unstable radionuclide will decay by the emission of positron particles. A positron is somehow similar to an electron, they have similar mass, albeit with a positive charge. If an electron is considered as a matter, then a positron is an antimatter particle to an electron. If an electron is denoted as a letter e with a negative sign (e^-), a positron has a symbol of β with a positive charge ($\beta +$). The positron emitters are classified based on their half-lives and mode of decay mechanism. The most commonly used PET radionuclides in molecular imaging, such as ^{11}C, and ^{18}F have very short half-lives and high branching ratios for $\beta +$ decay. While ^{89}Zr and ^{64}Cu have long radioactive half-lives and a low branching ratio for $\beta +$ decay (Conti and Eriksson 2016). A more detailed list of the physical property of positron emitters can be found in Fig. 3.1 (Nolting et al. 2012).

Isotope	Halflife (min)	Positron Energy (MeV)	Positron range (mm)	Production source
^{82}Rb	1.26	3.15	1.7	generator
^{15}O	2.03	1.70	1.5	cyclotron
^{13}N	9.97	1.19	1.4	cyclotron
^{11}C	20.3	0.96	1.1	cyclotron
^{18}F	109.8	0.64	1.0	cyclotron
^{64}Cu	768	0.66	n/a	cyclotron
^{68}Ga	67.72	1.90	2.9	generator

Fig. 3.1 Physical property of common positron emitters. Data obtained from Nolting et al. (2012)

When the unstable radionuclide, such as ^{18}F (fluorine-18) decays, it is converted into ^{18}O (oxygen-18), emitting a positron with a half-life $t_{1/2}$ of 110 min with an energy of 0.64 meV and a positron range of 1.0 mm. While ^{11}C (carbon-11) decays into ^{11}B (boron-11), emitting a much shorter half-life $t_{1/2}$ of 20 min. The question is, how do we know ^{18}F converts to ^{18}O and ^{11}C to ^{11}B and so on? During the positron decay process, a proton is converted into a neutron, so the next element will have one less number of protons and one more number of neutron. Regardless of this change, the mass number remains the same. In the $[^{18}_{9}F]$ case, the atomic number will change from 9 to 8. This means that $^{18}_{9}F$ is now converted to $^{18}_{8}O$ because only oxygen has the atomic number 8. Using the same analogy, $^{11}_{6}C$ converts to $^{11}_{5}B$ while emitting a positron.

Another notion worth discussing is the difference between conventional synthetic chemistry and PET or radiopharmaceutical production chemistry. Aside from the differences in the physical operations of these two areas, such as the scale of the reactions, and the reaction setup, maneuvering, characterization, and purification. All of these differences reflect the nature of dealing with short half-life radioisotopes. Two other tasks are required to perform and describe in PET chemistry, including radiochemical yield (RCY) and the specific activity. Further, radiochemical purity information is also useful data that need to be addressed. RCY refers to the total amount of radioactivity of the finally purified probe compared to the starting radioactivity used at the beginning of the labeling reaction. This suggests that ideal PET chemistry partly favors a short reaction time. And as the radiation decay progresses during the process, it is crucial to report whether the RCY is decay-corrected (d.c.) or non-decay-corrected (n.d.c.). If decay-corrected data is reported, specific time, such as when the decay-corrected is counted for, in the beginning of the reaction or at the end of bombardment (EOB).

The specific activity of a PET probe measures the amount of radioactivity (in Bq or mCi) per unit mass (in grams or milligrams) of the labeled compound. For example, at the end of the labeling reaction, 10 mg of a ligand contains 40 mCi (1480 MBq). Then, the specific activity should be reported as 4 mCi/mg (40/10) or 148 MBq/mg.

While molar activity, as suggested, measures the amount of radioactivity (MBq or mCi) per mol of the ligand. This information is crucial for imaging receptors. For instance, to study receptor density or specificity, it is necessary to perform the work using a probe with high specific activity; otherwise, if the probe has low specific activity, it means there is a large concentration of unlabeled materials, which will saturate the receptor before the reporter probe could bind. In this regard, the specific activity of the probe links to other terminologies that involve the production operation of radiopharmaceutical agents, i.e., carrier-added (c.a.) and no-carrier-added (n.c.a.). A carrier is usually an inactive material, and it can be isotopic or non-isotopic with the radionuclide, but chemically similar to the radionuclide (Organization 2008). A carrier can be added during processing and dispensing of a radiopharmaceutical preparation to enhance the chemical, physical, or biological properties of radiopharmaceutical preparation (Organization 2008). The term n.c.a. indicates that no dilution of the specific activity. For instance, in an n.c.a. production of an ^{18}F-labeled

PET tracer, no "cold" fluorine-19 has been added intentionally during radionuclide production or during the synthesis of the tracer. In contrast, in a c.a. production, "cold" fluorine-19 has been added intentionally (Edem et al. 2019). The addition of a carrier, for example (^{19}F-F$_2$), leads to an increased mass of the final radiotracer, consequently resulting in the receptor saturation and the reduced PET signal from specific binding (Cai et al. 2008).

Before delving more into the radiolabeling chemistry, it is worthwhile to get familiar with a number of essential nuclear chemistry terminologies usually described in the literature:

> **Annihilation radiation**: when a particle (electron) and its antiparticle (positron) collide to produce annihilation photons (511 -keV).
>
> **Autoradiography**: a photographic technique used to detect radioactive materials associated with a specific target on ex vivo tissue.
>
> **Becquerel**: unit of radioactivity rate one per second, abbreviated as Bq.
>
> **Betta emitter**: a radioactive atom, which changes into another atom by emitting a beta particle.
>
> **Carrier**: inactive material added to a radioactive material during production process to ensure that the radioactivity would maintain stable in all subsequent chemical and biological processes.
>
> **Carrier-free**: the production of radioactive isotope with high specific activity where no carrier is added.
>
> **Chemical yield**: this is also called radiochemical yield, meaning that a fraction of radiomaterial-associated compound compared to original activity achieved after a reaction.
>
> **Chromatogram**: a graphic information of the compound detected either by UV or gamma detector after purification to separate the labeled products.
>
> **Currie**: a unit of radioactivity, 1 Ci equals 3.7×10^{10} Bq.
>
> **Cyclotron**: a particle accelerator where the particles travel in a cylindrical vacuum chamber under the influence of a magnetic field and accelerated by a rapidly varying electric field produced by a high-frequency generator.
>
> **Decay (radioactive)**: a radioactive material loses energy by radiation.
>
> **EOB-End of bombardment**: nuclear bombardment reaction where a nucleus is bombarded by another nucleus or nuclear particle, where fission reaction occurs.
>
> **Fission reaction**: a nuclear reaction in which a nucleus splits due to an impact with another particle with the release of energy.
>
> **Gamma radiation**: electromagnetic radiation emitted during the process of particle annihilation.
>
> **Half-life**: the time required for the amount of radioisotope to reduce to one-half of its original value.
>
> **Isotopes**: nuclides that have the same atomic number but different mass number.
>
> **Molar activity**: the activity of one mole of the compound, expressed as Ci or Bq per mol (Ci/mol, Bq/mol).
>
> **Nuclide**: its also known as nuclear species, characterized by its mass number, atomic number, and nuclear energy state.

Radiopharmaceutical: a radioistopically labeled therapeutic/diagnostic ligand. **Specific activity**: activity of materials divided by the mass of that material, usually expressed as Ci of Bq per grams (Ci/g, Bq/g).

3.2 [^{18}F] Labeling

Fluorine-18 (^{18}F) is an attractive positron emitter compared to other counterparts. A brief statistical analysis of PubMed in the past ten years indicated that over 30% of literature covers ^{18}F making it the most commonly used probe in PET imaging, followed by ^{11}C, ^{64}Cu, ^{68}Ga, and ^{89}Zr. Although with a half-life of only 110 min, ^{18}F possesses a longer half-life than many. Next, ^{18}F decay emits a positron of relatively low energy, providing a low maximum distance traveled in tissue (2.4 mm). Further, the steric similarity to a hydrogen atom makes [^{18}F] an ideal isotope, which can potentially convert any active therapeutic agent into a PET probe without severely hindering the affinity for the molecular target (Jalilian et al. 2000; Mueller et al. 2007).

Fluorine-18 can be labeled to active biopharmaceutical ligands as [^{18}F]F$^-$ nucleophiles or [^{18}F]F$_2$ electrophiles. The nucleophilic labeling reaction is most favorable in PET chemistry since [^{18}F]fluorides ([^{18}F]F$^-$) can be obtained in high yield by convenient procedures that do not require the addition of carriers, so usually called n.c.a. reaction (Attina et al. 1982). This type of reaction usually has high molar activity (10^2GBq/μmol). Meanwhile, c.a. electrophilic [^{18}F]F$_2$ reactions have low molar activity (100–600 MBq/μmol) (Jacobson et al. 2015). Nevertheless, it is the only choice when direct labeling is impossible or more logistical burdens.

3.2.1 Production of Fluorine-18 Radioisotopes

Understanding how ^{18}F is produced helps to design better labeling conditions. To produce nucleophilic [^{18}F]F$^-$, stable isotope of H$_2$^{18}O is used as a target, the meV proton bombardment will convert ^{18}O to ^{18}F and a neutron as shown in Fig. 3.2.

Where p and n stand for proton and neutron, respectively. This expression is usually denoted as ^{18}O(p,n)^{18}F (Bailey et al. 2004). When the target is ^{18}O$_2$ gas, ^{18}F $-$ F$_2$ gas is obtained. ^{18}F $-$ F$_2$ is also prepared from deuteron-irradiation of Ne (Jacobson et al. 2015). The production method used is dependent on the desired

$$^{18}_{8}\text{O} \; + \; ^{1}_{1}\text{p} \longrightarrow \; ^{18}_{9}\text{F} \; + \; ^{1}_{0}\text{n} \; + \; \gamma$$

Fig. 3.2 Production of ^{18}F from ^{18}O-enriched water

subsequent chemical reactions; $^{18}F^-$ fluoride is produced for use as a nucleophile, while ^{18}F-fluorine is produced for use in electrophilic methods (Jacobson et al. 2015).

In the case of $[^{18}F]F^-$ production, to isolate the radionuclides, recover the ^{18}O-enriched water and remove impurities, the mixture of $[^{18}F]/H_2O$ in an aqueous solution is passed through an anion-exchange resin allowing the ^{18}O water to pass through while trapping the $[^{18}F]F^-$. Afterward, the $[^{18}F]F^-$ can then be released from the cartridge using either 0.01 M K_2CO_3 or 0.01 M Cs_2CO_3 (Schlyer et al. 1990), where potassium or cesium function as counterion for $[^{18}F]F^-$. Currently, the most popular approach for $[^{18}F]F^-$ radiosynthesis is to add a large counterion, such as tetrabutylammonium salts or the alternative approach of employing aminopolyether Kryptofix [2.2.2] (K_{222}), a phase-transfer crown ether macromolecule as a chelator of potassium, which has been found to drastically enhance the S_N2 nucleophilic substitution by $[^{18}F]$ F^- reactivity (Nolting et al. 2012). This act of sequestering the $[^{18}F]F^-$ away from the positively charged cations usually produces "naked fluoride anions." But the other significant role of K_{222} is to ensure the reactants, such as the precursors and the nucleophiles, must be in the same phase for the nucleophile–electrophile collision.

Currently, two methods are used to produce electrophilic $[^{18}F]F_2$ gas. One method focuses on deuteron bombardment of neon-20 gas target material to make ^{18}F via the $^{20}Ne(d,\alpha)^{18}F$ nuclear reaction, albeit this approach produces low molar activity $[^{18}F]F_2$ (12 Ci/mmol) (Bailey et al. 2004). Another method is more productive since $[^{18}F]F_2$ was produced from $[^{18}F]F^-$ after it was converted into $[^{18}F]CH_3F$, followed by mixing with carrier F_2 in an inert Ne matrix (Bergman and Solin 1997). This method generates electrophilic $[^{18}F]F_2$ with a molar radioactivity up to 55 GBq/μmol (1500 Ci/mmol).

Poor RCY and low specific activity due to the use of carrier-added method during the production of $[^{18}F]F_2$ gas, combined with complex maneuvering of gaseous $[^{18}F]F_2$ material, altogether make this approach less favorable compared to $[^{18}F]F^-$ counterpart. Nevertheless, innovative chemistry has been developed and showed the promising result, which will be discussed right below.

3.2.2   Methods of $[^{18}F]$labeling

A large number of methods have been reported in the past few decades on labeling bioactive ligands with ^{18}F. The following sections will discuss key experiments, organized into four main groups, such as (i) electrophilic substitution; (ii) nucleophilic substitution; (iii) via prosthetic groups; (iv) miscellaneous developments.

3.2.2.1   Electrophilic Substitution via $[^{18}F]F_2$

Human brains use glucose as a main source of energy, thus requiring continuous distribution of glucose to the brain crossing the blood–brain barrier. Disrupting glucose metabolism in the brain is the underlying mechanism related to many diseases. Thus, in vivo imaging glucose metabolism would a reveal dynamic condition of the brain. The 2-deoxy-2-[^{18}F] fluoro-D-glucose ([^{18}F]FDG) PET radioligand was designed by chemically mutated the hydroxyl groups on C-2 with ^{18}F. When [^{18}F]FDG enters the brain parenchyma, the probe will be phosphorylated by kinase, albeit with the absence of a hydroxyl group on C-2, the phosphorylated product would be intracellularly trapped at the site of metabolism, providing the metabolic information real time via the PET signal (Fowler and Ido 2002).

The first production of FDG PET agent, which opened up a new landscape about PET imaging in clinical practice and commercialization, went through [^{18}F]F$_2$ electrophilic radiofluorination reaction (Ido et al. 1977). The produced [^{18}F]F$_2$ was treated with 3,4,6-tri-O-acetyl-D-glucal (Fig. 3.3) in the presence of fluorotrichloromethane (Freon-11). After the reaction, Freon-11 was removed by a gentle stream of helium. This labeling produced a 3:1 mixture of the [^{18}F]-1,2-difluoroglucose isomer along with the undesired [^{18}F]-1,2-difluoromannose isomer. After purification by gas–liquid chromatography, [^{18}F]-1,2-difluoroglucose was treated with HCl to provide [^{18}F]FDG with 8% radiochemical yield, and the radiochemical purity was greater than 98%, with high molar activity (8750 mCi/mmol).

Aside from this [^{18}F]FDG labeling, it is worthwhile to note that direct fluorination of arenes using [^{18}F]F$_2$ mostly, if it happens, will be via aromatic electrophilic substitution S$_E$1 reaction. The detailed mechanism was discussed earlier in Chap. 1. The reaction occurs more readily if the ring is activated with electron-donating groups. Basically, these groups have unshared pair of electrons (RNH$_2$, OH, RO, RS), which can be delocalized into the ring, resulting in extending the π system through the resonance structures. More or less, this approach might lead to nonselective labeling since most of the activated groups favor labeling at ortho- or para-, and a small extent

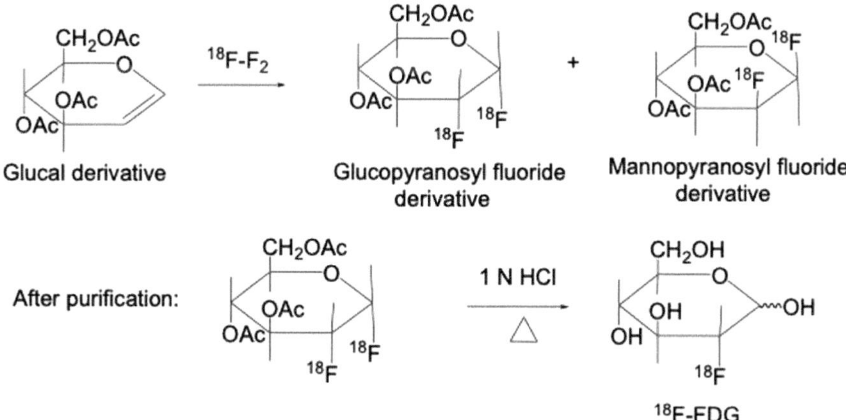

Fig. 3.3 [^{18}F]F$_2$ electrophilic reaction to generate first ^{18}F-FDG. Data obtained from Ido et al. (1977) with permission from the author

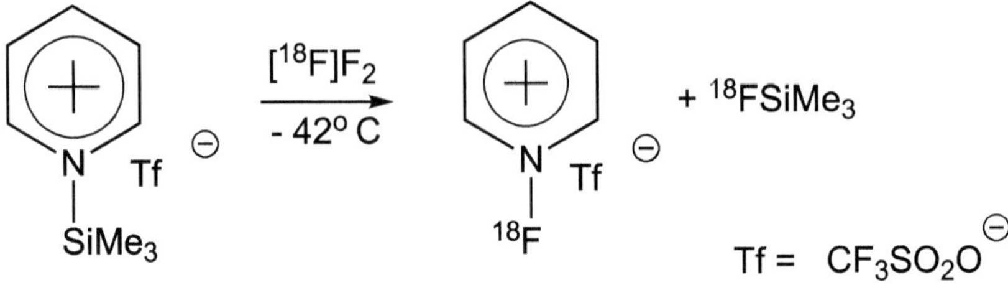

Fig. 3.4 Preparation of [^{18}F]fluoropyridinium triflate as an electrophilic ^{18}F-labeling reagent. Data adapted from Oberdorfer et al. (1988a)

of meta-position; thus, requiring extensive purification and characterization of the desired product and poor RCY.

^{18}F labeling via [^{18}F]F$_2$ was improved by generating new electrophilic ^{18}F-labeling reagents, with the capability to transfer ^{18}F to an sp^2-carbon within a reasonable time and under very mild conditions (Oberdorfer et al. 1988a). The idea of using electrophilic ^{18}F-labeling reagent is attributed to pioneering work trying to perform fluorination of 2-nitropropane with undecafluoropiperidine (Banks and Williamson 1964). This work suggests that fluorine of the N-F moiety can be readily attacked by a polarizable nucleophile such as a carbanion (Oberdorfer et al. 1988a). One typical experiment showed that N-[^{18}F]fluoropyridinium triflate can be prepared by cleavage of the N-Si(CH$_3$)$_3$ bond of the trimethylsilylpyridinium triflate using [^{18}F]F$_2$ (Fig. 3.4) (Oberdorfer et al. 1988a).

These [^{18}F]labeling reagents are robust for the selective transferring of ^{18}F to electron-rich substrates. The reaction was shown to be more selective and robust by cleavage of aryl-metal bonds of typically Ar-MR (M = Sn, Hg, Si, R = (CH$_3$)n) (Bergman and Solin 1997; Luxen et al. 1990; Namavari et al. 1992; Szajek et al. 1998). So far, a number of successfully developed fluorination reagents have been reported to enhance selectivity in electrophilic fluorination (Oberdorfer et al. 1988a, 1988b; Satyamurthy et al. 1990; Teare et al. 2007, 2010; Tredwell and Gouverneur 2012) (Fig. 3.5).

While it is widely recognized the limitations and logistic burdens of electrophilic [^{18}F]F$_2$ labeling, this and other progress in the development of ^{18}F-labeled electrophilic fluorinating reagents prove otherwise. Among the reagents shown in Fig. 3.5, probably, Selectfluor bis(triflate) is the most common reagent for electrophilic [18F]F$_2$ labeling to target drugs, given its high specificity and ease to prepare with commercially available starting material. In recent work, it has been demonstrated the reaction can be performed in very mild condition and robust, as simple as a "shake and mix" type of protocol (Teare et al. 2010). For example, Selectfluor bis(triflate) has shown to be a versatile ^{18}F transferring agent with promising reactivities toward metal substrates, including Ag, SnMe$_3$, OSiMe$_3$, and boronic esters (Fig. 3.6) (Tredwell and Gouverneur 2012; Stenhagen et al. 2013).

With the current state-of-the-art chemistry, electrophilic fluorination using ^{18}F transferring agents can reach molar activity up to 20 GBq/μmol (Teare et al.

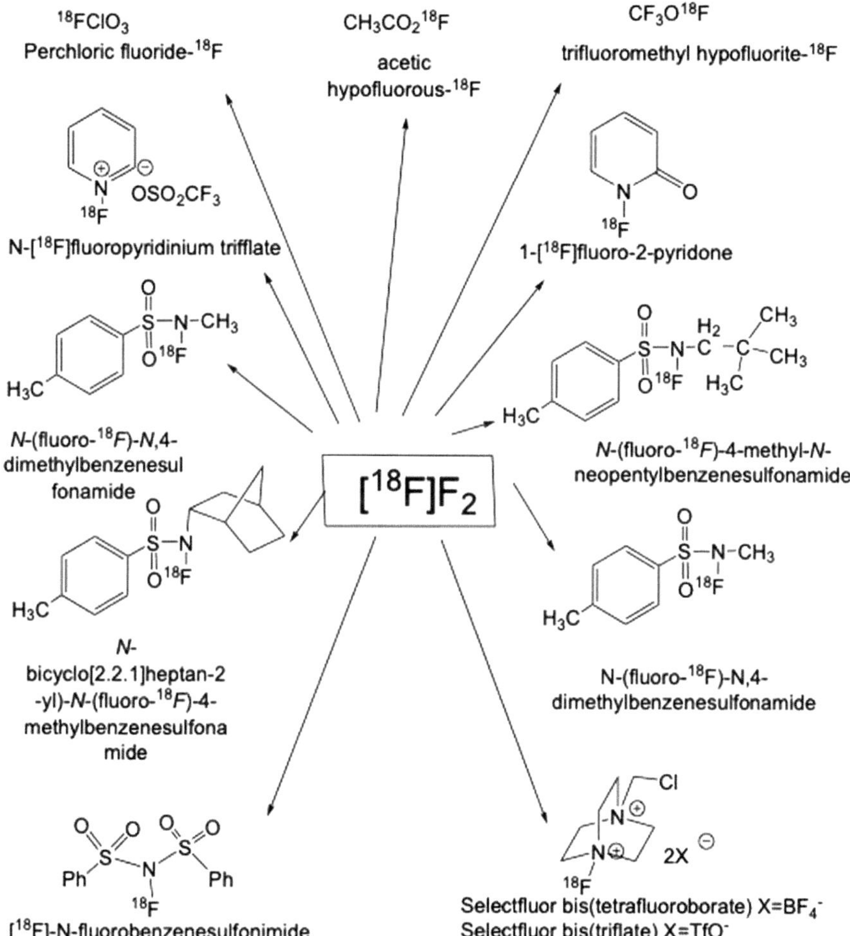

Fig. 3.5 A list of ^{18}F-labeled reagents for the synthesis of [^{18}F]arenes or for substrate with carbon moiety with high electron density. Data obtained from Teare et al. (2010) with permission from the author

2010). Although it is still far from [^{18}F]F$^-$ method (average of molar activity 300 gBq//μmol), it is predicted that the ability to label bioactive material with [^{18}F]F$_2$ via the transferring reagent as a choice not must be in a horizon. Altogether, this method will stand along with others to offer unlimited capabilities in how one should prepare a PET probe.

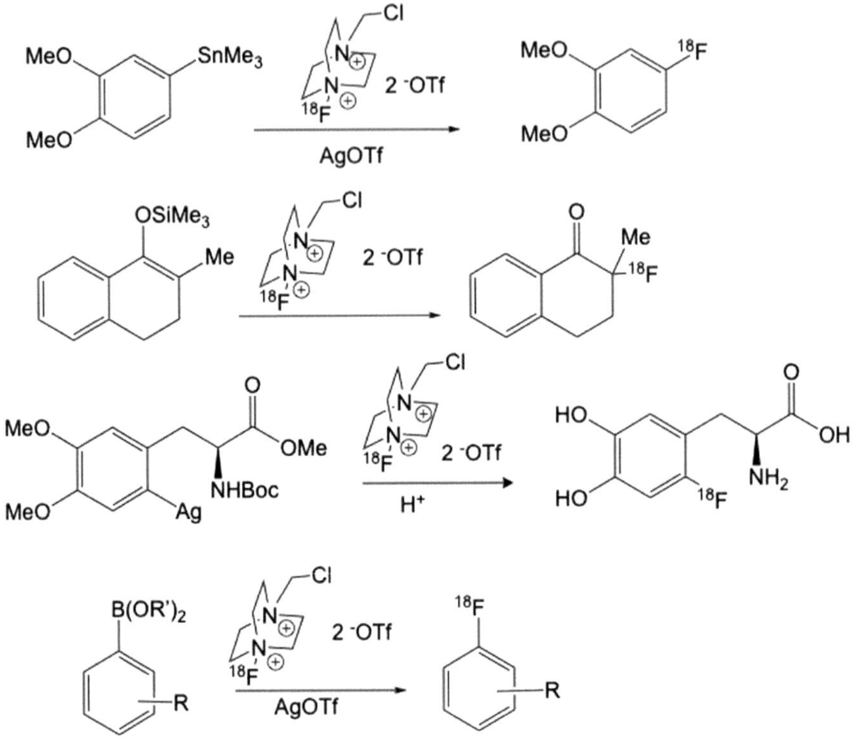

Fig. 3.6 Versatile roles of Selectfluor bis(triflate) as an alternative to $[^{18}F]F_2$ labeling via electrophilic fluorination. Data derived from Stenhagen et al. (2013)

3.2.2.2   Nucleophilic Substitution via $[^{18}F]F^-$

Most of the operations in a PET chemistry laboratory are focused on this type of reaction. It is more robust and more straightforward compared to electrophilic labeling, and thus, RCY and specific activity can be achieved with high yields. In a typical SN_2 reaction, virtually; every ligand, whether the aliphatic or aromatic-based structures, can be labeled effectively as far as they have appropriate leaving groups. The reaction mechanisms to label $[^{18}F]F^-$ can be classified into two categories. The first mechanism focuses on forming an aliphatic or sp^3-hybridized carbon bonded to fluorine, and the second is an aromatic or non-sp^3-hybridized carbon bonded to fluorine. Within these two distinct classes of reaction products, one can consider a wide range of products (Nolting et al. 2012). Both reactions rely on careful designing of the labeling conditions, among which the choice of a robust leaving group is critical given fluoride is not a strong nucleophile. This is a topic that merits detailed scrutinization depending on each individual case. Theoretically, nucleophilicity decreases going down the periodic table (Smith and March 2001a). In that case, fluoride is considered as one of the weakest nucleophiles among halides in the periodic table

Nucleophile	n	Nucleophile	n
SH$^-$	5.1	Br$^-$	3.5
CN$^-$	5.1	PhO$^-$	3.5
I$^-$	5.0	AcO$^-$	2.7
PhNH$_2$	4.5	Cl$^-$	2.7
OH$^-$	4.2	F$^-$	2.0
N$_3^-$	4.0	NO$_3^-$	1.0
Pyridine	3.6	H$_2$O	0.0

Fig. 3.7 Common nucleophiles and their strength. Data obtained from Wells (1963) with the permission from the American Chemical Society

($I^- > Br^- > Cl^- > F^-$). However, this order is solvent-dependent. In aprotic solvents, such as DMSO, DMF, and acetone, the order of nucleophilicity is $F^- > Cl^- > Br^- > I^-$. But overall, when comparing the nucleophilicity of F^- with others, it is a generally accepted that F^- is a weak nucleophile (Fig. 3.7) (Smith and March 2001b). Further, based on the soft–hard acid–base (HSAB) principle, fluoride is a hard base with a small and less polarized structure and reduced nucleophilicities.

With that in mind, designing of [^{18}F]F$^-$ labeling reaction conditions should focus on optimizing other key players. As mentioned above, it seems the nucleophilicity of F^- could be improved in aprotic solvents. Aside from using these solvents in the reaction, it is crucial to azeotropic removal of water from the mixture of Kryptofix [2.2.2] (K$_{222}$)/[^{18}F]F$^-$ by repeated additions and evaporations of solvents from the reaction vessel before introducing the precursor. In this working-against-time maneuvering, heating the reaction is nearly inevitable since for every ten degrees increased, the reaction rate will double, albeit pre-evaluation of the stability of labeling ligand is necessary to prevent decomposition. One of the alternatives to cut reaction time while achieving good radiochemical purity and specific activity is using the microwave module for heating during the labeling reaction. For instance, [^{18}F]fallypride was obtained in 60% labeling yield and 25% overall yield, compared with 25% labeling and 15% overall yield with all-conventional heating in the same module. The [^{18}F] preparation and radiosynthesis time were 25 min for the microwave versus 60 min for conventional heating (Ansari et al. 1304). With optimized conditions in regard to the use of ultrasonic energy to heat and good leaving groups, some labeling reaction can be achieved with over 96% RCY in merely one minute (Dolci et al. 1999). The other choice, which can decide the fate of the operation, relies on the design of the leaving groups.

Sulfonic Esters as Leaving Groups

Halides can be employed as leaving groups in [^{18}F]F$^-$ labeling reactions, albeit with modest RCY. On average, the RCY was reported somewhere between 10% and approximately 20%. As seen in Fig. 3.8, labeling of fluticasone propionate and methyltropane with halides as leaving groups provided merely 10 and 22% RCY,

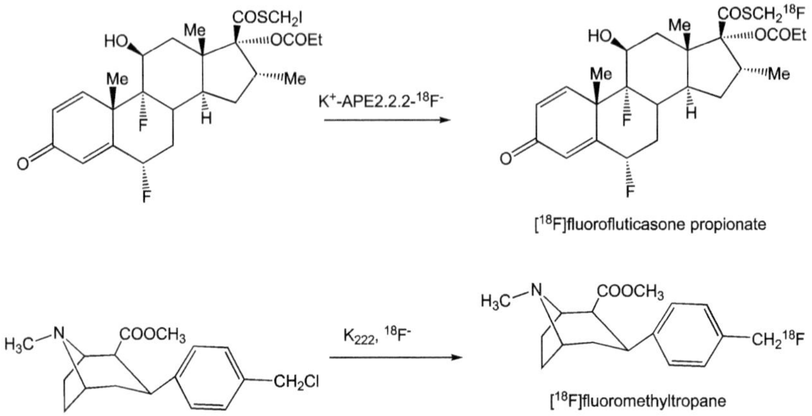

Fig. 3.8 Halides as leaving groups in [^{18}F]F$^-$ labeling reactions. Data obtained from Aigbirhio et al. (1997); Petric et al. (1999)

respectively (decay-corrected to EOB) (Aigbirhio et al. 1997; Petric et al. 1999). Therefore, it is rational to turn attention to sulfonic ester leaving groups, particularly when it comes to aliphatic nucleophilic fluorination. Typically, mesylates (R-OSO$_2$CH$_3$), tosylates (R-OSO$_2$C$_6$H$_4$CH$_3$), brosylates (R-OSO$_2$C$_6$H$_4$Br), nosylates (R-OSO$_2$C$_6$H$_4$NO$_2$) are much better than halide counterparts (Aigbirhio et al. 1997). If the condition allows, the fluorinated versions of these leaving groups are much appreciated, including triflates (R-OSO$_2$CF$_3$), tresylates (R-OSO$_2$CH$_2$CF$_3$), or nonaflates (R-OSO$_2$C$_4$F$_9$). Let's examine this current and most favorable method for developing [^{18}F]FDG via [^{18}F]F$^-$ labeling. In this work, triflate serves as a leaving group (Fig. 3.9). This no-carrier-added synthetic procedure was among the first attempts to facilitate the facile production of [^{18}F]FDG for clinical work (Hamacher et al. 1986). This work was more advantageous than others and became a gold standard for the synthesis of [^{18}F]FDG. The labeling chemistry was simple, using tetra-O-acetyl-2-triflate-β-mannose as the precursor. Along with the presence of the phase transfer catalyst, the experiment offered a mild and efficient nucleophilic fluorination. Further, by using a tetra-acetylated precursor, which can be rapidly removed under mild reaction conditions providing [^{18}F]FDG with RCY of approximately 50% in a synthesis time of 50 min from EOB.

Unlike conventional organic chemistry, PET labeling chemistry is a complex operation due to the short half-life of the positron emitters. Therefore, establishing an optimized procedure for each ligand is crucial for potential clinical work. Raising the temperature to increase the reaction rates is one of the best options. In this regard, the developed precursors need to be stable to survive such a condition. The data shown in Fig. 3.10 exemplify this notion. In this work, an analog of QNB was converted into a PET radioligand by labeling with [^{18}F]F$^-$ for in vivo imaging subtypes of the muscarinic acetylcholinergic receptor (mAChR). During the labeling, different leaving groups were evaluated using identical reaction conditions (Luo et al. 1998). In theory, triflate could be the most favorable leaving group. The presence of three

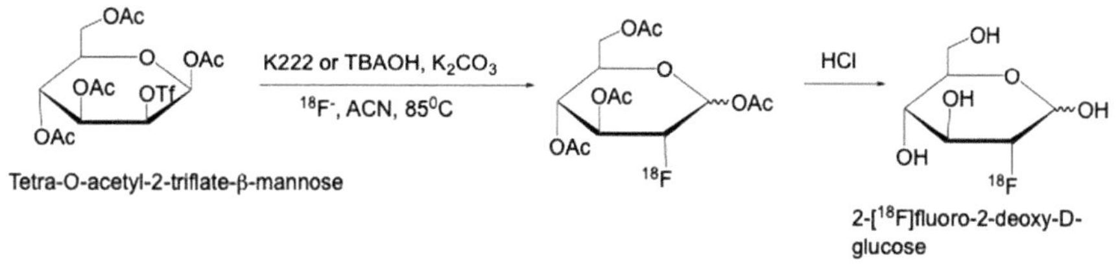

Fig. 3.9 Current approach for the synthesis of [^{18}F]FDG via [^{18}F]F$^-$ labeling. Data adapted from Hamacher et al. (1986)

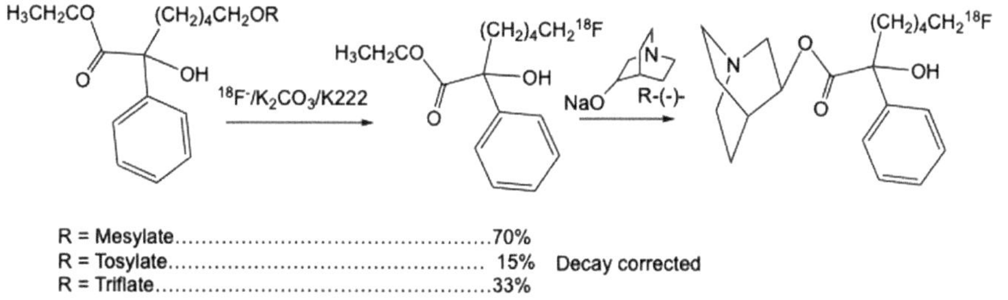

Fig. 3.10 Assessment of the effect of leaving groups. Data adapted from Luo et al. (1998)

fluorines, which are strong electron-withdrawing components that make triflate a very weak base and thus a better leaving group compared to others as shown in the experiment. It is estimated that triflate is several hundred times more reactive than mesylate. Unfortunately, the RCY is poor in this particular reaction, probably due to the intolerance of the precursor at elevated temperatures. And thus, only the ^{18}F-label intermediate prepared from the mesylate precursor was transesterified with the sodium salt of R-(-)-3-quinuclidinol to provide the desired QNB probe.

Aromatic Nucleophilic Substitution

Radiofluorination via the aromatic nucleophilic substitution (S_NAr fluorination) method to form the C (sp^2)-F bond probably constitutes the majority of literature because most bioactive ligands or drugs contain arenes and associated structures. In contrast to the aromatic electrophilic substitution mechanism discussed earlier, aromatic nucleophilic fluorination occurs on activated arenes with the presence of activated electron-withdrawing groups, with the order of NO_2 > CF_3 > CN > CHO > COR > COOR > COOH > Br > I > F > Me > NMe_2 > OH > NH_2, ortho and/or para to the leaving groups (Bailey et al. 2004; Kilbourn 1990; Angelini et al. 1985). The leaving groups are usually employed in these reactions, including, in reducing orders, NMe_3^+ > NO_2 > CN > F > Cl,Br, I > Oar > OR > SR > NH_2 (Bailey et al. 2004).

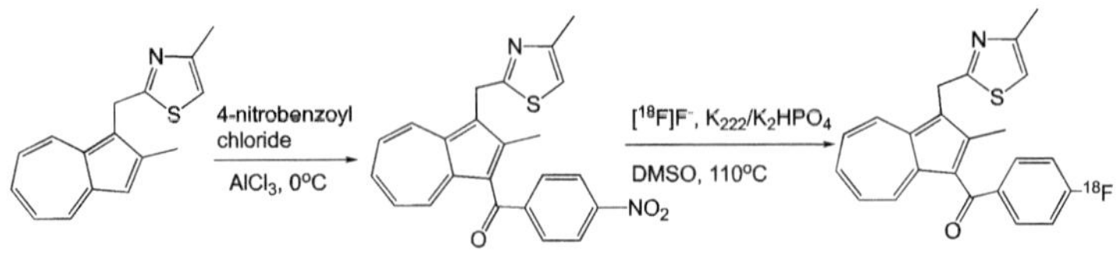

Fig. 3.11 Friedel–Crafts acylation for attaching aromatic nitro group as a precursor for ^{18}F labeling. Data obtained from Nolting et al. (2013)

Aromatic Nitro Moiety as a Leaving Group

The aromatic nitro compounds are versatile moieties that can be synthesized from a large number of available synthons. Further, their stability and ease of handling make them a favorable precursor for many productions of PET probes. Aside from building the desired compound through convergence synthesis to attach the nitro group in the compounds, direct nitration with fuming nitric acid in the presence of sulfuric acid is also a robust method to generate the nitrobenzene or substituted nitrobenzene molecules. Meanwhile, the aromatic nitro compounds can also be obtained via oxidation of the amine counterparts. With the availability of an extensive collection of aromatic nitro compounds, which serve as excellent synthons for convergent synthesis of PET probes. For example, in the development of COX-2 PET probe for imaging the role of inflammation in cancer (Fig. 3.11), the aromatic p-nitro group can be introduced to the main azulene structure via a simple Friedel–Crafts acylation reaction (Nolting et al. 2013). The reason para-nitro was chosen in this work since it is more nucleofugal than other positions. It has been shown in the past the order of nucleofugality in decreasing order: p-NO$_2$ > o-CN > o-NO$_2$ ~ p-CN > > m-NO$_2$ (Attina et al. 1982). The aromatic nitro intermediate served as a precursor for arene nucleophilic substitution using [^{18}F]F$^-$ to provide the desired product with RCY 3% (decay-corrected at EOS), with over 99% chemical and radiochemistry purities and a molar activity of 733 Ci/mmol (Fig. 3.11). Poor RCY was related to the stability of azulene under an elevated temperature. The precursor stability-high temperature-RCY relationship needs detailed consideration in the design. For example, in a different approach to developing the aromatic nitro group as a precursor to making phosphodiesterase 2A probe for neuroimaging, the 2-nitropyridine ring was incorporated onto the benzoimidazotriazine backbone via Suzuki coupling (Fig. 3.12). The labeling with [^{18}F]F$^-$ proved to be very successful, with up to 94% RCY during the early stage of the reaction when the temperature was approximately 100 °C (Ritawidya et al. 2019). When the temperature was increased to 120 °C, the RCY reduced to 93%. However, as the reaction progressed with the heating of about 150 °C, a significant reduction in RCY to 57% was recorded, indicating the decomposition of the [^{18}F]BIT1 under these conditions.

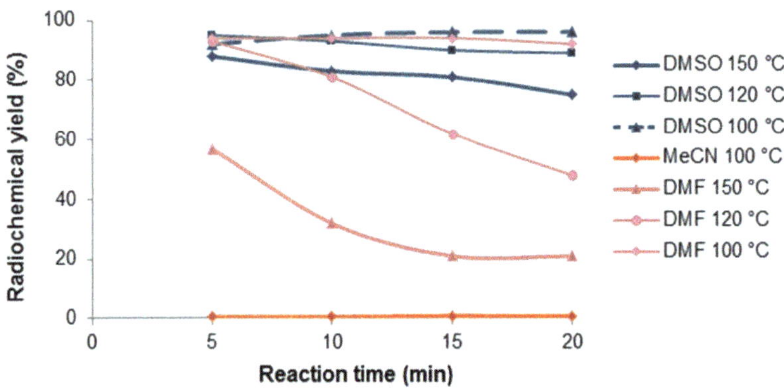

Fig. 3.12 Temperature-dependent labeling results for [^{18}F]BIT1. Data obtained from Ritawidya et al. (2019)

Trimethylammonium Salts as a Leaving Group

A decade ago, aromatic nucleophilic substitution via nitro precursors was considered the most popular method used in PET chemistry (Kilbourn and Haka 1988; Lemaire et al. 1987). One major issue with using nitro moiety as a leaving group involves using high temperatures. As shown in the examples above, poor RYC is due to decomposition of the material at elevated temperatures. Another practically difficult operation using nitro-substitute precursors is that separating the desired products from the precursors is a very time-consuming and sometimes difficult procedure (Haka et al. 1989). In that regard, trimethylammonium salts became a good candidate for the improvement for the following reasons: (i) it is one of the best-leaving groups for aromatic nucleophilic substitution; (ii) the reaction does not need high temperature, thus reducing the unprecedented decomposition of the labeled compounds; (iii) the salt precursor can be crystallized as solid or powder, which is much easier to handle and storage; (iv) the mobility of ammonium salt versus the label product is very much distinct chromatographically; thus, it is easier to monitor the progress of the reaction and purification.

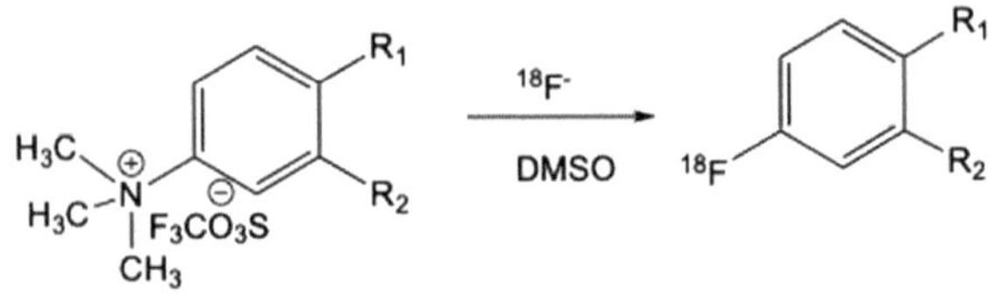

Fig. 3.13 Fluorination of activated aromatic ammonium salts with [^{18}F]F$^-$. Data adapted from Haka et al. (1989)

Fig. 3.14 [^{18}F]F$^-$ labeling in the presence of a labile ester group. Data obtained from Jalilian et al. (2000)

A number of different types of counterions have been used with trimethyl ammonium salts, such as halides, perchlorates, or triflate. However, the ease and robust reaction using methyl triflate as a methylating reagent to convert N,N-dimethylanilines to the corresponding N,N,N-trimethylammonium triflate salts, making triflate a preferable choice (Haka et al. 1989). The methylation reaction using methyl triflate is found to be tolerant to many substituents on the arenes. As shown in Fig. 3.13 that the trimethylammonium triflates produced better yields of ^{18}F-labeled aromatic products. The reaction occurs at low temperatures that and is thus applicable for an extensive repertoire of substituted aromatic compounds. The mild labeling condition afforded by activated aryltrimethylammonium triflate salts had broadened the scope of synthesis of several pharmaceutical active compounds. This development resolves the main issue that many compounds face, particularly those with labile groups, such as esters, which usually could not survive harsh labeling reaction conditions at high temperature. One typical example to demonstrate the power of this technology is through the synthesis of a benzodiazepine PET radioligand for neuroimaging. In this no-carrier-added ^{18}F-labeling experiment, trimethylanilinium triflate precursor was treated with the azeotropically dried mixture of Kryptofix [2.2.2] and potassium fluoride in DMSO at 90 °C (Fig. 3.14) (Jalilian et al. 2000). The total synthesis and purification of the desired probe were achieved in 15 min with an RCY of approximately 75% (decay-corrected) with the molar activity of 3000 Ci/mmol and a radiochemical purity of greater than 95% (Jalilian et al. 2000). There was apparent hydrolysis of the ester functional groups when the labeling occurred with concomitant presence of high temperature and an excess amount of base. Thus, the reaction should be optimized depending on each ligand in terms of temperature, exposure time, solvent, and Kryptofix/base ratio to obtain the best labeling conditions. It is found that extended heating during labeling alone would reduce the synthetic yield due to precursor decomposition or ester cleavage of the precursor and/or product.

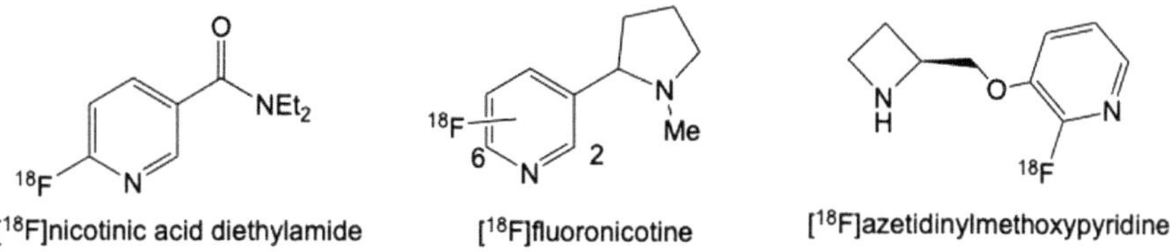

Fig. 3.15 Nucleophilic aromatic substitution on heterocyclic molecules with [^{18}F]fluoride. Data obtained from Dolci et al. (1999)

Heteroaromatic Systems

With all of the leaving group strategies discussed so far (halides, NO_2, N^+Me_3) for the homoaromatic nucleophilic ^{18}F labeling (S_NAr fluorination) can be applied without much modification for nucleophilic substitution with [^{18}F]F^- of heteroaromatic molecules. One of the limitations of dealing with S_NAr fluorination of heteroaromatic molecules is the need to elevate the reaction temperature (usually in the range of 130–180°C). As the result, the RCY is low due to the decomposition of the substrates and/or products.

Pyridine is an ideal substrate to explore the labeling conditions of the heterocyclic system. One of the first reported synthesis of 2-[^{18}F]fluoropyridine to generate [^{18}F]nicotinic acid diethylamide was achieved via S_NAr fluorination with 40% RCY starting with 2-chloropyridine analog and $^{18}F^-$ as cesium salt, in acetamide at 200°C (Fig. 3.15) (Dolci et al. 1999). A very similar labeling condition used $^{18}F^-$ as cesium salt, 2- and 6-[^{18}F]fluoronicotines were prepared from the corresponding 2- or 6-bromonicotine precursors to afford the products with RCY approximately 30–40%. Again, this labeling reaction requires the temperature up to 210 °C in DMSO (Ballinger et al. 1984). Another similar labeling was performed during the preparation of a PET radioligand targeting nicotinic acetylcholine receptor. The [^{18}F]azetidinylmethoxypyridine (Fig. 3.15) was synthesized by nucleophilic aromatic iodo-to-fluoro substitution in DMSO by heating the labeling reaction up to 150 °C for 20 min to provide 10% decay-corrected RCY.

A more comprehensive and comparable study was performed on the synthesis of 2-[^{18}F]fluoropyridine utilizing a whole range of leaving groups, including -Cl, -Br, -I, -NO_2, -N^+Me_3 (Fig. 3.16) (Dolci et al. 1999). Similar to the aromatic system, the yield of 2-[^{18}F]fluoropyridine was higher when nitro or trimethylammonium was used as leaving groups. The labeling was particularly high for the trimethylammonium case, both when heating the mixture in a conventional method or by microwave-assisted reactions. Notably, the data showed poor reaction yields with halides as leaving groups.

Altogether, the examples above attest to the difficulty to deal with S_NAr fluorination of heteroarenes. Later, the reaction condition was improved by replacing costly materials like cesium fluoride (CsF) by potassium fluoride (KF)/tetrabutylammonium chloride (Bu$_4$NCl) (See et al. 2020). KF, in general, does not solubilize in aprotic solvents very well. Thus, it must be used along with the phase transfer catalyst like

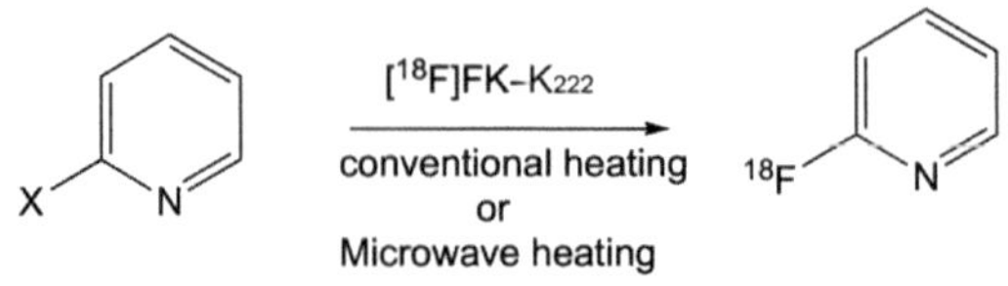

$$X\text{-pyridine} \xrightarrow[\substack{\text{conventional heating} \\ \text{or} \\ \text{Microwave heating}}]{[^{18}F]FK\text{–}K_{222}} {}^{18}F\text{-pyridine}$$

X = Cl, Br, I, NO$_2$, N$^+$Me$_3$.CF$_3$SO$_3^-$

Temperature		Substituent		Time of reaction (min)	
		X	5	10	20
120°C		Cl, Br or I	0	0	0
		NO$_2$	11	76	82
		N$^+$Me$_3$	81	87	91
150°C		Cl	1	3	23
		Br	1	16	25
		I	0	0	1
		NO$_2$	52	85	92
		N$^+$Me$_3$	89	89	90
180°C		Cl	11	28	57
		Br	56	60	87
		I	2	5	19
		NO$_2$	77	88	89
		N$^+$Me$_3$	88	91	92
	50		Time of reaction (min)		
			1	2	4
		Cl	0	1	26
		Br	0	1	68
		I	0	1	8
Microwave		NO$_2$	3	67	76
Power		N$^+$Me$_3$	77	94	96
(Watt)	100	Cl	1	22	n/a
		Br	11	71	n/a
		I	1	14	n/a
		NO$_2$	59	88	n/a
		N$^+$Me$_3$	96	90	n/a

Fig. 3.16 Comparisons of the effect of leaving groups for 2-[^{18}F]pyridine fluorination. Indicated yields are the average of three independent runs. Data adapted from Dolci et al. (1999)

Bu$_4$NCl to provide good labeling yield. The formation of the anhydrous Bu$_4$NF enabling effective fluorination, at moderately high temperature.

3.2.2.3 Diaryliodonium Salts as Electrophilic Arylating Reagents for ^{18}F$^-$

Hypervalent iodine compounds, such as diaryliodonium salts (Ar$_2$I$^+$X$^-$) and aryliodonium ylides (ArI$^+$R$^-$), have emerged as useful precursors for labeling homoarenes and heterorarenes with no-carrier-added cyclotron-produced [^{18}F]F$^-$ ion in the past several years (Pike 2018). It was tantamount like a versatile reagent, which induces radiofluorination of electron-deficient, as well as electron-rich (non-activated or deactivated) arenes, with an unrestricted choice of position on the rings, responding to the unmet need in critical PET radiopharmaceutical production. These salts are stable, suitable for storage for long shelf time; they can be used under mild conditions without the need for inert conditions. Another advantage of using these agents is the ease of preparation, low toxicity, and ease to handle (Uyanik et al. 2010). The main mechanism of diaryliodonium salts in this particular chemistry is purely electrophilic transferring of the aryl groups to the electron-rich substrate, such as [^{18}F]F$^-$ ion. This method fills the gap in labeling arenes. A large number of literature have demonstrated these reactions occur by a mechanism that is distinct from classical S$_N$Ar reaction (Telu et al. 2011). And it broadens the scope of S$_N$Ar reaction on electron-rich rings and/or at the meta-position and extend the preparation of many useful bioactive PET agents, which otherwise more challenging to achieve. One of the first experiments demonstrated that diaryliodonium salts served as precursors for facile labeling arenes with [^{18}F]F$^-$ in the presence of K$_{222}$, including those arenes without activated with electron-withdrawing substituents (Fig. 3.17) (Pike 2018). The hypervalent iodine (III) compounds have a three-center-four-electron bond. They are also called λ^3-iodanes, which possess a distorted trigonal bipyramidal geometry, where the central iodine atom has 10 valence electrons (Zhdankin and Stang 2008). One aryl ring occupies an axial position and the other an equatorial position. These rings can exchange position via a fast rotation in a process known as Berry pseudo-rotation (Fig. 3.18) (Pike 2018).

From the chemical structure, it is apparent that diaryliodonium salts can be prepared as symmetric and asymmetric products. Regardless of what approach, the general method to make diaryliodonium salts typically involves 2–3 steps with first oxidation of an aryl iodide to + 3 oxidation state (iodine (III)), followed by

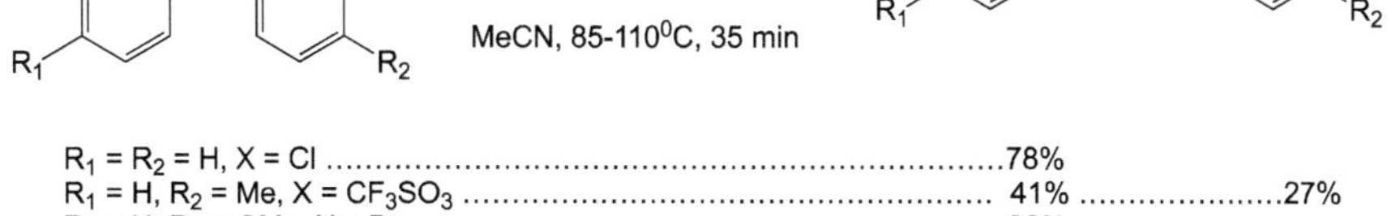

R$_1$ = R$_2$ = H, X = Cl 78%
R$_1$ = H, R$_2$ = Me, X = CF$_3$SO$_3$ .... 41% 27%
R$_1$ = H, R$_2$ = OMe, X = Br 88%
R$_1$ = R$_2$ = OMe, X = CF$_3$CO$_2$ 73%

Fig. 3.17 Diaryliodonium salts a precursor in arene radiofluorination. Data obtained from Pike (2018) with permission from the author

Fig. 3.18 Configuration of diaryliodonium salt. Data obtained from Pike (2018)

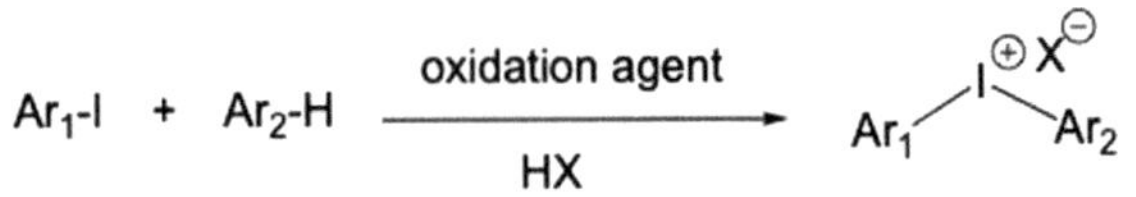

Fig. 3.19 Generalized method for the synthesis of diaryliodonium salts. Data adapted from Bielawski and Olofsson (2007)

ligand exchange with an arene to obtain diaryliodonium salt (Ochiai et al. 1997). A simple one-pot synthesis of diaryliodonium salts involves the use of aryl iodide with a commercially available oxidant in the presence of arene and a suitable acid, the anion of which would end up in the iodonium salt (Fig. 3.19) (Bielawski and Olofsson 2007).

In practical experiments, mCPBA could be used as an oxidation reagent along with boron trifluoride etherate as a Lewis acid (Bielawski et al. 2008) or toluenesulfonic acid (tosic acid) (Lindstedt et al. 2017) or in the presence of trifluoromethanesulfonic acid (triflic acid) (Bielawski and Olofsson 2007) (Fig. 3.20). More recently, diaryliodonium salts were prepared using Oxone in the presence of sulfuric acid (Soldatova et al. 2018). Although not much improvement from the other established procedure, the simplicity, and the use of inexpensive and available materials, along with typically good yields of iodonium salts underlying the significance of this new method. The role of the counterion has been shown to play a crucial role in the development and reactivity of diaryliodonium salts. The most commonly seen counterions, including TFA, Cl, PF_6, Br, OTf, BF_4, and OTs. They can be achieved by synthesis or via anion-exchange process, but not necessarily from the use of Lewis acid as precursors during the experiment. The choice of counterion dictates by how easily the products can be obtained. For instance, boron trifluoride etherate salt poses more challenging for purification; meanwhile, triflate counterion facilitates isolation of the desired products, and thus, there is a penchant to use triflate salt in the literature (Lindstedt et al. 2017). Figure 3.21 describes a procedure for an in situ anion exchange from diaryliodonium tosylate to diaryliodonium triflate.

The reaction mechanism of diaryliodonium salts in regard to ^{18}F radiofluorination or nucleophiles, in general, is a subject that merits several investigations. One of the earlier mechanistic studies showed that no radical intermediates were involved in the reaction pathway. The study demonstrated that arylation of β-keto ester enolates with diaryliodonium salt resulted in anticipated products, but no detectable aryl radical in the process (Ochiai et al. 2003). Currently, the most accepted mechanism of $^{18}F^-$ labeling via diaryliodonium salt reaction is postulated as a two-step reaction via an initial ligand exchange on the iodine (III) with the nucleophile, followed by ligand

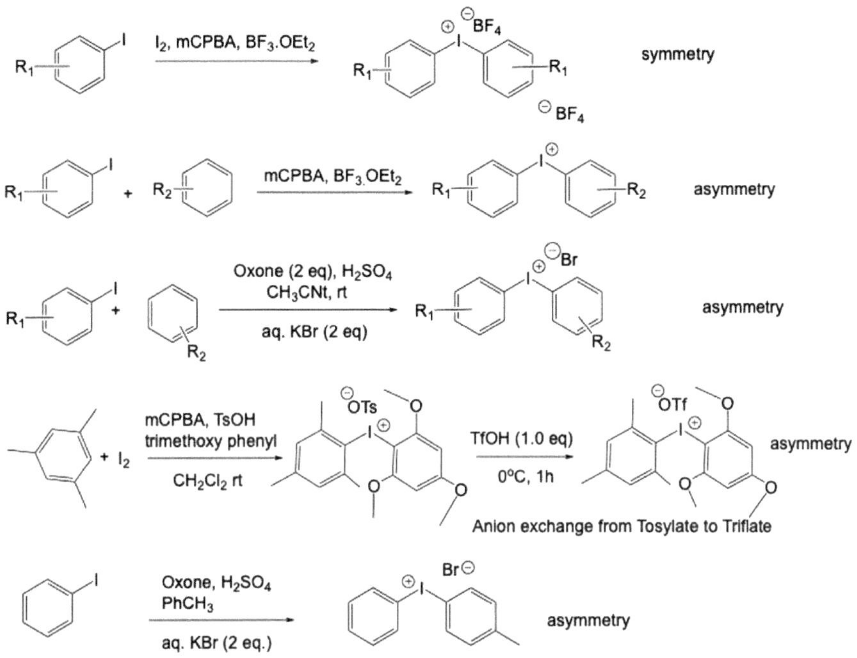

Fig. 3.20 Different routes of synthesis of substituted diaryliodonium salts. Data adapted from Bielawski and Olofsson (2007), Bielawski et al. (2008), Lindstedt et al. (2017), Soldatova et al. (2018)

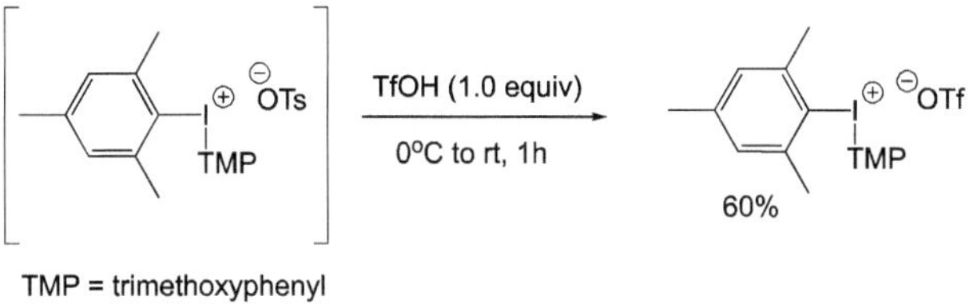

Fig. 3.21 In situ anion exchange from tosylate to triflate. Data obtained from Lindstedt et al. (2017) with permission from the American Chemical Society

coupling (Fig. 3.22) (Ochiai et al. 2003; Chun et al. 2010). In the ligand exchange step, the ^{18}F$^-$ would attack the intrinsically electrophilic iodine center, displacing the counterion X$^-$ group to form an intermediate in the transition state between iodine, the ^{18}F$^-$ and the equatorial arene. This process generated two isoforms of diaryliodane fluorides, which exist as an equilibrium mixture through rapid pseudo-rotation on iodine (III) (Ochiai et al. 1990). Next, the ^{18}F$^-$ is transferred to the equatorial arene through a reductive elimination process (Pike 2018). Another potential explanation of the robustness of iodonium salt as an electrophilic carbon transfer reagent. Similar to the conventional aromatic nucleophilic substitution discussed above. The

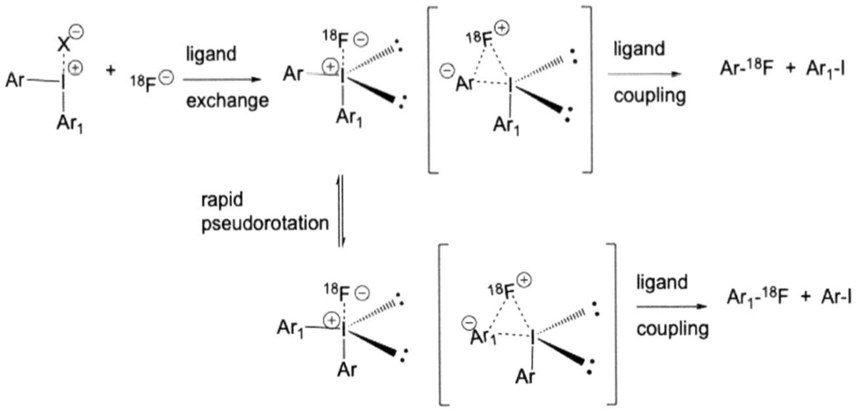

Fig. 3.22 Mechanism of $^{18}F^-$ labeling using a diaryliodonium salt. Data obtained from Chun et al. (2010) with permission from the author

$^{18}F^-$ anion will displace iodoarene (ArI^+), which is considered as a "hyperleaving group." It has been shown that iodoarene is a remarkable nucleofuge moiety, better than many leaving groups, including triflate. The nucleofugality of ArI^+ is about 10^6 better than triflate (Okuyama et al. 1995).

For the design of successful radiofluorination reaction, asymmetric diaryliodonium salts are more effective than their symmetric counterparts. First, the quite distinguished physical property of the iodoarene leaving group facilitates purification to separate from the desired radiofluorination product. Second, labeling selectivity can be achieved using asymmetric salts, in which the arene associated with the iodoarene leaving group can be designed as electron-rich to reduce the chance for fluorination. For instance, when the iodonium salt is comprised of an arene and an alkyl (R)- or alkoxy (OCH_3)-substituted arene, ^{18}F-labeled arene was the only fluorinated product (Shah et al. 1998). Furthermore, if the iodonium salt contains a phenyl group and fluoro-arene, both the [^{18}F]fluorobenzene and [^{18}F]halofluorobenzene will be formed, albeit the latter is a dominant product due to the mild electron-withdrawing group of halides renders more electron-deficient arenes. And most interestingly, iodonium compounds with bulky aryl rings tend to undergo nucleophilic substituents on the bulky ring (Grushin et al. 1992). This phenomenon is in contrast to other methods and thus could be exploited as a useful tool for regioselective labeling.

To date, numerous $^{18}F^-$ labeling designs and optimization have been developed using diaryliodonium salts as a major synthon for various imaging applications (Yusubov et al. 2013). For instance, an innovative design of a version of the thioflavin-based compound for imaging β-amyloid plaques was developed with ^{18}F labeling as an alternative to the less robust [^{11}C] counterpart (Fig. 3.23), using diaryliodonium tosylate precursors to provide 95% RCY and molar activities of 85–118 GBq/μmol (Lee et al. 2011). Notably, this diaryliodonium-mediated-$^{18}F^-$ labeling was performed in the presence of the active aromatic nitro group, demonstrating the superior activity of this chemistry. Another important neuroimaging

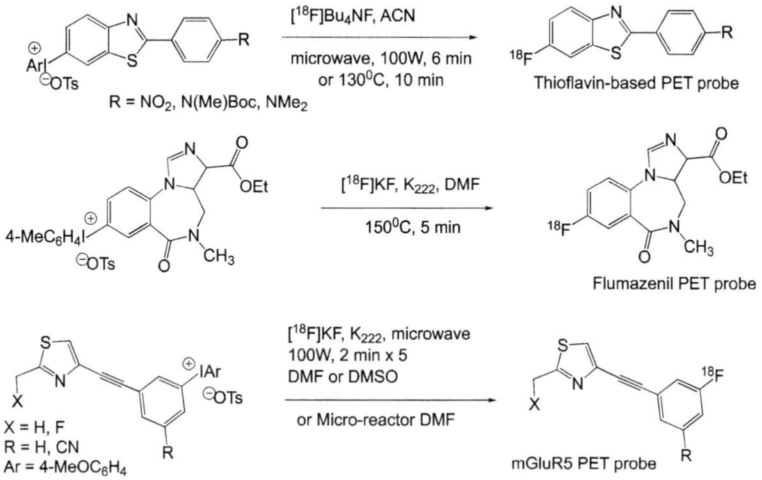

Fig. 3.23 Representative of PET probes developed utilizing diaryliodonium salts. Data adapted from Lee et al. (2011), Moon et al. (2011), Telu et al. (2011)

probe, such as [^{18}F]flumazenil radioligand was developed for the assessment of central benzodiazepine receptor concentration in the brain. In this work, the asymmetric diaryliodonium tosylate precursor was conveniently labeled with less than 60 min to provide RCY 64% with a specific activity of 370–450 GBq/μmol (Moon et al. 2011). In another work, notably, the aromatic nucleophilic substitution at meta-position was achieved via nucleophilic radiofluorination of diaryliodonium salts; otherwise, it is challenging to obtain by other methods. In this work, 3-fluoro-1-[(thiazol-4-yl)ethynyl]benzenes, an important class of high-affinity metabotropic glutamate subtype 5 receptor (mGluR5) was labeled with F$^-$. The work demonstrated the beauty and benefit of diaryliodonium chemistry where the highly electron-rich 4-methoxyphenyl ring to direct the radiofluorination to the opposite aryl ring, particularly when the target ring is also a mild electron-rich entity (Telu et al. 2011). This mGluR5 probe was achieved with 42% RCY, and it benefits molecular imaging of brain mGluR5 in preclinical and clinical works.

3.2.2.4 Aryliodonium Ylides

If the diaryliodonium salt method is designed for ^{18}F$^-$ labeling for electron-deficient aromatic systems, in contrast, aryliodonium ylides are for electron-rich aromatic analogs. Aryliodonium ylides are attractive ^{18}F$^-$ transferring agents, not only for robust labeling chemistry but also in the process of how to prepare the hypervalent iodine reagents. Aryliodonium ylides are denoted as $ArI^+-CR_2^-$, which is indicative of a zwitterionic nature in the ylidic C-I bond, where R represents electron-withdrawing groups, such as carbonyl, cyano, nitro, or sulfonyl moieties; Aryliodonium ylides represent an iodonium class that can be used for radiofluorination of non-activated arenes (Yusubov et al. 2013).

The first preparation of stable iodonium ylide by the reaction of dimedone and (difluoroiodo) benzene was reported in 1957 (Zhdankin and Stang 2008). Nevertheless, iodonium ylides-based precursors for $^{18}F^-$ labeling were first reported from UCLA group not too long ago (Satyamurthy and Barrio 2010). This radiosynthesis of n.c.a ^{18}F-labeled arenes using iodonium ylide derivatives quickly became well recognized due to its regiospecific labeling capability. Iodonium ylides are pretty stable and isolable compared to oxonium ylides. Nevertheless, they are less stable than iodonium salts discussed above, and thus, in situ generations of the ylides or conducting the coupling reaction at low temperatures is mandatory. Stable iodonium ylides can be obtained by treating iodosoarenes or iodosoarenediacetates with 1,3-dicarbonyl analogs or -disulfone derivatives (Kirmse 2005). One of the most practical ways involves the treatment of diacetoxyiodobenzene with malonate esters in the presence of a strong base (Fig. 3.24) (Goudreau et al. 2009). Other methods for preparing iodonium ylides have been reported using (diacetoxyiodo)arenes and Meldrum's acid or its derivatives in the presence of a mild base (Chun et al. 2010).

More stable iodonium ylides were reported based on the chemical structure of dimedonide analogs (Zhu et al. 2012). Basically, the o-alkoxyphenyliodonium dimedonide ylides were obtained by the reaction between diacetates and dimedone under basic conditions. In most cases, diacetates could be synthesized in a one-step reaction by treating iodobenzene or substituted versions with freshly prepared peracetic acid (AcOOH-AcOH). The iodonium dimedonide ylides are very stable at room temperature throughout exposure in several months without signs of degradation. Another advantage of using dimedone-based iodonium ylides is related to better solubility in common organic solvents.

In regard to the application in nucleophilic radiofluorination, as zwitterion analogs, iodonium ylides would favor regiospecific substitution reaction, in which of the nucleophile, such as $^{18}F^-$ would be labeled on the aromatic ring in most of the cases. The ^{18}F-labeling of arenes was limited to electron-deficient arenes for many years, but the iodonium ylides provided access for direct labeling of electron-rich arenes (Petersen et al. 2017). For instance, the derivatives of 5-HT$_{2A}$ receptor agonist, shown in Fig. 3.25, were labeled with $^{18}F^-$ via iodonium ylide precursors, followed by the deprotection of the Boc group in TFA/CH$_2$Cl$_2$ to obtain the intermediate. Then, which went through reductive amination to afford the desired labeling product. This reaction led to 8–15% decay-corrected RCY (Petersen et al. 2017).

As far as the complex nature of ^{18}F labeling of electron-rich arenes, the RCYs of most of the ^{18}F labeling via iodonium ylide method are not extraordinary; however, if we compared the same type of reactions using other approaches, such as NO$_2$

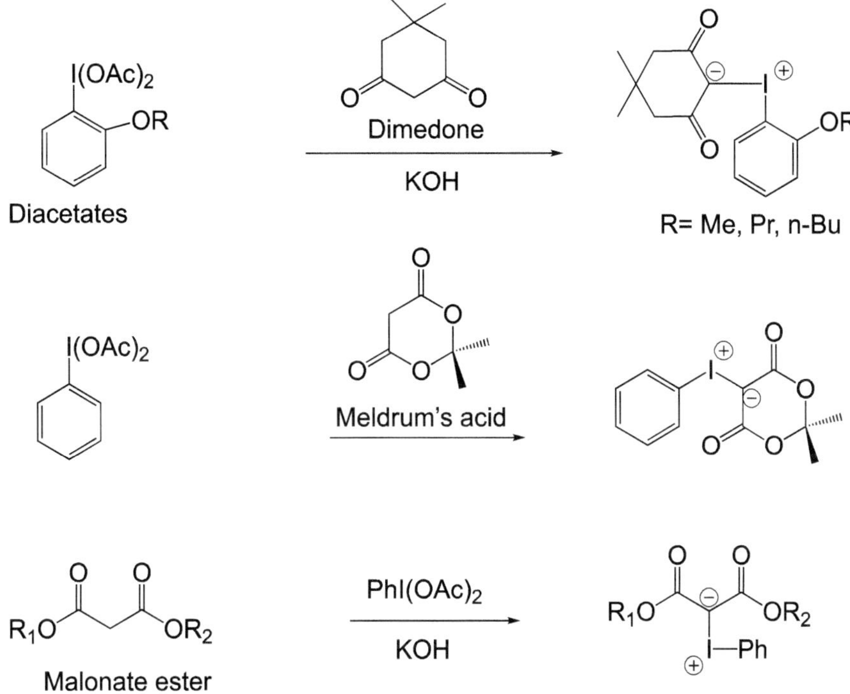

Fig. 3.24 Examples to demonstrate how to prepare stable iodonium ylides. Data adapted from Satyamurthy and Barrio (2010), Yusubov et al. (2016)

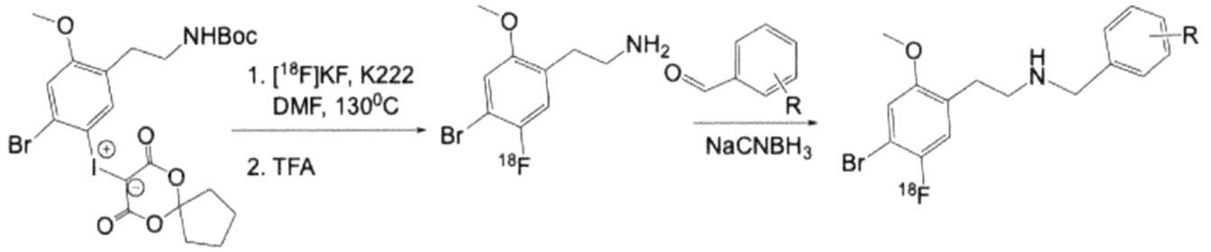

Fig. 3.25 Reaction of iodonium ylides with ^{18}F$^-$ in a typical PET labeling. Data adapted from Petersen et al. (2017)

or halides as leaving groups, then, in general, iodonium ylides offer RCY 4–five-fold better than others (Stephenson et al. 2015). One particular reaction exemplifies this notion-the radiosynthesis of ^{18}F-3-fluoro-5-[(pyridin-3-yl)ethynyl] benzonitrile ([^{18}F]F-PEB), a metabotropic glutamate receptor subtype 5 (mGlu$_5$) antagonist used in neuroimaging. Radiosynthesis of ^{18}F-FPEB using traditional S$_N$Ar fluorination reaction with ^{18}F$^-$ was challenging because the nucleophilic displacement of common leaving groups, such as Cl, Br, or NO$_2$ by ^{18}F$^-$ at the meta-position was not ideal. Thus, high temperature is required, and as a consequence, several impurities

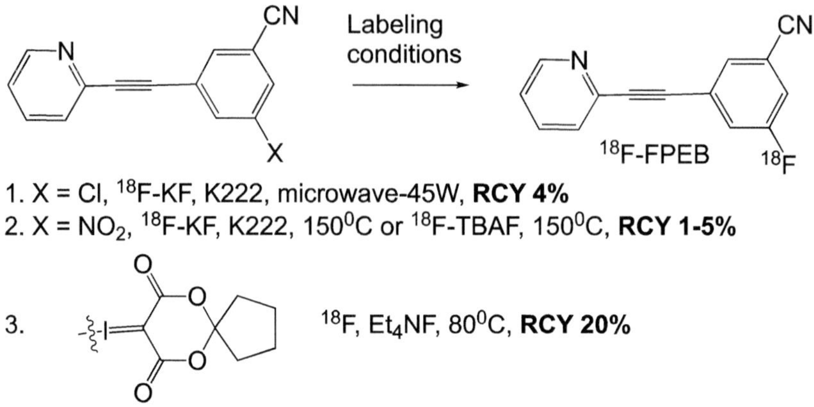
CN
N
Labeling
conditions
X
CN
N
18F-FPEB
18F
1. X = Cl, 18F-KF, K222, microwave-45W, RCY 4%
2. X = NO2, 18F-KF, K222, 150⁰C or 18F-TBAF, 150⁰C, RCY 1-5%
3.
O
O
I
O
O
18F, Et4NF, 80⁰C, RCY 20%

Fig. 3.26 Comparison of different labeling methods for the clinical production of ^{18}F-FPEB. Data adapted from Stephenson et al. (2015)

were generated in the process. In general, the RCY of a typical ^{18}F-FPEB reaction is about 1–5% (Fig. 3.26) (Stephenson et al. 2015). While using the iodonium ylide-based precursor, the resulted RCY was reported 20%, representing a tenfold increase over the other methods using NO_2 as a precursor (Liang et al. 2014). The labeling reaction time was cut 30% shorter with a twofold increase in molar activity (18 Ci/μmol). Further, the simplicity of the reaction facilitates the incorporation of the process into automation to support routine radiopharmaceutical production.

Not every fluorination labeling using aryliodonium ylides would end up at the aromatic ring, depending on the reaction conditions and the use of starting materials. Particularly, in the presence of protic acids, such as HF or HCl, it is apparent that fluorination or chlorination of iodonium ylides with these compounds provides the corresponding halogenated products derived from the C-protonation of the ylides, followed by displacement with fluoride ion (Gondo and Kitamura 2012). The observation derived from the fluorination and chlorination reactions of dibenzoylmethane with (diacetoxyiodo)benzene to form the iodonium ylide, which formed 2-fluoro dibenzoylmethane upon exposure to HF reagent. Mechanistic studies demonstrated that in the presence of acids, the protonation occurs first when exposed to the iodonium ylide (Fig. 3.27). Among the resonance isoforms, only the C-protonation leads to the fluorinated product, in which the nucleophile (halides) displaces the iodophenyl group. Although no ^{18}F labeling has been reported for this observation, more reactions need to be performed to confirm and generalize this mechanism for future work.

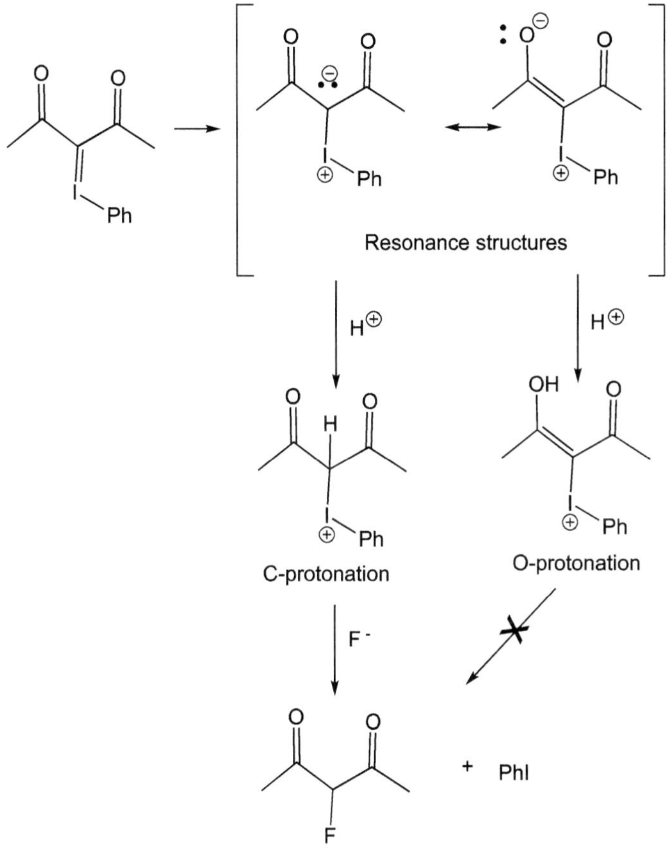

Fig. 3.27 Mechanism to explain C-protonation is the key intermediate from where displacement of the iodophenyl group by a halide ion. Data obtained from Gondo and Kitamura (2012)

3.3 Methods of [^{11}C] Labeling

With a half-life of merely 20 min, [^{11}C]carbon is one of the most diverse radionuclides used in today's PET chemistry since [^{11}C]carbon can be produced and rapidly transformed into a large variety of valuable synthons suitable for incorporation into many synthetic pathways. [^{11}C]carbon labeling becomes attractive to clinical radiopharmaceutical production because it generates the probes with an identical structure compared to the mother compound. The production of molecular probes with an identical chemical structure to the ligand retains the key profiles, including biodistribution, bioavailability, binding affinity, pharmacokinetics, and toxicology.

Several nuclear reactions are available for the production of [^{11}C]carbon (Wolf and Redvanly 1977). But the most effective and convenient procedure is to generate [^{11}C]carbon via the ^{14}N(p, α)^{11}C nuclear reaction. The reaction involves high-energy proton bombardment of a cyclotron target containing nitrogen gas along

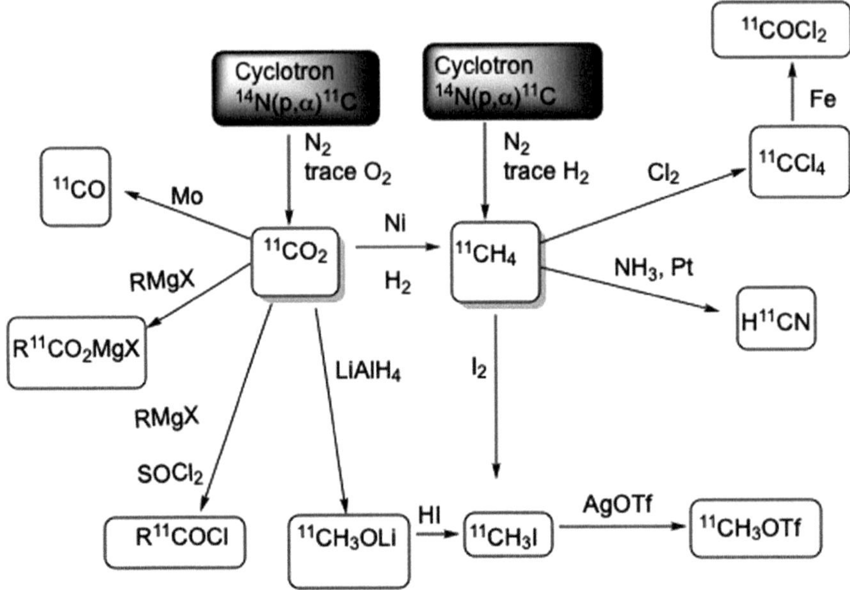

Fig. 3.28 Innovative chemistries to convert cyclotron [^{11}C]carbon into useful synthons for labeling purpose. Data derived from Dahl et al. (2017) with approval from Springer Nature

with the minute amount of oxygen and hydrogen gases to provide [^{11}C]carbon in the form of [^{11}C]carbon dioxide ($^{11}CO_2$) and [^{11}C]methane ($^{11}CH_4$), respectively (Dahl et al. 2017). Sometimes, these active synthons can be used as labeling agents for ^{11}C labeling (as described below for the case of $^{11}CO_2$). However, in most cases, these primary synthons will be converted into more reactive species before making [^{11}C]carbon labeling in realization.

Based on these cyclotron-generated products, [^{11}C]carbon labeling methods are tailored utilizing innovative chemistries to transform them into useful synthons for labeling any needed bioactive radiopharmaceutical agents (Fig. 3.28). Particularly, the [^{11}C]CO_2 could serve as a source of carbonyl due to its electrophilicity. For instance, [^{11}C]acetate and [^{11}C]palmitate were synthesized from [^{11}C]CO_2 after reacting with the corresponding Grignard reagents (Runkle et al. 2011). Recently, the [^{11}C]CO_2 fixation method proves to be versatile method, which deserves detailed discussion.

3.3.1   Direct [^{11}C]carboxylation

Carbonyl groups exist in many, if not most of the bioactive materials. This is the reason why [^{11}C]CO_2 has been employed as a synthon for direct incorporation of [^{11}C]CO_2 into radiopharmaceuticals, despite its low chemical reactivity (Luzi et al.

2020). Several methods have been developed in the past decades for the generation of [^{11}C]ureas, [^{11}C]carbamates, [^{11}C]amides, [^{11}C]esters, [^{11}C]carboxylic acids, [^{11}C]acid chlorides, [^{11}C]oxazolidinone, and more. Grignard reactions involve significantly in this ^{11}C-carbonylation protocol since it is very convenient for the conversion of alkyl or aryl magnesium halides into [^{11}C]carboxylic acids (Eriksson et al. 2021). In addition to Grignard synthesis, organolithium reactions were incorporated in ^{11}C labeling chemistry. However, the strength of these reactive nucleophiles for [^{11}C]CO$_2$ is also their shortcoming. They are notoriously sensitive to water, thus making the preparation of the intermediate, storage, and reaction setup more demanding (Fig. 3.29).

Another significant development utilizes more stable boronic esters as substrates for the incorporation of [^{11}C]CO$_2$ catalyzed by Cu(I) (Riss et al. 2012). The work showed that CuI is the best choice for this reaction. Particularly, using CuI along with K$_{222}$, and KF under homogenous conditions resulted in significant improvement in carboxylation yields (Fig. 3.30). Given the availability of a large repertoire of boronic substrates, this type of reaction contributes to diversifying capability to label a range of functional groups. The [^{11}C]carboxylic acids obtained from direct conversion of boronic esters can rapidly be converted into [^{11}C]esters or [^{11}C]amides.

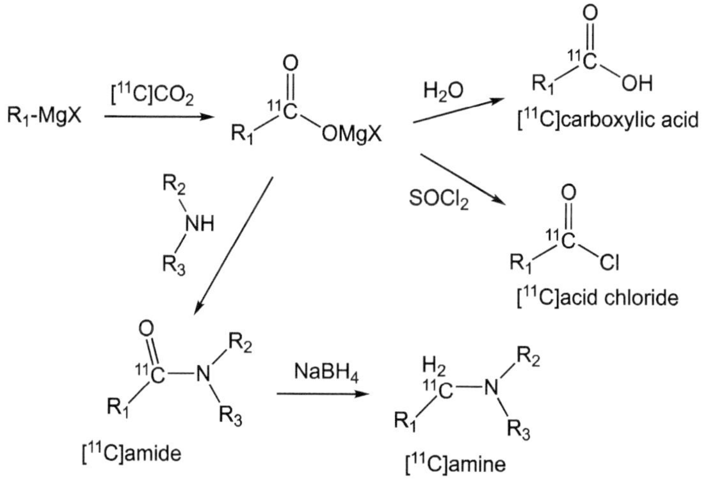

Fig. 3.29 Direct [^{11}C]CO$_2$ labeling using Grignard reaction. Data obtained from Taddei and Gee (2018) with permission from John Wiley and Sons

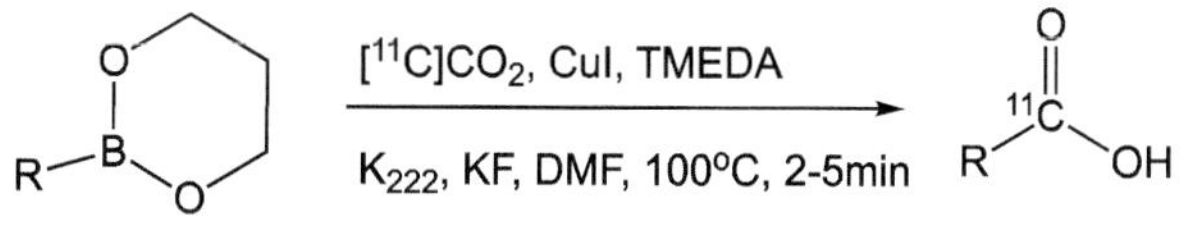

Fig. 3.30 Cu(I) catalyzed [^{11}C]CO$_2$ incorporation via boronic esters. Data obtained from Riss et al. (2012) with permission from the author

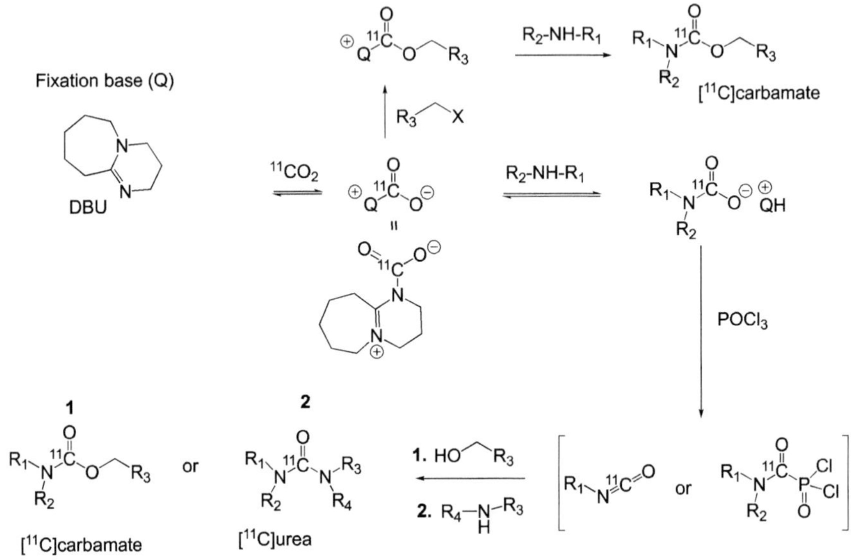

Fig. 3.31 Synthesis of [^{11}C]ureas and [^{11}C]carbamates by [^{11}C]CO$_2$ fixation. Data obtained from Dahl et al. (2017) with permission from Springer Nature

Poor solubility of CO$_2$ in the reaction mixture is the culprit behind sluggish reaction outcome and low yield. One of the strategies to enhance the activity of CO$_2$ involved bubbling of 1,8-diazabicyclo[5.4.0]undec-7-ene (DBU) or 2-*tert*-butylimino-2-diethylamino-1,3-dimethyl-perhydro-1,3,2-diazaphosphorine (BEMP) in the reaction. These bases act as organomediators by activation of CO$_2$ prior to the covalent bond formation (Dahl et al. 2017). This chemical fixation of CO$_2$ became the main procedure for ^{11}C labeling of pharmaceutical radioactive substances, including ureas, carbamates, oxazolidinones, and carboxylic acids (Fig. 3.31).

With this design, [^{11}C]methyl carbamates can be synthesized from primary or secondary amine in a one-pot reaction. Generally, the labeled products can be obtained in merely 1 min at room temperature (Fig. 3.32) (Wilson et al. 2010). It is observed that increasing reaction time does not improve the reaction yield. Overall, this procedure is impeccable for short half-life [^{11}C]carbon radionuclide. The RCYs of [^{11}C]-methylcarbamates were excellent for both primary and secondary aliphatic amines (~75–90%). However, if the same reactions occurred with weak nucleophilic amines, such as in the case of aniline or 4-nitroaniline, the reactions suffered poor yields (~3–40%).

Aside from the methods for synthesizing symmetric and asymmetric [^{11}C]urea shown in Fig. 3.31, another robust and efficient method for generating these ureas was developed by adapting Mitsunobu reaction with the [^{11}C]CO$_2$ fixation process (Dheere et al. 2015). The reaction for [^{11}C]urea synthesis was hypothesized to occur in three steps. First, the reaction commences with incorporating [^{11}C]CO$_2$

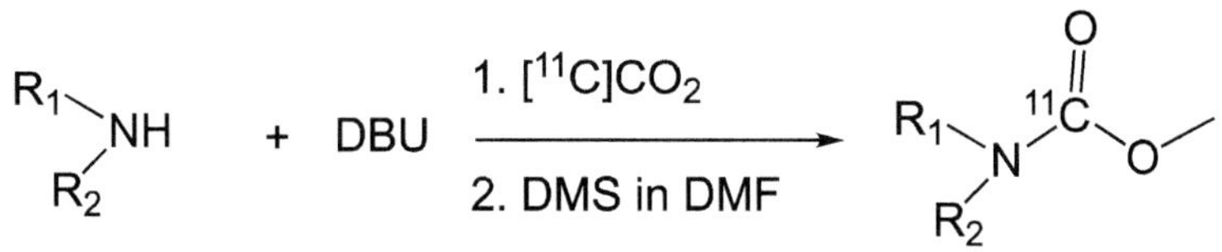

Fig. 3.32 [^{11}C]CO$_2$ direct incorporation for the synthesis of [^{11}C-carbonyl]-methylcarbamates using DBU. DMS = dimethyl sulfate, which serves as a methylating reagent. Data adapted from Wilson et al. (2010)

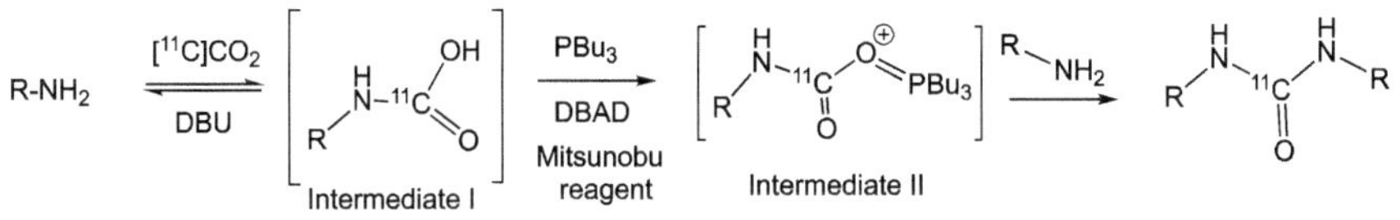

Fig. 3.33 Proposed mechanism for the synthesis of symmetrical [^{11}C] urea. Data adapted from Dheere et al. (2015)

onto primary amine in the presence of DBU to form intermediate I. Then, followed by the reaction of this intermediate with Mitsunobu reagents (DBAD: di-*tert*-butyl azodicarboxylate) to convert to intermediate II. Finally, nucleophilic attack from another amine on II to form the urea (Fig. 3.33) (Dheere et al. 2015). The [^{11}C] urea derivatives were obtained with a good RCY of over 70%. This simple and robust methodology provides a convenient way to prepare ^{11}C-labeled ureas previously inaccessible by other methods and enabling their utilization for in vivo applications. A similar approach could be used for the synthesis of asymmetric [^{11}C]ureas.

3.3.2   *[^{11}C]methylation*

One of most common methods for [^{11}C]-labeling employs [^{11}C]methyl iodide ([^{11}C]CH$_3$I) or [^{11}C]methyl triflate ([^{11}C]CH$_3$OTf) as active synthons. In general, [^{11}C]methylation chemistry is simple and it enables several innovative modifications to improve the operation. Particularly, the engagement of automation and other innovative devices, such as synthesis on solid-phase cartridge (Boudjemeline et al. 2017; Singleton et al. 2019) or in loops (steel, PTFE or PEEK) methods (Iwata et al. 2001; Studenov et al. 2004), all help to reduce reaction time and improve RCY. These developments enabled reproducible production of [^{11}C]-label radioligands with high specificity are the real impetus behind the clinical translations.

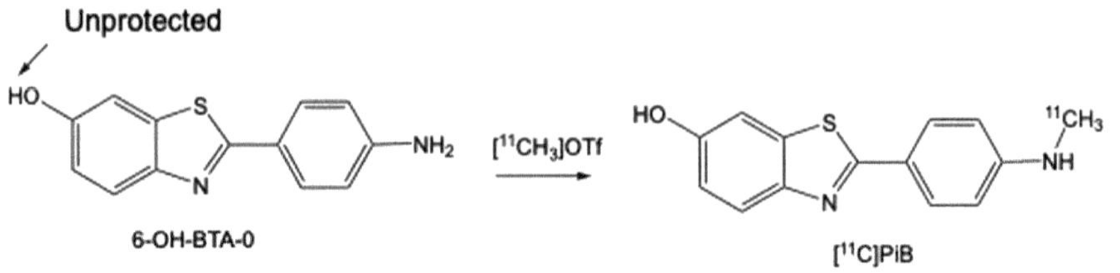

Unprotected

HO

NH$_2$ [^{11}CH$_3$]OTf

6-OH-BTA-0

HO

^{11}CH$_3$

NH

[^{11}C]PiB

Fig. 3.34 Radiosynthesis of [^{11}C]PiB from the unprotected 6-OH-BTA-0 using [^{11}C]CH$_3$OTf. Data adapted from Boudjemeline et al. (2017)

3.3.2.1 Solid-Phase [^{11}C]methylation

Toward this endeavor, a recent highly efficient solid-phase supported the radiosynthesis of [^{11}C]PiB by [^{11}C]methylation of the 6-OH-BTA-0 (Fig. 3.34) on a disposable cartridge, without the need to protect the hydroxyl group, warrants some attention (Boudjemeline et al. 2017). It is apparent that this one-step synthesis offered several advantages compared to the original radiosynthesis of [^{11}C]PiB, which required additional protection of the hydroxyl group with O-methoxymethyl (MOM) and deprotection after radioisotope labeling with [^{11}C]CH$_3$I. In this new and automated approach, [^{11}C]methylation was performed by passing gaseous [^{11}C]CH$_3$OTf through the tC18 cartridge preloaded with 6-OH-BTA-0 in acetone. The reaction lasted 2–3 min at room temperature, then the impurities were flushed out to the waste container using 12.5–15% ethanol solution, and the product [^{11}C]PiB was eluted into the final vial through a sterile filter using 50% ethanol followed by sterile phosphate buffer. The advantage of this approach is that no HPLC purification involved all three steps, production, purification, and formulation, using only the tC18 cartridge. The [^{11}C]PiB derived from this method has been used for human injection with the overall reaction time within 10 min starting from [^{11}C]CH$_3$OTf with 22% isolated yield (not decay-corrected) and molar activity of 190 GBq/μmol. The remarkable short reaction time thanks to the labeling using unprotected 6-OH-BTA-0.

3.3.2.2 Loop Chemistry-Based [^{11}C]methylation

This is another innovative [^{11}C]labeling operation, with respect to [^{11}C]methylation that has been used routinely in the clinical settings, worth further discussion. By far, this is the predecessor of the later development called microfluidics. These thin tubing loops (0.75–1 mm ID) typically attached to the HPLC injector (Iwata et al. 2001; Wilson et al. 2000) are used in place of the reaction, a perfect idea for tracer chemistry. This method proved to be an ideal alternative to the bubbling method for tracer development and offered a convenient way for direct transferring the reaction mixture to an HPLC column (Iwata et al. 2001). The overall goal of this method is to improve labeling procedure, RYC, and specific activity of the radiopharmaceutical compounds and to reduce reaction time. The other advantage of loop chemistry involves the use of small amount of solvents and reagents, enabling the preparations of highly concentrated precursor solutions, facilitating the improved methylation outcome. It has been reported that lowering the precursor concentration below some thresholds may lead to a drop in [^{11}C]methylation yields (Studenov et al. 2004). Recently, loop chemistry has been implemented for fully automated production of [^{11}C]Raclopride and [^{11}C]DASB employing only ethanol from start to end. And thus eliminating other organic solvents, simplifying QC, and thus facilitating routine clinical production of these radiopharmaceutical probes with better RCY and specific activities compared to other approaches (Shao et al. 2013). In these experiments, the precursor was dissolved in ethanol (100 μL) and loaded into the 2 mL HPLC loop and purging air with nitrogen gas for 20 s at a rate of 10 ml/min before passing the [^{11}C]MeOTf through the loop for 3 min at the rate of 40 ml/min, followed by semi-preparative HPLC. The radioactive peak indicated the product, was collected and diluted in saline before storing in the vial by passing through a sterile filter. For [^{11}C]Raclopride and [^{11}C]DASB, the RCY was 3.7 and 3%, molar activity of 770 TBq/mmol and 560 TBq/mmol, respectively.

Like the bubbling method, loop chemistry is critically solvent-dependent due to the encountering small surface area and the use of a remarkably minute amount of solvent. Since the process now is different from the bubbling mechanism, many solvents that work well in bubbling reactions might behave differently in loop chemistry. For instance, acetone is commonly used in [^{11}C]methylations via [^{11}C]CH$_3$OTf. However, acetone seems to be too volatile for loop syntheses; this is compounded by reduced volume to be used compared to the bubbling method. This observation was proved to be true when other versions of higher boiling point ketone, such as methylethylketone, diethylketone, dipropylketone or cyclohexanone. Among these ketones, cyclohexanone, which has the highest boiling point, provides notably the best results (Iwata et al. 2001). Using this solvent modification, [^{11}C]Raclopride was prepared with over 40% RCY based on [^{11}C]CH$_3$OTf decay-corrected within 40 min from EOB using the simple loop method.

While it is clear that the amount of gas used to deliver [^{11}C] source is critical for successful synthesis, this is related to the characteristics of the loops. It has been demonstrated that different loop materials profoundly impact the reaction yield. Comparison of the three loop types, including Tefzel, PEEK, and stainless steels, revealed that the loss of trapped radioactivity is much faster for Tefzel loops compared to PEEK or stainless steel counterparts and leads to a decrease of the product radiochemical yield (Studenov et al. 2004).

3.3.2.3 [^{11}C]methylation Using Inorganic Bases

[^{11}C]methylation of arylamines like anilines or pyrroles with [^{11}C]CH$_3$OTf should proceed without issues. In some cases, when dealing with low or moderate reactivity toward [^{11}C]CH$_3$OTf, conventional methods for reaction optimization would include elevating the reaction temperature or using a catalytic amount of base. However, the sluggish reactions may happen with [^{11}C]CH$_3$I or if the methylation occurs with arylamines, where electron-withdrawing groups further reduce the ring's electronic effect. In most cases, the forceful use of harsh conditions, such as high temperature combined with strong bases would lead to the decomposition of either the starting materials or the products, or both. Given the abundance and prevalence of anilines and pyrroles and related bioisosteres in medicinal chemistry for drug development, precarious and shortcoming of [^{11}C]methylation chemistry prevents translating many promising candidates into useful PET probes. Therefore, it is utterly critical to improve this important labeling method. One of the innovative chemistries reported not too long ago involved the integration of ultrasound with inorganic bases to generate agitated solid inorganic bases in dimethylformamide to overcome the poor reaction activity of arylamines, enabling rapid [^{11}C]methylation labeling reaction (<10 min) at room temperature (Cai et al. 2012). Ultrasound and solid bases play a critical role in the successful reaction. It has been demonstrated that ultrasound effectively accelerates the chemical reaction at the solid–liquid interfaces, such as the use of solid bases in this case. In stark contrast to conventional chemistry, where bringing all reactants into one homogenous phase is crucial for the completion of the reaction, heterogenous liquid–solid is a unique domain for sonochemistry. The ultrasonochemical effect generates bubbles in a liquid medium when collapse will liberate considerable energy in terms of heat in 10^{-6} s, causing tremendous agitation at the interphase between the reactants (Einhorn et al. 1989). This method has been shown to be very successful for [^{11}C]methylation of several pyridine derivatives. However, it is worthwhile to note whether anilines or substituted anilines, besides the desired product, there is an accompany of a side product of N-methylformanilide, since dimethylformamide is the source of formylation of anilines. As shown in Fig. 3.35, during the [^{11}C]methylation of a version of PiB compounds for imaging beta-amyloid plaques using Li$_3$N or KOH, N-methylformanilide side product was actually a major product. However, [^{11}C]-methylformanilide can be converted to the desired product by hydrolysis with 1 M KOH at room temperature for 5 min. Based on this mechanism, the conclusion is that the N-formyl compounds are the progenitors of the labeled [^{11}C]N-methylarylamines.

3.3.2.4 Palladium-Mediated Cross-Coupling of [^{11}C]CH$_3$I

So far, the discussion focuses on N-methylation. Now, for the [^{11}C]-methylation via C–C bond formation or on the C-aromatic system, the contribution of transition metal in this type of reaction is indispensable. Transition metal-catalyzed cross-coupling reactions between organometals with organic halides to form carbon–carbon bonds serve as excellent means for [^{11}C]methylation using [^{11}C]CH$_3$I. These reactions include Stille coupling using stannanes, Suzuki reaction using organoboronic acid, Hiyama coupling with organosilanes, and Heck reaction with alkenes; all use palladium as a catalyst. The translation of such knowledge from conventional organic chemistry for [^{11}C]-methylation via transition metal was first demonstrated in 1995, and this breakthrough work formed the foundation for using palladium-catalyzed cross-coupling in the reaction with CH$_3$I (Andersson et al.

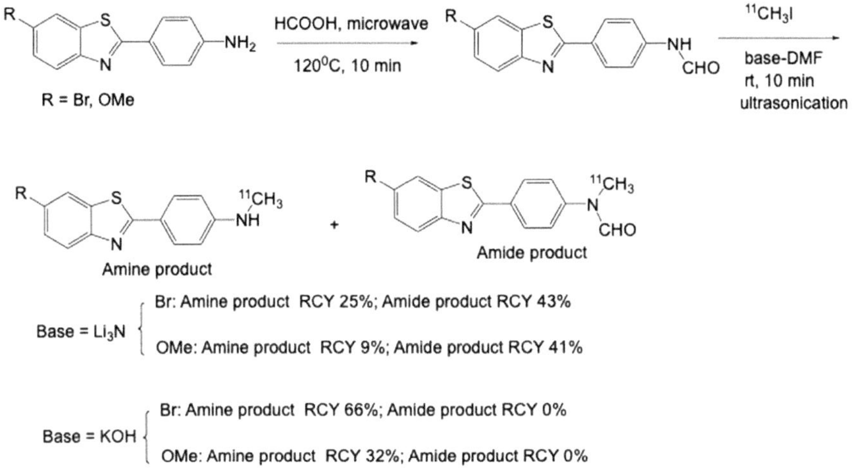

Fig. 3.35 [^{11}C]-N-methylation with [^{11}C]methyl iodide by activation of arylamine as N-formyl group using solid inorganic bases. Data obtained from Cai et al. (2012) with permission from the author

1995). The reason why this chemistry is so significant in the field is that alkyl halide with an sp^3-hybridized carbon (C(sp^3)-X) acts as a weak electrophile for transition–metal-catalyzed cross-coupling compared to the corresponding vinyl or aryl halides with sp^2-hybridized carbon atoms (C(sp^2)-X) (Doi 2015). For reactions that involve transition metals, the electron-rich C(sp^3)-X would undergo oxidative addition to a low-valent transition metal complex at a much lower rate than that of C(sp^2)-X, as found in, for example, vinyl or aryl halides (Frisch and Beller 2005). As shown in Fig. 3.36 (Andersson et al. 1995), [^{11}C]methylation using [^{11}C]CH$_3$I proceeded smoothly via Stille and Suzuki coupling reactions to form carbon–carbon bond formation across many key substrates to label acrylic, alkenylic, and alkylic positions. Surprisingly, the reaction operation was convenient and simple. For detailed information, the [^{11}C]CH$_3$I was transferred by a stream of nitrogen gas into a vessel containing tetrakis(triphenylphosphine)palladium [Pd(PPh$_3$)$_4$] in an appropriate solvent, followed by the addition of the organostannane or organoborane substrate. The reaction was then heated to 90 °C for 4 min. And the labeled product was purified by semi-preparative HPLC. The RCY was calculated and decay-corrected based on [^{11}C]CH$_3$I. It is worth noting that conventional cross-coupling reactions usually require extended reaction time, several hours to the least. It turns out transition metal-catalyzed cross-coupling reactions using [^{11}C]methyl halides have a unique advantage, that is due to the intrinsic characteristics of tracer chemistry in the PET labeling procedure.

The use of large excess amounts of Pd^0 and stannyl/boryl substrate, which will facilitate effective oxidative addition reaction of tracer [^{11}C]methyl halides and trans-metalation step. Altogether, these are the driving forces behind the high RCY and speedy labeling reaction time (Fig. 3.37) (Doi 2015).

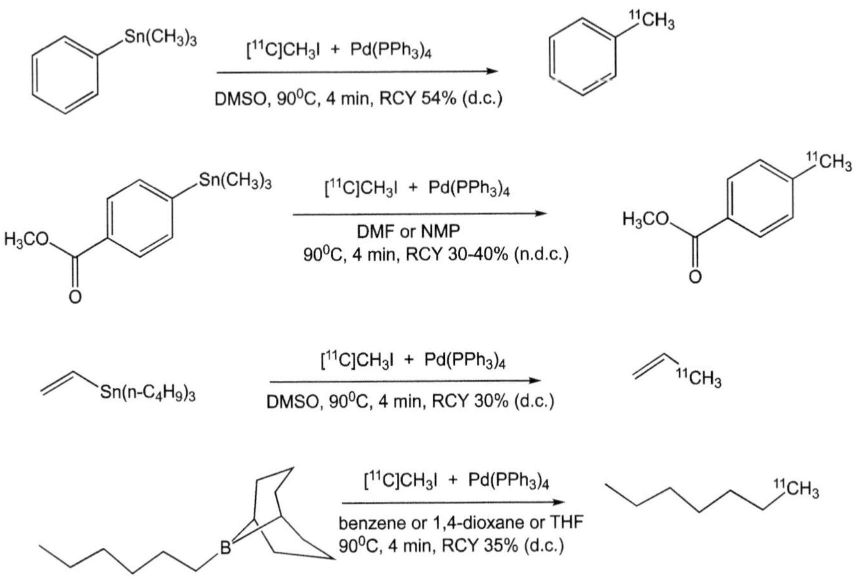

Fig. 3.36 Palladium-mediated cross-coupling reaction of $[^{11}C]CH_3I$ with organotin and organoboron compounds. Data obtained from Andersson et al. (1995) with permission from the author

Besides optimizing the reaction yield and labeling efficiency, the toxicity of the participating materials needs further scrutinization to ensure future clinical translation of the radioligands. That balance is going to inevitably be more demanding for any type of reaction using transition metals. For example, both Stille and Suzuki cross-coupling reactions have great implications for the generation of C–C bond, not limited to only aryls, but also including alkyls, alkenyls, and alkynyls. Stille coupling reaction has robust reaction rates. However, as mentioned above PET labeling chemistry is different from conventional synthetic organic chemistry. The reaction is actually pseudo-first-order, given the use of large excess of the precursor/substrate, and thus, the concentration of the precursor remains constant during the labeling process (Anderson et al. 2013). Thus, the higher reactivity in a Stille reaction does not play a key role in here (Anderson et al. 2013). Aside from the fact that the stannyl substrates are air and moisture stable, the reactions are also tolerant to the functional groups. In contrast to the Suzuki reaction, the Stille coupling reaction does not need a base, and thus, the reaction is suitable for labeling any base-sensitive compounds (Anderson et al. 2013). However, a major limitation of Stille coupling, particularly if the work is related to clinical translation, is its toxicity. Last but not least, Stille coupling is unflavored in alkyl-alkyl cross-reaction (Doi 2015). In that respect, Pd-mediated cross-coupling of $[^{11}C]$methyl iodide with organoborane via Suzuki coupling is more favorable than Stille coupling, given that the former uses organobronic acids, which are safer than stannanes. In fact, boron is more than safe; it has been known by scientists for many years that trace mineral boron is a micronutrient with diverse and

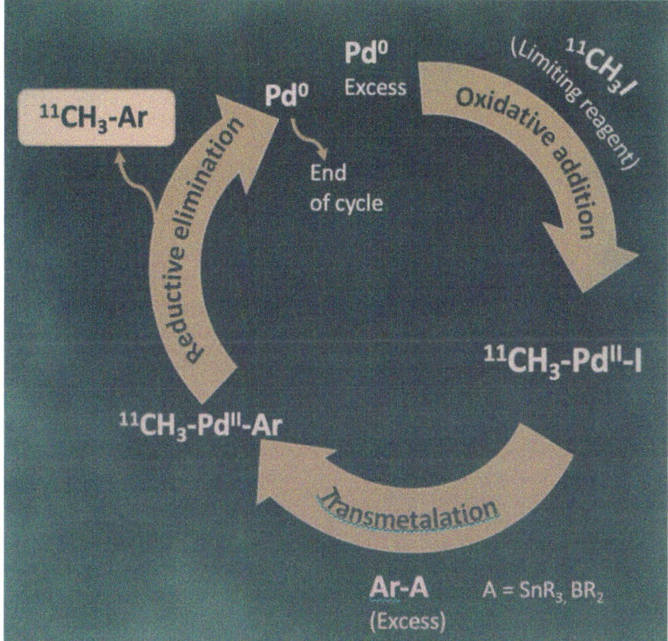

Fig. 3.37 Proposed mechanism to explain the improved reaction time. The presence of excess amount of Pd catalyst and stannyl or boryl substrate can accelerate the transmetalation process. Data reproduced from Doi (2015)

vitally important roles in metabolism, and thus it is necessary for plant, animal, and human health (Pizzorno 2015). But foremost, the reliability of the reaction along with the mild condition that can be used orthogonally to other labile function groups, combined with stability and simple preparation of the trialkylborane precursors, the reaction needs no inert air, altogether making Suzuki reaction a versatile reaction for [^{11}C]methylation of bioactive ligands.

Figure 3.38 shows a typical Suzuki reaction for the preparation [ω-^{11}C]palmitic acid for imaging fatty acid metabolism (Hostetler et al. 1998). The versatility of Suzuki coupling enables to work with different types of substrates, combined with the stability and simple preparation of the borane precursors making application of the Suzuki coupling an attractive strategy for the synthesis of tail-labeled fatty acids through the incorporation of [^{11}C]CH$_3$ (Hostetler et al. 1998). Despite different substrates, the labeling condition was identical: [^{11}C]methyl iodide was collected in THF at 0°C, then Pd(PPh$_3$)$_4$, borane substrate (furan or ester), and NaOH were added in succession, and the temperature was heated to reflux for 4 min to provide the labeling product. The *t*-butyl ester was removed with neat TFA at 90°C for 1 min to provide the desired product. For other derivative, ozonolysis of furan followed treatment with peracetic acid to provide the desired product. Both reactions afforded high RCY and purity.

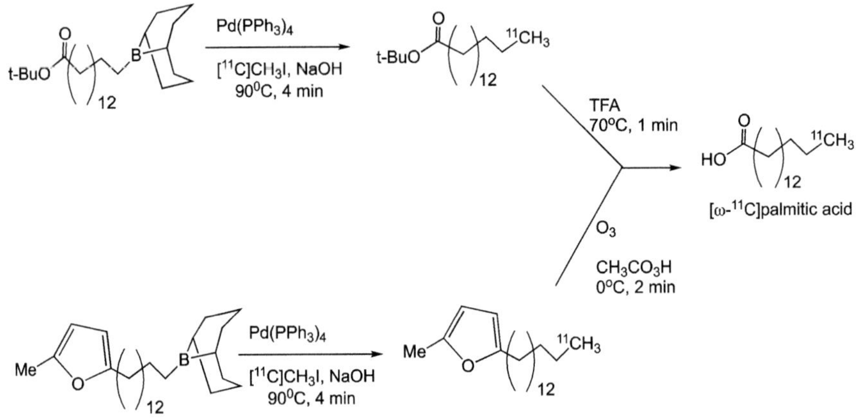

Fig. 3.38 Suzuki reaction for the synthesis of a [^{11}C]palmitic acid probe. Data obtained from Hostetler et al. (1998) with permission from the American Chemical Society

Another advantage of using the Suzuki reaction is to label aromatic systems with [^{11}C]CH$_3$ moiety in the presence of unprotected, free amines, including primary, secondary, and tertiary amines or other functionalized groups without competing alkylation at these amino groups (Anderson et al. 2013). The work eliminates protection/deprotection operations, which usually require additional time and cost. This approach offers a unique platform for developing radiopharmaceutical probes since many biological-relevant compounds or drug-like molecules contain amino-laden arenes. In principle, one would think that competing alkylation of the nitrogen by [^{11}C]CH$_3$I would dominate the reaction. However, based on the prior proposed reaction mechanism (Fig. 3.37), the Suzuki reaction between [^{11}C]CH$_3$I and arylboronic acid is a stepwise process, first started with oxidative addition of [^{11}C]CH$_3$I, followed by transmetalation and finally reductive elimination in the presence of boronic substrate. This led to a hypothesis that N-^{11}C-methylation of amines, in this condition, would not happen. Experimental data have proved that this theory holds true. For example, a variety of [^{11}C]CH$_3$ labeled aromatic amines were synthesized; since no protection/deprotection was involved in the process, the labeled products were isolated with high RCY (56–92%) and molar activity > 4000 Ci/mmol in less than 20 min following the production of [^{11}C]methyl iodide (Anderson et al. 2013). Representative data is shown in Fig. 3.39.

Recently, direct aryl and heteroaryl acetylation using [^{11}C]CH$_3$I mediated by (carbonyl)cobalt was successfully developed for the generation of the labeled aryl methyl ketones, a versatile synthetic synthon for the synthesis of many pharmacologically active molecules (Dahl et al. 2016). This type of reaction is fast and simple compared to other established work. In the past, [^{11}C]labeled aryl methyl ketones can be obtained by palladium-mediated [^{11}C]carbonylation reactions using [^{11}C]CO (Rahman et al. 2004; Karimi et al. 2005; Dahl et al. 2013) or acetyl chloride (Arai et al. 2009a, 2009b; Arai 2012). However, these reactions are demanding,

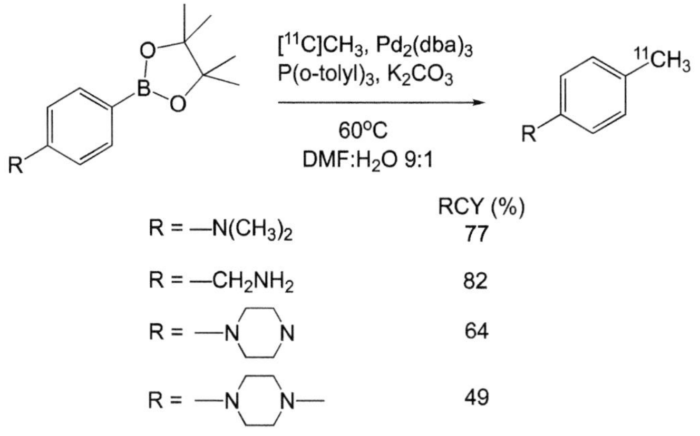

Fig. 3.39 [^{11}C]CH$_3$ labeling via Suzuki reaction with unprotected aromatic amines. Data obtained from Anderson et al. (2013) with permission from Tetrahedron Letters

the problem stemmed from the multistep radiosynthesis starting with [^{11}C]CO$_2$. Further, labeling with [^{11}C]CO$_2$ usually confers low specific activity compared to those obtained from [^{11}C]CH$_3$I. And for imaging neuroreceptors, it is crucial to achieve high specific activity to avoid blocking the receptors with non-radioactive carrier molecules. Altogether, there is an unmet need to develop of a more robust method to improve the work. In this cobalt-catalyzed reaction, it is found that the diaryl ketone often posed as a major product. However, by treating the reaction with a large excess of aryl halide to [^{11}C]CH$_3$I, a reasonably accepted amount of desired [^{11}C]-labeled aryl methyl ketone material could be achieved. The optimized reaction conditions include the delivery of [^{11}C]CH$_3$I into a vessel containing a solution of aryl halide in acetonitrile at room temperature. The solution was transferred to a second vessel containing Co$_2$(CO)$_8$ resided in a microwave cavity. The resultant mixture was let to react at 130 °C for 1 min (Dahl et al. 2016). This reaction condition is applicable for the radiosynthesis of a variety of functionalized aryl methyl ketones. As shown in Fig. 3.40, the aryl chloride starting materials that have the electron-withdrawing groups, such as Br, carbonyl, cyano, and haloalkyl (entries 4–7) at para-position offered good RCYs. The same holds true for the derivatives with electron-donating groups, including hydroxyl, alkoxyl, and amine (entries 1–3). Further, heterocyclic systems are also a good candidate for this reaction. The [^{11}C]acetylation of 2-chlorothiophene provided 31% radiochemical conversion into the desired aryl methyl ketone product. While [^{11}C]2-acetylpyridine was formed from the halide substrates with acceptable yield from 16–22%.

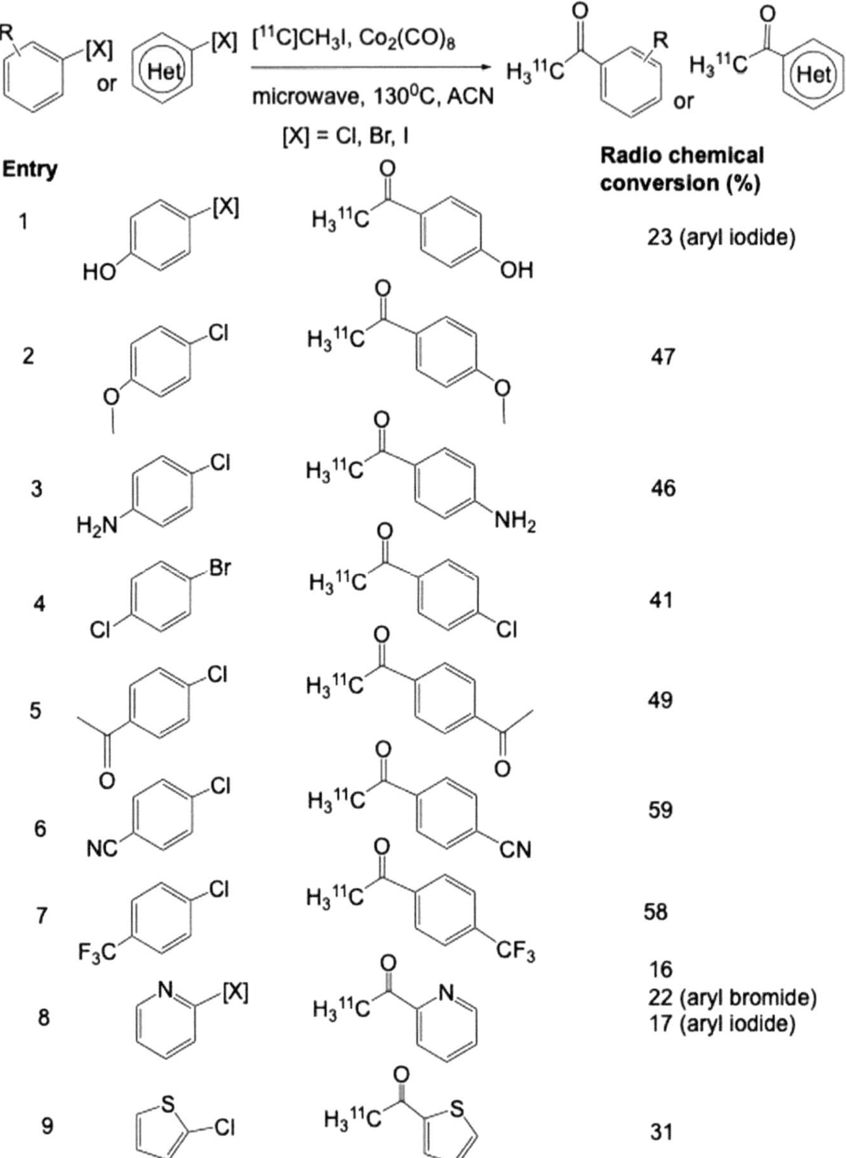

Fig. 3.40 (Carbonyl)cobalt-mediated [^{11}C]acetylation of different versions of aryl halides with [^{11}C]CH$_3$I. Data obtained from Dahl et al. (2016) with permission from the author

3.3.3 [^{11}C]labeling via Carbonylation Using [^{11}C]CO

[^{11}C]CO was one of the first [^{11}C] labeled compounds to be used in tracer experiments in human (Kihlberg and Langstrom 1999; Tobias et al. 1945). The development of PET tracers using [^{11}C]CO as a precursor for labeling carbonyl moiety is indispensable to radiopharmaceutical chemistry, given the prodigious contributions and ubiquity of carbonyl analogs in drug development, including amides, ketones, aldehydes, carboxylic acids, and esters. In many cases, the participation of the carbonyl groups in the metabolic pathway, sometimes, makes it even more advantageous than methyl groups as labeling position (Kihlberg and Langstrom 1999). The proclivity to bring more [^{11}C]CO applications in PET chemistry due to the convenient synthesis of transition metal-mediated carbonylation reactions, which is already very much established in conventional organic chemistry (Brennfuhrer et al. 2009; Kealey et al. 2014; Rahman 2015). Two synthetic methods can be adapted to [^{11}C]carbonylation; these are through transition metal- or radical-catalyzed reactions. In the former case, transition metals in groups 9 and 10 of the periodic table play a key role in synthesizing several functional groups found in bioactive materials. For group-10 metals, the proposed mechanism for carbonylation of organic halides via cross-coupling reactions started first with oxidative addition of zerovalent metal with an organic halide, followed by insertion of carbon monoxide to form the carbonyl intermediate with electrophilic characters suitable for a nucleophilic attack. Finally, a reductive elimination step to confer the final product and release the zerovalent metal ready for catalyzing the next cycle (Fig. 3.41) (Rahman 2015).

Further, group 9 transition metal, such as rhodium, can also catalyze to mediate C–C bond-forming reactions. One of the typical experiments of rhodium-catalyzed carbonylation using [^{11}C]CO was reported for the synthesis of [^{11}C]urea and [^{11}C]carbamates using phenyl azide and [^{11}C]CO with 1,2-bis(diphenylphosphino)ethane-bound Rh(I) complex (Doi et al. 2004). These reactions are thought to go through the formation of [^{11}C]isocyanate intermediate or an [^{11}C]isocyanate-coordinated Rh complex before forming the products (Fig. 3.42).

Free-radical carbonylation is well-described in conventional chemistry. Specifically, long-chain organohalides can be converted into aldehydes in the presence of Bu$_3$SnH and AIBN in high-pressure condition using an autoclave (Ryu et al. 1990). The general mechanism involves the conversion of an alkyl halide into a radical via Bu$_3$Sn$^•$. The carbon radical would trap carbon monoxide, followed by hydrogen abstraction from tin hydride (Bu$_3$SnH) to form an aldehyde. The application of free-radical carbonylation for the synthesis of ^{11}C-labeled aliphatic carboxylic acids, esters, and amides has been reported in the literature using different sources of radical generation (Rahman 2015). For instance, [carboxyl-^{11}C]carboxylic acids and esters were prepared from alkyl iodides via photoinitiated radical reactions using [^{11}C]CO (Itsenko and Langstrom 2005a, 2005b).

The current limitation in PET chemistry is the lack of methods for robust production of [^{11}C]CO with good concentration. This is compounded by the poor reactivity

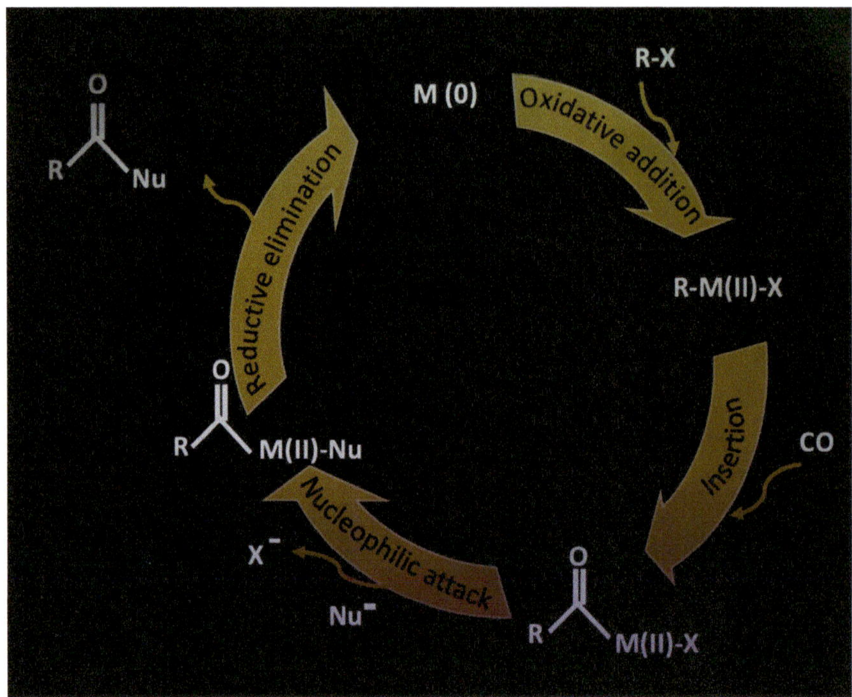

Fig. 3.41 Proposed mechanism of transition metal-mediated carbonylation. Data reproduced from Rahman (2015)

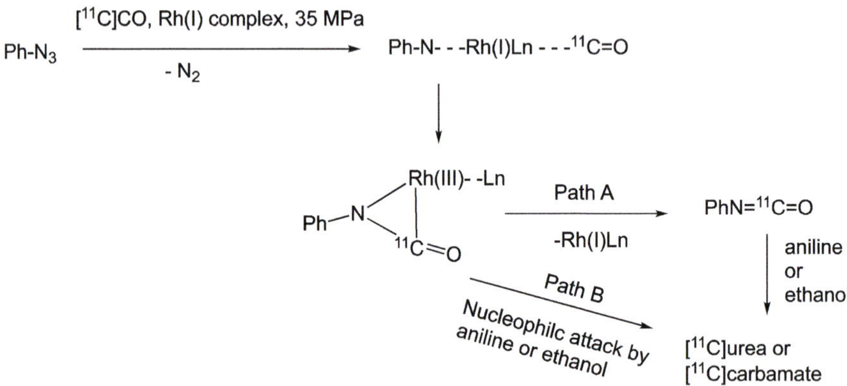

Fig. 3.42 Proposed mechanism for rhodium-catalyzed [^{11}C]carbonylation. Data obtained from Doi et al. (2004) with permission from the author

and solubility of carbon monoxide, and thus, it is challenging to trap [^{11}C]CO in low-volume reaction vessels.

Reports in the past demonstrated that delivery of [^{11}C]CO by conventional bubbling method provided poor availability in the reaction vials. Typically, it has been reported that 1–10% [^{11}C]CO remaining in the vial after delivered by [^{11}C]CO-He gas stream in THF and 4.5–6.2% in 1,2-dimethoxyethane/water (Lidstrom et al. 1997; Al-Qahtani and Pike 2000). One of the approaches to overcoming this drawback involves trapping [^{11}C]CO in the micro-autoclave at high pressure. This approach seems to be very successful with the synthesis of [^{11}C]labeled amides. Aside from using very high pressurized conditions (35 MPa), the reactions are simple, with no apparent complications found in the process. A typical run would involve all the reagents being dissolved in 1,4-dioxane, and the resulting mixture was heated with [^{11}C]carbon monoxide at 130–150 °C. Using this innovative method, the trapping efficiency was nearly quantitative, and thus the labeled amide products got molar activity in the range of 400–1300 GBq/μmol (Fig. 3.43) (Kihlberg and Langstrom 1999). In conclusion, the use of [^{11}C]CO in metal-mediated (such as palladium or other metal in group 10 of the periodic table) carbonylative coupling of aryl halides with amines has been demonstrated as a versatile approach to the production of amide ^{11}C-labeled in the carbonyl moiety (Kihlberg and Langstrom 1999).

Later on, another method for enhanced trapping [^{11}C]CO at atmospheric pressure using reversible BH$_3$.THF complexing with [^{11}C]CO for the palladium-mediated carbonylation of alkyl halides has been reported (Fig. 3.44) (Audrain et al. 2004). The first step of this operation involves the conversion of [^{11}C]CO into the [^{11}C]BH$_3$.CO complex and trapping it at a very low temperature. Another reaction condition worth mentioning is the choice of a base, which is crucial in this reaction, specifically for the amidation step. However, no correlation is found between the strength of the bases and the reaction outcome. For example, using a strong base like 1,8-diazabicyclo[5.4.0]undec-7-ene (DBU) afforded only traces of the product. In contrast, much weaker base like pyridine or intermediate bases like 2,2,6,6-tetramethylpiperidine (TMP) and benzylamine afforded radiochemical yields of 36, 8, and 20%, respectively. Overall, triethylamine (TEA) seems to be the best base, which provides 47% RCY. Furthermore, the polarity of solvents seems to have no role in the process of carbonylation, albeit THF proved to be the best solvent compared to DMF or 1,2-dichloroethane.

The reaction setup seemed to be laborious, starting with the trapping of [^{11}C]CO at −192 °C in the presence of molecular sieves. Then, the [^{11}C]CO was transferred to the first vial containing BH$_3$.THF mixture to form the [^{11}C]BH$_3$.CO complex at room temperature. The mixture was then transferred to a second vial, which was cooled at −60 °C in order to condense THF, and finally, the mixture was transferred to the third vial containing the reactants in solution cooled at −78 °C. After trapping there for a few minutes, the reaction mixture was heated in the range of 50–140 °C, depending on the substrate, for a few minutes before processing to purify the desired products. The feasibility of this technique was proved for the successful synthesis of amides and lactones.

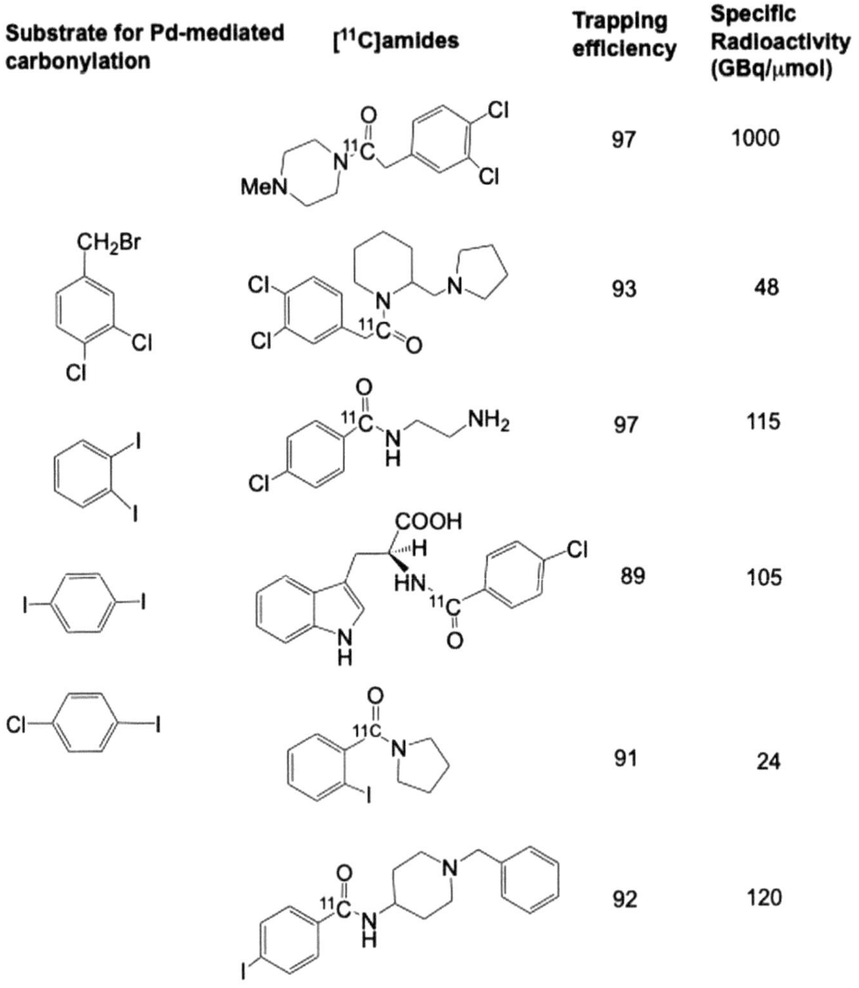

Substrate for Pd-mediated carbonylation	[^{11}C]amides	Trapping efficiency	Specific Radioactivity (GBq/μmol)
		97	1000
CH$_2$Br (3,4-dichlorobenzyl bromide)		93	48
1,2-diiodobenzene		97	115
1,4-diiodobenzene		89	105
1-chloro-4-iodobenzene		91	24
		92	120

Fig. 3.43 Palladium-mediated carbonylation via [^{11}C]CO under high pressure. Data obtained from Kihlberg and Langstrom (1999) with permission from the American Chemical Society

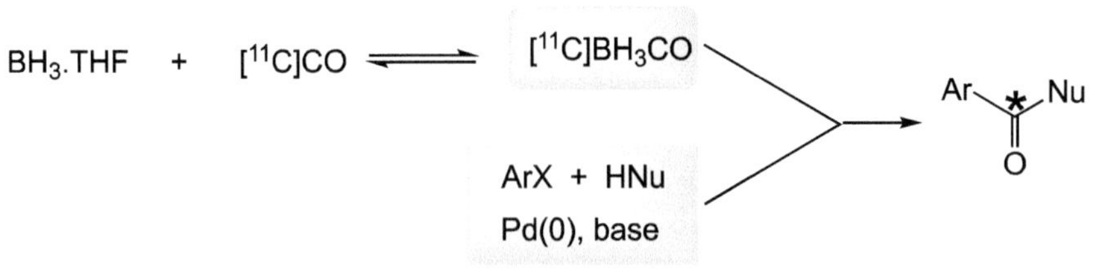

$$BH_3 \cdot THF \; + \; [^{11}C]CO \; \rightleftharpoons \; [^{11}C]BH_3CO$$

ArX + HNu

Pd(0), base

$$\rightarrow \; Ar\overset{*}{C}(=O)Nu$$

Fig. 3.44 [^{11}C]carbonylation reaction at atmospheric pressure using [^{11}C]carbon monoxide-boron complex. Data adapted from Audrain et al. (2004)

Another notable method uses metals, such as copper(I) tris(pyrazolyl)borate complexed with carbon monoxide for trapping [^{11}C]CO. The complex has been used in palladium-medicated [^{11}C]carbonylation reactions (Kealey et al. 2009). Before this discovery, the first well-characterized stable copper-carbon monoxide complex was reported in 1969 (Scott et al. 1969). The presence of the carbonyl in the association was characterized by a single υ(CO) band at 2093 cm^{-1}. However, the complex is unstable, it is air-sensitive, and thus, such a system is unfit for trapping [^{11}C]CO since CO dissociation is likely in high dilution condition in an inert carrier gas stream. To overcome this issue, attention has been focused on tris(pyrazolyl)borate, which has a high affinity for copper(I) and can apply to the trapping of carbon monoxide due to the electron-releasing property of the system may help to strengthen the copper-carbonyl bond. Particularly, potassium tris(3,5-dimethylpyrazolyl)borate (K[Tp*] is very effective in trapping [^{11}C]CO when treated with CuCl in THF (Fig. 3.45)(Kealey et al. 2009). And most importantly, [^{11}C]CO can be released by the addition of a competing phosphine donor, which has a strong affinity for copper(I) (Bruce and Ostazewski 1973) to the mixture. Overall, this copper complex significantly improved the solubility of [^{11}C]CO at room temperature and pressure. The method has led to the successful development of [^{11}C]-labeled amides or ureas using the corresponding palladium catalyst.

In a recent development, [^{11}C]CO labeling reaction was optimized to improve RCY and reduce burden during the operation. Basically, [^{11}C]CO was transferred to a reaction vial with great trapping capability using noble gas, such as Xenon. The process is free of using high-pressure conditions and chemical-trapping additives (Eriksson et al. 2012). This method credits the high solubility of xenon gas in

Fig. 3.45 **a** Effective trapping of [^{11}C]CO via the formation of Cu[Tp*]^{11}CO. Addition of triphenylphosphine triggers the release of [^{11}C]CO from the complex. **b** Synthesis of [^{11}C]amide via carbonylation reaction between aryl halide and benzylamine using Cu[Tp*]^{11}CO as a source of [^{11}C]CO. Data obtained from Kealey et al. (2009) with permission from the author

Gas	T(K)				
	263.15	273.15	283.15	293.15	303.15
He	0.746	0.820	0.894	0.965	1.042
Ne	0.933	1.037	1.166	1.261	1.382
Ar	10.55	9.999	9.595	9.245	8.958
Kr	31.55	29.44	26.91	25.23	23.59
Xe	167.0	139.2	117.6	101.6	88.74
H_2	2.316	2.403	2.537	2.635	2.746
D_2	2.284	2.412	2.582	2.767	2.893
N_2	4.548	4.563	4.586	4.589	4.635
O_2	8.196	8.065	7.879	7.743	7.637
CH_4	22.44	20.86	19.67	18.24	17.23
C_2H_4	143.7	119.6	101.4	87.49	76.73
C_2H_6	202.8	165.2	136.5	114.7	97.68
CF_4	6.318	6.190	6.055	5.857	5.711
SF_6	37.26	32.81	28.84	25.75	23.43
CO_2	135.9	112.1	93.44	78.99	68.53

Fig. 3.46 Solubility ($10^4 \chi_2$) common gases in 1-butanol at low pressure (101.33 kPa) at different temperatures. χ_2, mole fraction. Data obtained from Pardo et al. (1995) with permission from Elsevier

organic solvents, such as THF, an ideal solvent reported in several prior carbonylation reactions, for the immersion of the delivered $[^{11}C]CO$ into the solvent without using chemical-trapping additives. Xenon has been considered an ideal gas with high solubility in THF across different temperatures, among other common gases that had been tried for the delivery of $[^{11}C]CO$ (Fig. 3.46) (Gibanel et al. 1993). This explains greater trapping activity in the reaction vessel when using xenon (26%) versus 16% when using helium. With simplified reaction maneuverings and avoiding the use of trapping chemicals, this xenon-assisted method offers a better opportunity to explore more reaction labeling of critical radiopharmaceutical precursors using $[^{11}C]CO$. A number of contemporaneous labeling experiments using this method have been demonstrated with moderate outcomes, including the $[^{11}C]$carbonylation of amide ($[^{11}C]$carbonyl-N-benzylbenzamide), urea ($[^{11}C]$carbonyltriclocarban), and ester ($[^{11}C]$carbonylmethyl nicotinate), provided with fair RCY of 71, 42, and 29% (dc), respectively (Eriksson et al. 2012).

3.4 Labeling via Prosthetic Chelators

Now we shift our discussion to a different labeling technique. So far, we mentioned radiolabeling of small organic molecules with well-defined chemical structures by exposing the precursors to the reaction conditions with scant concerns about elevated pHs, temperatures, and chemical additives. For active biologics, such as peptides, aptamers, DNAs, oligomers, proteins, or antibodies, among many others, the labeling conditions are more stringent than for small molecules. These macromolecules not only have defined structures, but their modes of action solely depend on maintaining appropriate three-dimensional conformations. Unfortunately, these precise structural orientations, required for the macromolecules to recognize their specific targets in a lock-and-key arrangement, are sensitive to pH and temperature. These new challenges have necessitated the development of a mild labeling strategy using prosthetic groups. These nonspecific analogs have no role whatsoever in the study; rather, they can be used to label with positron emitters as any conventional methods described above. Further, they possess an amine- or thiol-activated group for conjugation to the biologics.

3.4.1 Labeling Monoclonal Antibodies with $[^{18}F]F^-$

Over a hundred years ago, Nobel laureate Paul Ehrlich had an idea of targeting specific receptors for chemotherapy in a term he coined, "magic bullet." Further, he also pioneered the idea of using one's own immune system to treat diseases. His vision, later on, was realized with the development of monoclonal antibodies using hybridoma technology (Kohler and Milstein 1975). What makes Paul Ehrlich's work so special is that he also pioneered the foundation of imaging technology. He was the first who used fluorescence dyes for therapy as well as for visualization of specific biological structures (Strebhardt and Ullrich 2008)". The emergence of PET imaging combined with visualization of antibodies via PET imaging at the beginning of the twenty-first century is again a testimony of his visionary thinking in biomedical research.

The average size of an antibody is approximately 150KDa, making them unable to penetrate imaging targets effectively and thus resulting in a poor/signal-to-noise ratio (Shan 2004). One of the approaches to overcome slow distribution due to the long half-life of intrinsic antibodies involved in the development of engineered $[^{18}F]$-labeled antibody fragments (Fab) for imaging applications. A number of engineered Ab fragments have been reported, with modifications that differ in antigen-binding sites and size, including Fab (~50KDa), F(ab')$_2$(~110 KDa) fragments, single-chain Fv (scFv, ~ 25 KDa) with one antigen-binding site called monovalent, bivalent scFv dimmers (~50 KDa), and scFv-fusion proteins or sometimes called minibodies (~80 KDa) (Olafsen and Wu 2010).

In the early days, a bifunctional linker such as N-succinimidyl 4-$[^{18}F]$fluorobenzoate ($[^{18}F]$SFB) was developed with n.c.a for linking radionuclides to biologically active materials. This prosthetic linker was accomplished, starting with 4-formyl-N,N,N-trimethylanilinium triflate (Fig. 3.47) (Vaidyanathan and Zalutsky 1992). The $[^{18}F]F^-$ labeling was achieved to substitute the quaternary amine salt at 120° for 25 min to afford 4-$[^{18}F]$fluorobenzaldehyde in a 40–80% yield, followed by an oxidation reaction to convert the aldehyde group into the corresponding carboxylic acid using potassium permanganate. Finally, the carboxylic group was activated as an N-hydroxysuccinimide ester using dicyclohexylcarbodiimide (DCC). The overall RCY for the preparation of the $[^{18}F]$SFB was about 25%, with the molar activity at

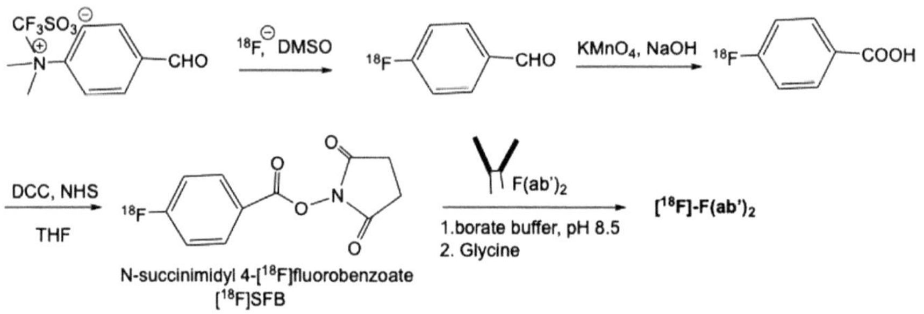

$CF_3SO_3^{\ominus}$

$^{18}F^-$, DMSO → KMnO$_4$, NaOH → DCC, NHS / THF

N-succinimidyl 4-[^{18}F]fluorobenzoate
[^{18}F]SFB

F(ab')$_2$
1. borate buffer, pH 8.5
2. Glycine
→ [^{18}F]-F(ab')$_2$

Fig. 3.47 Development of [^{18}F]SFB for labeling antibody. Data obtained from Vaidyanathan and Zalutsky (1992) with permission from Elsevier

the end of synthesis is 300 Ci/mmol. Esterification via DCC intermediate is a slow process that explains the reason related to moderate-to-low RCY of [^{18}F]SFB.

Regardless of that, the [^{18}F]SFB succinimidyl ester continues to serve as mainstream for Ab labeling. In a typical reaction, [^{18}F]SFB can be conjugated to F(ab')$_2$ fragments or any kind of engineered antibodies via the ε-amino group of lysine. The reaction time varies, depending on the condition setup for the experiment. In general, if the labeling is carried out at room temperature with the antibody concentration in the range of 3–4 mg/mL, the labeling time is about 15–20 min. The borate buffer offers the best condition to maintain the Ab integrity. While it is intuitive to use high pH for the labeling reaction since the pKa of the ε-amino group of lysine is approximately 10–11. However, the stability and configuration of antibodies restrict the pH to not more than 8.5 (Maerle et al. 2019). At this pH8.3, both the N terminal α-amino group (pKa 8.95) and the ε-amino group of lysine (pKa 10.53) are active and referred to as reactive amino groups (Madler et al. 2009; Mattson et al. 1993). The coupling reaction can be quenched using glycine (0.2 M) before purification against lower molecular weight using the Sephadex G-25 column. The specific activity of the [^{18}F]-labeled engineered antibody fragment is approximately 5 mCi/mg using this protocol.

To overcome the long activation time for in situ generation of the succinimide ester, another bifunctional prosthetic linker was developed for antibody labeling, in which the [^{18}F]F$^-$ labeling occurs in the presence of the succinimide ester. The chemistry started with the synthesis of the nosylate derivatives of N-succinimidyl methylbenzoate (Fig. 3.48) (Lang and Eckelman 1994). Since the linker has two activated sites, susceptible to nucleophilic attacks, and thus labeling conditions like high temperature and strong base are discouraged to prevent hydrolysis of these activated moieties. This necessitates the use of a very reactive leaving group such as the nosylate, which is tenfold more reactive than tosylate, to ensure the nucleophilic substitution can occur at room temperature. Further, this strategy also reduces unspecific labeling at the succinimide ester to form the undesired product of acid fluoride. For the first time, [^{18}F]labeling was performed at room temperature to provide the [^{18}F]-N-succinimidyl 4-(fluoromethyl)benzoate with 34% yield, and the undesired acid fluoride 27% based on radio TLC analysis. If the temperature was raised gently to 5 °C, the yield was 25%, albeit reducing the undesired acid fluoride to less than 5%. The optimal labeling was achieved using carbonate as a counterion, and acetone as a solvent. Over the course of 30–35 min [^{18}F]labeling, the HPLC-isolated [^{18}F] labeled succinimide ester was stable after 4 h. Protein was labeled with the [^{18}F] ester to provide an overall radiochemical yield of 10% (EOS) in 65 min. This linker was the first to be used in labeling antibody fragments and evaluated the biodistribution of [^{18}F]-labeled Fv fragment in vivo. The anti-Tac disulfide-stabilized Fv fragment was engineered from murine monoclonal antibodies that recognized the alpha subunit of interleukin 2 (IL-2α) receptor (Choi et al. 1995). The data showed that blood clearance of the probe was rapid, only < 10% retained in the blood by 15 min. The probe accumulates in the tumor with 4.2% injected dose/g between 45 and 90 min, and the major clear-out route was in the kidneys.

Further improved development of [^{18}F]SFB, mainly aiming to reduce the reaction time, got more momentum with the emergence of new technology. The integration of microwave-assisted synthesis has cut short the reaction procedure from roughly 2 h to 40 min in a conveniently one-pot synthesis (Olafsen et al. 2012). As shown in Fig. 3.49 the microwave-assisted synthesis happened in each step of the process, starting from [^{18}F]fluorination reaction to the deprotection of the ester and activation of the carboxylic acid. Each step took approximately 1–2 min and thus reducing the overall reaction time. The whole labeling process required only 30–45 min to provide the RCY of 13–24% and radiochemical purity of 95%. It is worth mentioning that the reaction time was significantly reduced thanks to N,N,N',N'-tetramethyl-O-(N-succinimidyl)uranium tetrafluoroborate (TSTU) instead of using DCC/NHS. The TSTU is already one step ahead of the O-acylisourea intermediate if using DCC/NHS. And in general, to effectively activate the carboxylic group into O-acylisourea sometimes, takes 18 h. The prosthetic bifunctional linker successfully labeled HER2 cys-diabody (Cys-Db) via the ε-amino group of lysine residues on the antibody fragment. In vivo imaging demonstrated that the antibody fragments cleared rapidly from the blood 2 h post-injection. PET imaging data using a tumor xenograft mouse model of HER2-positive tumor confirmed that the tumor is detected readily 1–4 h post-injection. From 4 to 6 h, the signal was stronger in the tumor than other organs, except the kidneys, which are confirmed as the clearance route.

3.4.2 Labeling Monoclonal Antibodies with ^{64}Cu

So far, we have discussed short half-life positron emitters for imaging antibodies, such as [^{18}F]-labeled antibody fragments. The advantage of using the [^{18}F] tag is the ease to access to the materials, and [^{18}F]-labeled antibodies have been used in clinical trials thanks to their safety; thus, the road to clinical translation is within reach. However, [^{18}F] radionuclides might not be truly ideal for use with biologics, like antibodies, which have a long half-life and slow pharmacokinetics. With a half-life of 12.7 h, ^{64}Cu is a great candidate for the work.

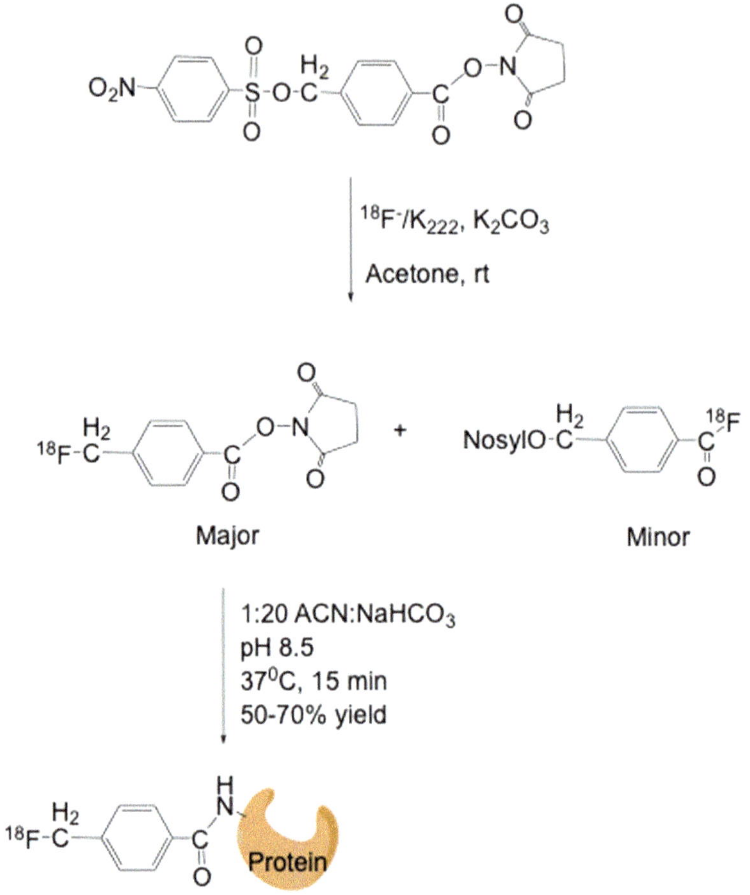

Fig. 3.48 Optimization of the reaction condition for labeling an activated bifunctional prosthetic linker. Data obtained from Lang and Eckelman (1994) with permission from Elsevier

A number of different methods have been reported for the production of ^{64}Cu, including the use of ^{64}Zn(n,p) ^{64}Cu reaction, albeit this method provides poor yield while high contamination of ^{67}Cu (Nayak and Brechbiel 2009). Lately, ^{64}Ni(p,n) ^{64}Cu reaction was developed and showed to produce high-quality ^{64}Cu with sufficient quantities and qualities for therapeutic applications (Obata et al. 2003).

To generate an imaging probe, metals, like ^{64}Cu, must be trapped in tight coordination via the chelators. These bonds are generated through dative interactions with the unshared pairs of electrons on each oxygen and nitrogen atom of the respective chelator (Hermanson 2008). In the early days, the activated EDTA chelator served as a bifunctional linker for labeling peptides and antibodies with ^{64}Cu (Fig. 3.50). Among the radioisotopes studied so far, such as ^{111}In and ^{57}Co, it seems EDTA can

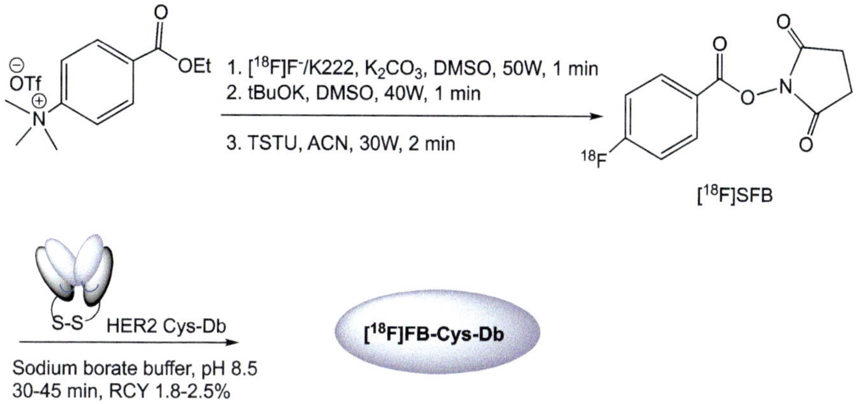

Fig. 3.49 Microwave-assisted synthesis of [^{18}F]SFB bifunctional linker for labeling antibody fragments for in vivo application. The animal was injected with 82 µCi of [^{18}F]FB-anti-HER2-Cys-Db starting 2 h post-injection, the MCF-7/HER2 tumor was visualized via microPET imaging (arrow). Data adapted from Olafsen et al. (2012) with permission from the author

establish firm coordination to ^{64}Cu. However, it is reported that TETA (1,4,8,11-tetraazacyclotetradecane-N,N',N'',N'''-tetraacetic acid)-copper association is better than EDTA-copper counterpart (Cole et al. 1986). Besides these chelators, different versions of 1,4,7,10-tetraazacyclododecane-N,N',N'',N'''-tetraacetic acid (DOTA) succinimidyl ester or maleimide were employed for labeling antibodies on amino or thiol groups, respectively. Another well-recognized chelator, such as diethylenetriaminepentaacetic acid (DTPA), has been used to label ^{64}Cu and other metal cations. It contains three tertiary amines and five sterically unconstrained carboxylic acids that can be used for coordinating metal ions and serve as a functionalization handle (Deblonde et al. 2018; Gut and Holland 2019). As a chelator-linker where the amino group of antibodies can react with the anhydride moiety on DTPA to create a multivalent and metal chelating through a strong coordination complex for in vivo applications. In fact, several ^{111}In-labeled antibodies associated with DTPA chelator have received clinical approval for SPECT imaging. These include ^{111}In-Satumomab for diagnosing colorectal and ovarian carcinomas (Corman et al. 1994), and ^{111}In-Capromab for prostate cancer therapy (Boros and Holland 2018).

Another noteworthy mention is that the hydrophobicity of macrocyclic molecules sometimes may hinder antibody labeling. To overcome this drawback, the succinimidyl ester in DOTA was incorporated with a polar group, such as a sulfonate moiety, which enables antibody bioconjugation in aqueous conditions (Fig. 3.50). The beauty of this antibody-chelate conjugation chemistry is that after labeling the physicochemical property of the complex is largely dictated by the large-sized antibody. Further, the hydrophobicity of macrocyclic chelators significantly reduces after they coordinate to ^{64}Cu.

Although each macrocyclic chelator has different strengths or kinetics in how they capture ^{64}Cu, they all share the same mechanism. Basically, the coordination with

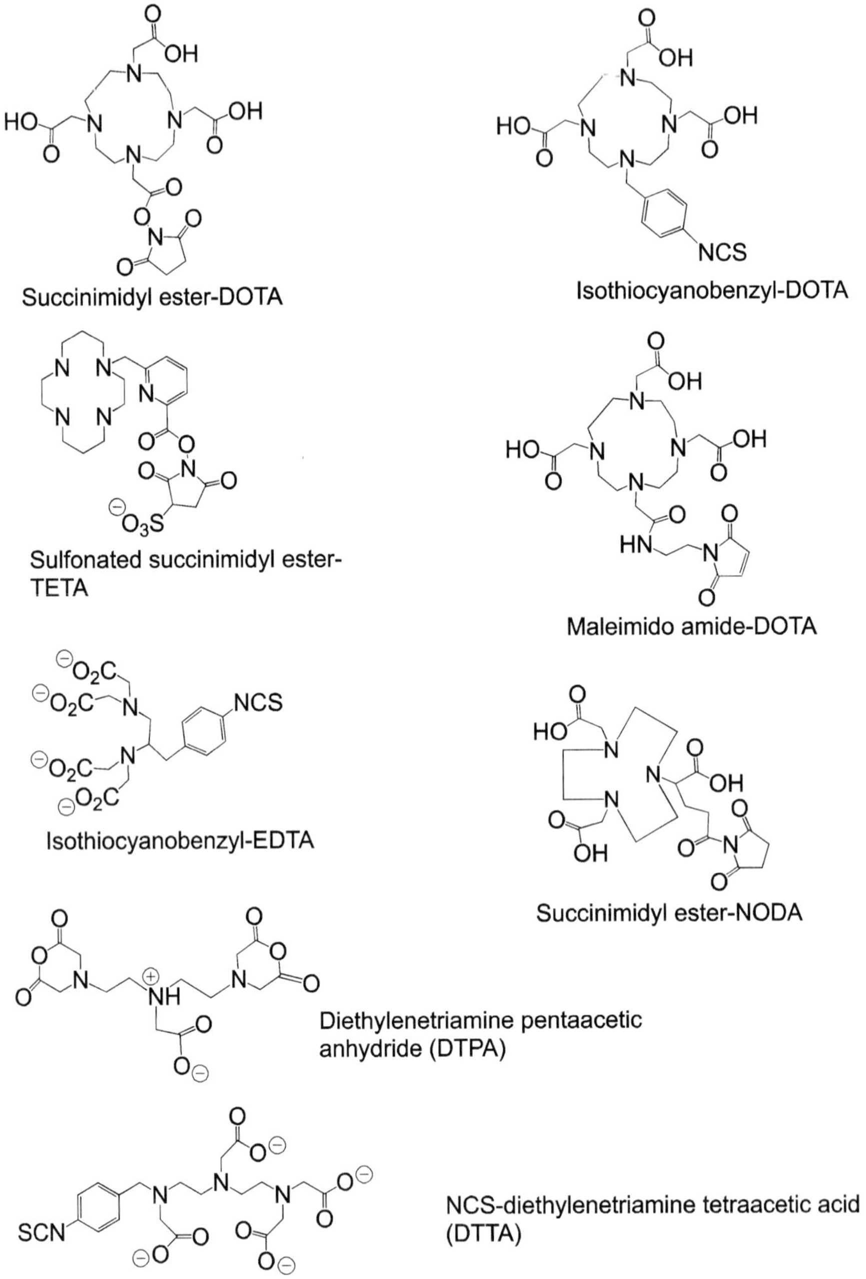

Fig. 3.50 Activated macrocyclic compounds used for labeling peptides, proteins, and antibodies with ^{64}Cu. Partially obtained from Anderson et al. (2008) with permission from the author

metal must go through the polyaminocarboxylate network, wherein the interaction happens via the unshared pair of electrons on each oxygen and nitrogen atom. Most importantly, the association with ^{64}Cu should be thermodynamically stable to prevent the immature release of ^{64}Cu and exchange with other ligands in vivo, leading to apparent toxicity. Therefore, designing effective chelators for ^{64}Cu labeling depend not only on the metal involved, but the process must also emphasize the structure of the chelator (Margerum et al. 1978). As a matter of fact, the choice of chelator becomes a topic of medical interest when it comes to using along with ^{64}Cu for theragnostic purposes. For instance, if a typical 10 mg of radioactive metal chelate-antibody is injected intravenously into a patient, this dose equals approximately 20 nM concentration in the blood circulation. With this diluted material, transmetal exchange may occur if the ^{64}Cu coordination with the chelator is labile since many highly concentrated metal cations in the blood can displace ^{64}Cu, such as albumin, Ca^{2+}, Mg^{2+}, transferrin, etc. Therefore, if the intention is to develop the ^{64}Cu-chelator for clinical work, the rate of dissociation in human serum at physiological temperature and pH should be assessed. For example, it has been reported that the rate of loss of copper from DOTA was approximately 0.0043 per day (Meares et al. 1990). In a typical experiment, the ε-amino group of lysine usually serves as a handle for attaching to the chelator via the succinimide ester. In the event, if the leaving groups on the chelator are not robust, or the chelator is equipped with the maleimide moiety, it is necessary to start with a strong nucleophile on the antibodies. In this case, the amino group can be converted to a thiol group using Traut's reagent (Fig. 3.51).

After bioconjugation of DOTA onto antibody, the ^{64}Cu can be associated with the chelator, most ideally at pH 5–5.5, using ammonium citrate. In case if the antibody

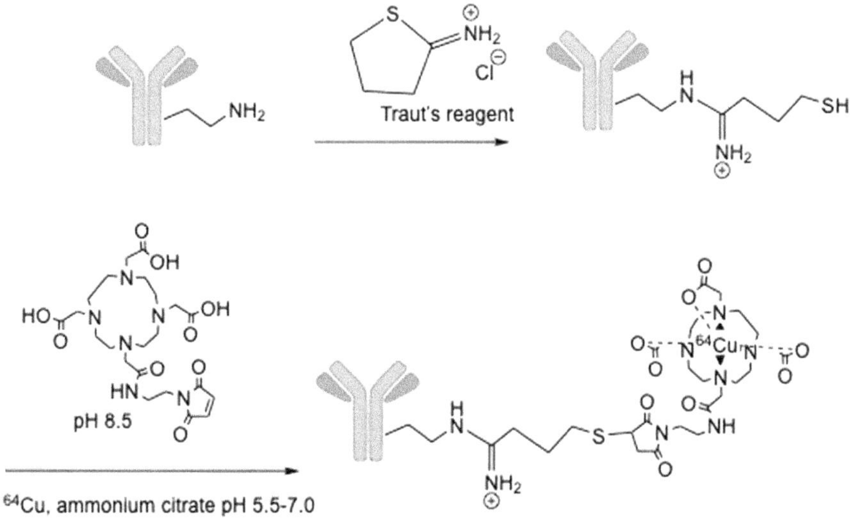

Fig. 3.51 Modification of amino group on antibody for labeling with cyclen ring using Traut's reagent

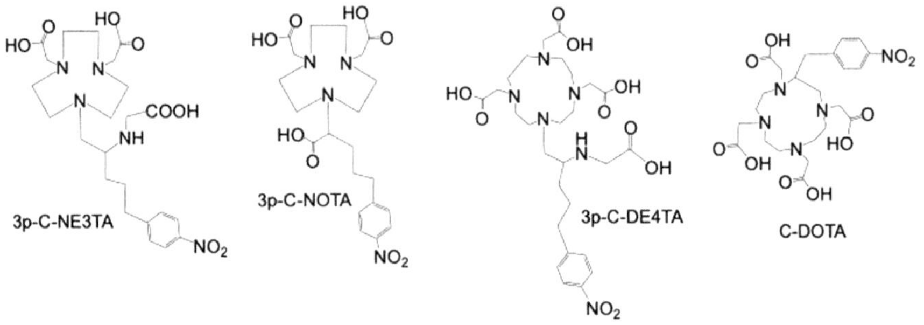

Fig. 3.52 Bifunctional chelators for ^{64}Cu. Data obtained from Wu et al. (2016) with permission from the author

stability is a concern at low pH, then the labeling buffer could be adjusted to 7. The reaction can be performed at room temperature, albeit at a longer time, while the reaction can be achieved with a high yield at 90 °C for 1 h. In a recent study, it has been shown that cyclen-based rings are excellent in coordination with ^{64}Cu. The ^{64}Cu-associated hexa-, hepta-, and octadentate cyclens remained stable, without any loss of ^{64}Cu for two days in human serum (Wu et al. 2016). Further analysis confirmed that the smaller macrocyclic rings, such as 3p-C-NE3TA and 3p-C-NOTA are more stable regarding the ^{64}Cu association than larger cyclen counterparts, like C-DOTA and C-DE4TA (Fig. 3.52). In vivo study also indicated that ^{64}Cu-3p-C-NOTA and ^{64}Cu-3p-C-NE3TA cleared well in vivo while ^{64}Cu-3p-C-DE4TA dissociated in vivo and substantial renal and liver retention, corroborating with previous observations (McCabe et al. 2012). Since the chelation is stable, the ^{64}Cu-NOTA complex is prone to resist transmetalation in vivo compared to ^{64}Cu-DOTA (Nedrow et al. 2014; Wadas et al. 2010).

It is not very surprising regarding the stability of hexadentate NOTA chelates over octadentate DOTA counterparts. Although it seems DOTAs should provide a stronger complex with ^{64}Cu, given they have more number of coordinations than NOTA. A large number of established knowledge in coordination chemistry has already demonstrated that increasing the ring size of a chelator would lead to a decreased complex stability (Hancock and Martell 1989). This observation was confirmed in the early days when scientists compared the entropy effects associated with the more extended connecting link between the donor atoms of ligands that form large chelator rings versus smaller versions. Other evidence also pointed out that the drop in coordination stability with increasing chelator ring size is due to unfavorable enthalpy, or in other words, weak thermodynamics of complex formation (Martell and Smith 1974). While many other criteria contribute to the labile ligand–metal coordination, one of the general notions that command attention in probe design is that the larger the cyclen chelators, the more difficult it is to bring together the coordination with metals and destabilize the complex (Hancock and Martell 1989).

Another critical aspect in the design of a strong chelator-^{64}Cu complex is the prevention of steric strain of the coordination. Molecular mechanics should be employed in the process to predict the orientation of the metal ion along the line with ligand coordinates from the chelator with respect to steric strain. Thus, a number of factors need to be considered for the design, apart from the size of the chelator discussed above, such as the specificity of the ligands for the metal ion, spatial arrangement of the ligands via high-ordered chelating structures, and size of the metals.

To understand what ligands are suitable for the metal to be chelated, or in this case of Cu^{2+}, we should borrow the knowledge from Pearson's hard–soft Lewis acid–base (HSAB) principle (Pearson 1963), which can provide handy information for probe design. Accordingly, hard Lewis acids and bases have small ionic radii and are non-polarizable because the electrons are held tightly closed to the nucleus, and have opposite charges. While soft Lewis acids and bases are characterized by large ionic radii, and with a more diffuse distribution of the electrons. Aside from these distinct groups, many of other compounds are in the middle range, classified as borderline molecules. According to the HSAB principle, hard acids like to react with hard bases, and soft acids prefer soft bases. Based on Fig. 3.53 (Hancock and Martell 1989), Cu^{2+} is either hard or soft acid, but it is in the borderline, and thus it is best suited with chelating reagents equipped with borderline bases as ligands, such as anilines or ligands with pyridyl donor groups might be a solution for generating strong and selective coordination for Cu^{2+}. Particularly, the pyridyl ligands can induce rigidity to the coordination complex due to the rigidity of the aromatic ring system (Martell and Smith 1974).

Strong ligand–metal complex stabilization can also be achieved by designing high-ordered and rigid chelators. One of the practical ways to impart structural rigidity involves bridging adjacent secondary amino groups via alkylation reactions. For instance, when the DOTA cyclen was reinforced with an ethylene cross-bridge, this leads to an increase in complex stability for Ni(II) (Fig. 3.54) (Ramasubbu and Wainwright 1982).

	Hard	**Borderline**	**Soft**
Acids	H^+, Li^+, Na^+, K^+ Be^{2+}, Mg^{2+}, Ca^{2+}, Ba^{2+} BF_3, BCl_3, $B(OR)_3$ Al^{3+}, Sc^{3+}, Ga^{3+}, In^{3+} Cr^{3+}, Mn^{3+}, Fe^{3+}, Co^{3+}	Cu^{2+}, Fe^{2+}, Co^{2+}, Ni^{2+}, Zn^{2+}, Sn^{2+}, Ir^{3+}, Ru^{3+}, $B(CH_3)_3$, BBr_3	Cu^+, Ag^+, Au^+, Hg^+, Pd^{2+}, Cd^{2+}, Hg^{2+}, Tl^+, BH_3, $Tl(CH_3)_3$
Bases	F^-, Cl^-, OH^-, RO^-, NO_3^-, ClO_4^-, ROH, NH_3, RNH_2, SO_4^{2-}	N_2, N_3^-, NO_2^-, Br, SO_3^{2-}, $C_6H_5NH_2$, C_5H_5N	CN^-, I^-, RS^-, SCN, CO, R_2S, RSH, $S_2O_3^{2-}$, RNC, C_6H_6

Fig. 3.53 Classifications of acids and bases based on Pearson's HSAB principle. Data obtained from Hancock and Martell (1989) with permission from the American Chemical Society

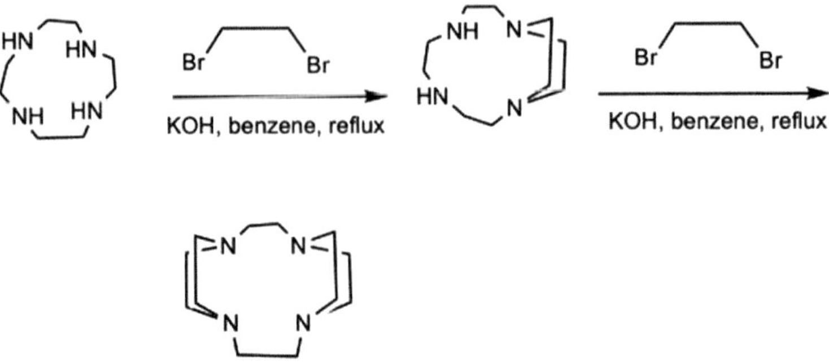

Fig. 3.54 Rigidized cyclen with alkylation to convert secondary amine to tertiary amine to improve metal–ligand complex stability. Data adapted from Ramasubbu and Wainwright (1982)

Cross-bridged macrocyclic chelators are another example of a contemporary approach to improving metal–ligand complex stability. This innovative development proved to be more advantageous in the chelating of ^{64}Cu but markedly improved in vivo stability compared to DOTA and TETA (Cai and Anderson 2014). In vivo evidence showed that structurally reinforced chelators are a determinant in stabilizing the complex stability and reducing metal loss via transmetalation. The ethylene cross-bridged (CB)-Te2A has a stable complex with ^{64}Cu, thus less prone to go through transchelation in vivo compared to ^{64}Cu-DOTA (Fig. 3.55) (Boswell et al. 2004). Cross-bridged chelators offer stability and reduce transmetalation in vivo; however, at the expense of slow complex formation, requiring stringent labeling conditions, including high temperature, most cases are incompatible with biologics. A number of modifications have been reported to address this shortcoming. Particularly, a family of propylene cross-bridged tetraaza macrocyclic chelators (PCB-TE2A) was reported recently the capability to form the coordination labeling with Cu^{2+} in mild reaction conditions. The chelators can be potentially modified with active functional groups, which enables coupling with biologics prior to complexation with metal ions without concerns of protein decomposition. Further, they also have effective complex stability both in vitro and in vivo (Yoo et al. 2012).

Next, the size of the metal ions matters in the design of imaging probe with the intention for enhanced complex stability. The chelating complex stability logK can be assessed by comparing a number of metal ions of different sizes. It is apparent

Fig. 3.55 Cyclam-based cross-bridged chelators

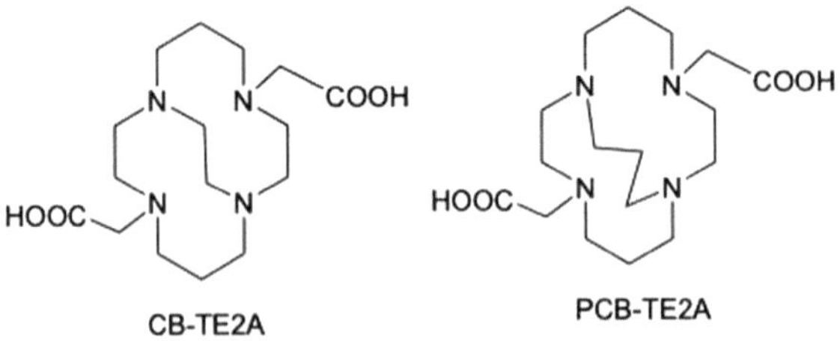

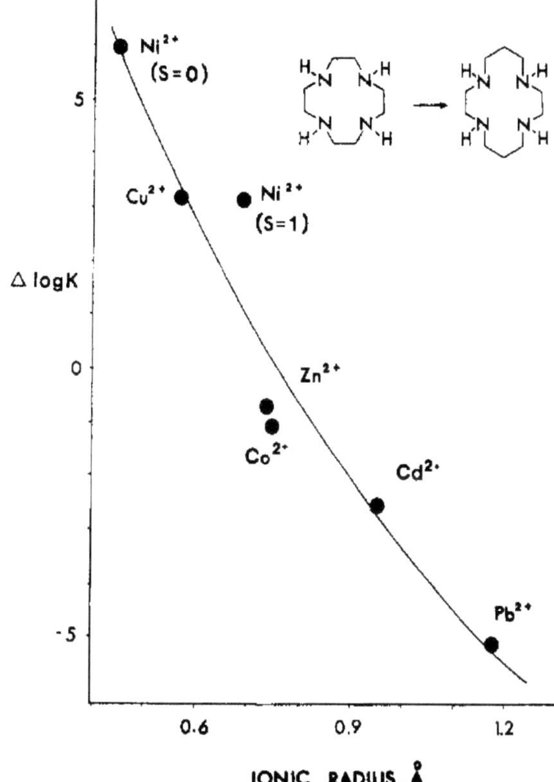

Fig. 3.56 Effect of increasing chelator size on the complex stability as a function of metal ion radius. Data obtained from Hancock and Martell (1989) with permission from the American Chemical Society

that there is a nonreciprocal relationship between the size of metal ions and logK. Figure 3.56 (Hancock and Martell 1989) shows that the complex stability logK reduces significantly as the metal radius increases when passing from a small chelator to a larger counterpart.

In this regard, it is likely of advantage for ^{64}Cu, which is considered to possess a small size (radius ~ 87 pm) when it comes to chelating with a smaller chelator. Altogether, the take-home lesson drawn from Fig. 3.56 indicates that decreasing the ring size of the chelator along with the use of small metal ions like ^{64}Cu would enhance the complex stability than using larger metal ions.

3.4.3 Labeling Peptides or Monoclonal Antibodies with ^{68}Ga

As shown in Fig. 3.53, gallium is considered as a hard Lewis acid, and it prefers to form octahedral complexes with hard Lewis bases, such as nitrogen or oxygen atoms. So as one would guess, the DOTA, NOTA chelators are the ideal structures for complexing with gallium. Although the half-life is shorter than ^{18}F and ^{64}Cu,

^{68}Ga earns the same popularity in PET imaging due to the ease and low cost of the ^{68}Ga radioisotope production. Basically, ^{68}Ga can be made available in the lab using a compact benchtop generator, which can be deployed in any research and clinical site without the need of a cyclotron, reducing the burden from teamwork coordination and enhancing flexibility. In a typical generator production, ^{68}Ga radionuclide can be produced by ^{68}Ge/^{68}Ga radionuclide generators via ^{69}Ga(p,2n)^{68}Ge, in which a proton beam is irradiated into ^{69}Ga target (Kambali and Wibowo 1198). The resulting ^{68}Ge decays by electron capture to ^{68}Ga with a half-life of approximately 68 min. And the end product ^{68}Ga^{3+} chloride can be associated with DOTA or NOTA chelators. It sounds simple as it seems; however, a number of technological innovations must be derived in order to provide reproducible and high-grade materials for radiopharmaceutical productions. Three major issues that need improvement (i) the ^{68}Ga^{3+} eluted from the matrix are too diluted, unsuitable for peptide/protein labeling with the desired radiochemical yield; (ii) the eluent is too acidic, a condition not favored for chelating, and it may destabilize the peptides; (iii) metallic impurities. In fact, one of the major issues related to the purification from impurities during the nuclear reaction. The produced parent isotope ^{68}Ge is trapped in titanium dioxide (TiO$_2$) or tin dioxide (SnO$_2$) adsorbents in the column matrix to decay to ^{68}Ga, which is then eluted in the mobile phase as ^{68}Ga^{3+} (^{68}GaCl$_3$) in a few milliliters of 0.1 M or 1.0 M hydrochloric acid (Talip et al. 2020). From there, ^{68}Ga^{3+} can be used directly for reactions with DOTA or NOTA to form the prosthetic group for peptide/antibody labeling. However, during this process, significantly detected impurities due to the presence of undesired metals such as Fe(II)/Fe(III), Mn(II), and Zn(II) may adversely affect the labeling process; therefore, intensive post-processing to ensure fast, reproducible production of ^{68}GaCl$_3$ with high specific activity must be pursued, particularly, when clinical interest for ^{68}Ga-label biologics are on the rise, and thus adhering to strict safety regulations are mandatory (Meyer et al. 2004).

A number of approaches have been developed to improve the purity of the ^{68}Ga^{3+} in the eluent, including fractionation of the initial generator eluate (Breeman et al. 2005). Basically, the assumption is that two-thirds of the total ^{68}Ga^{3+} activity elutes in 1–2 mL of the total eluate volume. This approach helps to concentrate the radioisotopes to increase the yield, but mainly impossible to eliminate ^{68}Ge and other metallic impurities. Other improved procedures involve chromatographic purification using anion- and cation-exchange chromatography. In anion-exchange chromatography, ^{68}Ga^{3+} ions were converted to [GaCl$_4$]$^-$ complex, which can be adsorbed quantitatively on a strong anion-exchange resin. Then, the metallic cations and other impurities can be washed off the column (Meyer et al. 2004). Using this procedure, other metal cationic impurities were rigorously excluded. The total elution time from the microcolumn was less than one minute, with the desorption efficiency ranging from 85–90%. The resulting pH of ^{68}Ga^{3+} eluted solution is approximately 2; thus, it is necessary to supplement with 1 M HEPES to adjust the pH to 3.5–4, which is optimum for the complexation of ^{68}Ga^{3+} with DOTA cage. Some limitations of this method include the inability to directly load ^{68}Ga^{3+} on the anion-exchange resin. And some residuals of Zn(II) and Fe(III) were found in the eluent (Zhernosekov et al. 2007).

For cationic-exchange chromatography, the cation-exchange resin was used along with the treatment of acetone and hydrochloric acid with an increasing gradient, where the concentration of hydrochloric acid was carefully enhanced from 0.1 N to 0.15 N or 0.2 N for selective removal of impurities from the resin while $^{68}Ga^{3+}$ remains adhering on the column (Zhernosekov et al. 2007). $^{68}Ga^{3+}$ was then eluted with acidified acetone using 0.05 N hydrochloric acid.

In general, in order to develop a peptide-based probe, the chelators can be conjugated into the peptide backbone via the N terminal amines via solid-phase peptide chemistry.

As a rule of thumb, all of the active function groups from the peptide or the chelator should be protected; and orthogonal to Fmoc chemistry. For instance, the carboxylic groups from the chelator can be protected as a *tert*-butyl ester (Fig. 3.57). The coupling could be performed using diisopropylcarbodiimide and an anti-racemization reagent, such as triazole 1-hydroxy-benzotriazole (HOBt). After attaching the chelator to the peptide sequence, the resin, the peptide side chain protection groups, and *tert*-butyl ester from the chelator can be removed using trifluoroacetic acid (TFA). The peptide product is precipitated in ether and purified using HPLC. After characterization and constituted in slightly acidic conditions, the chelator-peptide product is used for complexing with $^{68}Ga^{3+}$ to provide the labeled product.

It is worthwhile to mention that aside from using a chelator where the carboxylic groups are protected with t-butyl groups, an allyl protecting group can also be employed. And these groups can be easily deprotected using a catalytic amount of tetrakis(triphenylphosphine)-palladium (Pd(PPh$_3$)$_4$.

In the event that the peptides are unstable due to exposure to acidic conditions, different reaction plans should be designed. Further, this case is mostly applicable to labeling proteins or antibodies. In solid-phase chemistry, a small chelator molecule can be attached to the water-soluble peptide in the organic solvents. After detaching the peptide from the resin, the solubility of the conjugated product is dictated by the

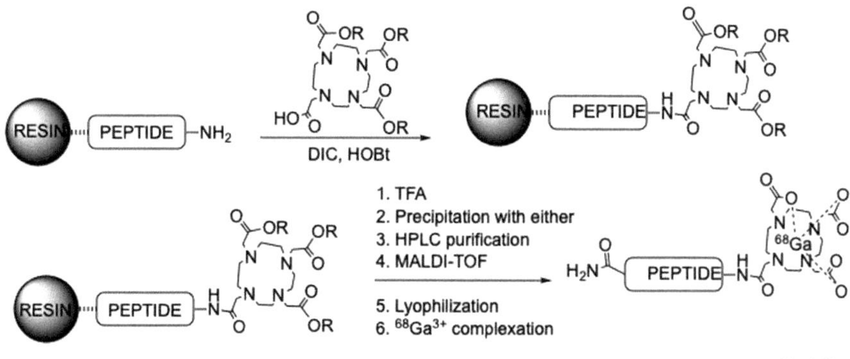

Fig. 3.57 Development of ^{68}Ga-peptide probe using solid-phase peptide chemistry

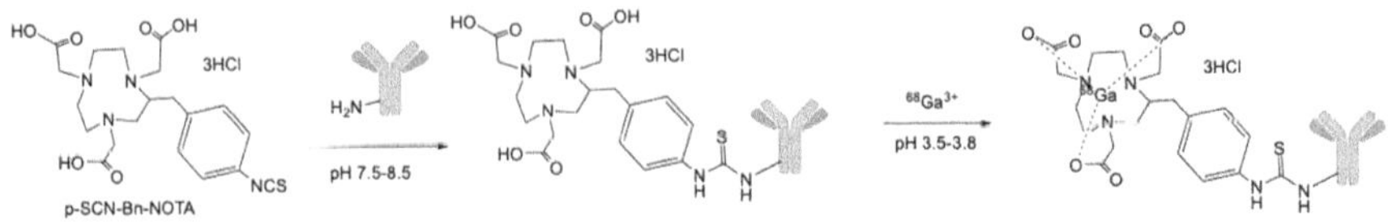

Fig. 3.58 Labeling of acidic-sensitive materials with ^{68}Ga

peptide. In most cases, the final product is very water-soluble. However, conjugation of a chelator directly to proteins or antibodies may be daunting due to incompatible solubility. To bring all materials in the same phase to foster effective conjugation, the chelator should have sulfonate group(s) to enhance water solubility, the same holds true for the succinimide moiety. Some commercial chelators available in the salt forms (Fig. 3.58) can also contribute to labeling antibodies.

3.5   Conclusion

This chapter just covered the radio-chemistry for labeling PET probes using a number of popular isotopes. Mastering the knowledge helps with the laboratory work and helps to design better labeling procedures. For example, given the burden of relying on chemical synthesis and purification prior to preclinical and, most importantly, clinical works, a number of efforts have been developed to transform this demanding operation into a more reproducible, reliable, and effortless task via the labeling kits. And hopefully, this lecture will in some way contribute to that effort. Now, let's move on to another lecture on the design of MR imaging probes.

References

M.E. Phelps, Molecular imaging with positron emission tomography. Annu. Rev. Nucl. Part. Sci. **52**, 303–338 (2002)

M. Conti, L. Eriksson, Physics of pure and non-pure positron emitters for PET: a review and a discussion. EJNMMI Phys. **3**, 8 (2016)

D.D. Nolting, M.L. Nickels, N. Guo, W. Pham, Molecular imaging probe development: a chemistry perspective. Am. J. Nucl. Med. Mol. Imaging. **2**, 273–306 (2012)

W.H. Organization, *Radiopharmaceuticals, The International Pharmacopoeia*, 4th edn. (2008)

P.E. Edem, E.J.L. Steen, A. Kjar, M.M. Herth, Fluorine-18 radiolabeling strategies-advantages and disadvantages of currently applied labeling methods, in *Late-Stage Fluorination of Bioactive Molecules and Biologically-Relevant Substrates, Chapter 2* (2019), pp. 29–103

L. Cai, S. Lu, P.V. W., Chemistry with [18F]fluoride ion. Eur. J. Org. Chem. 2853–2873 (2008)

A.R. Jalilian, S.A. Tabatabai, A. Shafiee, H. Afarideh, R. Najafi, M. Bineshmarvasti, One-step, no-carrier-added synthesis of a 18F-labeled benzodiazepine receptor ligand. J. Label. Compd. Radiopharm. **43**, 545–555 (2000)

K. Mueller, C. Faeh, F. Diederich, Fluorine in pharmaceuticals: looking beyond intuition. Science **317**, 1881–1886 (2007)

M. Attina, F. Cacace, A.P. Wolf, Labeled aryl fluorides from the nucleophilic displacement of activated nitro groups by 18F-F. J. Label. Compd. Radiopharm. **20**, 501–514 (1982)

O. Jacobson, D.O. Kiesewetter, X. Chen, Fluorine-18 radiochemistry, labeling strategies and synthetic routes. Bioconjug. Chem. **26**, 1–18 (2015)

D.L. Bailey, D.W. Townsend, P.E. Valk, M.N. Maisey, N.S. Mason, C.A. Mathis, Positron emission tomography: basic sciences and clinical practice, in *Springer Science & Business Media-Radiohalogens for PET Imaging* (2004), p. 219

D.J. Schlyer, M.A. Bastos, D. Alexoff, A.P. Wolf, Separation of [18F]fluoride from [18O]water using anion exchange resin. Int. J. Rad. Appl. Instrum. A. **41**, 531–533 (1990)

J. Bergman, O. Solin, Fluorine-18-labeled fluorine gas for synthesis of tracer molecules. Nucl. Med. Biol. **24**, 677–683 (1997)

J.S. Fowler, T. Ido, Initial and subsequent approach for the synthesis of 18FDG. Semin. Nucl. Med. **32**, 6–12 (2002)

T. Ido, C.N. Wan, V. Casella, J.S. Fowler, A.P. Wolf, Labeled 2-deoxy-D-glucose analogs. 18F-labeled 2-deoxy-2-fluoro-D-glucose, 2-deoxy-2-fluoro-D-mannose and 14C-2-deoxy-2-fluoro-D-glucose. J. Label. Compd. Radiopharm. **14**, 175–183 (1977)

F. Oberdorfer, E. Hofmann, W. Maier-Borst, Preparation of 18F-labelled N-fluoropyridinium triflate. J. Label. Compd. Radiopharm. **25**, 999–1005 (1988a)

R.E. Banks, G.E. Williamson, Chem. Ind. 1864 (1964)

A. Luxen, M. Perlmutter, G.T. Bida, G. Van Moffaert, J.S. Cook, N. Satyamurthy, M.E. Phelps, J.R. Barrio, Remote, semiautomated production of 6-[18F]fluoro-L-dopa for human studies with PET. Int. J. Rad. Appl. Instrum. A **41**, 275–281 (1990)

M. Namavari, A. Bishop, N. Satyamurthy, G. Bida, J.R. Barrio, Regioselective radiofluorodestannylation with [18F]F2 and [18F]CH3COOF: a high yield synthesis of 6-[18F]Fluoro-L-dopa. Int. J. Rad. Appl. Instrum. A **43**, 989–996 (1992)

L.P. Szajek, M.A. Channing, W.C. Eckelman, Appl. Radiat. Isot. **49**, 795–804 (1998)

F. Oberdorfer, E. Hofmann, W. Maier-Borst, Appl. Radiat. Isot. **39**, 685–688 (1988b)

N. Satyamurthy, G.T. Bida, M.E. Phelps, J.R. Barrio, N-[18F]fluoro-N-alkylsulfonamides: novel reagents for mild and regioselective radiofluorination. Int. J. Rad. Appl. Instrum. A **41**, 733–738 (1990)

H. Teare, E.G. Robins, E. Arstad, S.K. Luthra, V. Gouverneur, Synthesis and reactivity of [18F]-N-fluorobenzenesulfonimide. Chem. Commun. (Camb) 2330–2332 (2007)

H. Teare, E.G. Robins, A. Kirjavainen, S. Forsback, G. Sandford, O. Solin, S.K. Luthra, V. Gouverneur, Radiosynthesis and evaluation of [18F]Selectfluor bis(triflate). Angew. Chem. Int. Ed. Engl. **49**, 6821–6824 (2010)

M. Tredwell, V. Gouverneur, 18F labeling of arenes. Angew. Chem. Int. Ed. Engl. **51**, 11426–11437 (2012)

I.S. Stenhagen, A.K. Kirjavainen, S.J. Forsback, C.G. Jorgensen, E.G. Robins, S.K. Luthra, O. Solin, V. Gouverneur, [18F]fluorination of an arylboronic ester using [18F]selectfluor bis(triflate): application to 6-[18F]fluoro-L-DOPA. Chem. Commun. (camb) **49**, 1386–1388 (2013)

M.B. Smith, J. March, *March's Advanced Organic Chemistry*, 5th edn (2001a), p. 439

M.B. Smith, J. March, *March's Advanced Organic Chemistry*, 5th edn (2001b), p. 444

P.R. Wells, Linear free energy relationships. Chem. Rev. **63**, 171–219 (1963)

M. Ansari, R. Kessler, R. Baldwin, Microwave labeling of [18F]fallypride with a commercial synthesis module. J. Nucl. Med. (Radiopharmacy Posters), 1304 (poster number) (2008)

L. Dolci, F. Dolle, S. Jubeau, F. Vaufrey, C. Crouzel, 2-[18F]fluoropyridines by no-carrier-added nucleophilic aromatic substitution with [18F]FK-K222-A comparative study. J. Label. Compd. Radiopharm. **42**, 975–985 (1999)

F.I. Aigbirhio, R.M. Carr, V.W. Pike, C.J. Steel, D.R. Sutherland, Automated radiosynthesis of no-carrier-added [S-fluoromethyl-18F]fluticasone propionate as a radiotracer for lung deposition studies with PET. J. Label. Compd. Radiopharm. **39**, 567–584 (1997)

A. Petric, J.R. Barrio, M. Namavari, S.C. Huang, N. Satyamurthy, Synthesis of 3beta-(4-[18F]fluoromethylphenyl)- and 3beta-(2-[18F] fluoromethylphenyl)tropane-2beta-carboxylic acid methyl esters: new ligands for mapping brain dopamine transporter with positron emission tomography. Nucl. Med. Biol. **26**, 529–535 (1999)

K. Hamacher, H.H. Coenen, G. Stocklin, Efficient stereospecific synthesis of no-carrier-added 2-[18F]-fluoro-2-deoxy-D-glucose using aminopolyether supported nucleophilic substitution. J. Nucl. Med. **27**, 235–238 (1986)

H. Luo, A.L. Beets, M.J. McAllister, M. Greenbaum, D.W. McPherson, J. Knapp, F.F., Resolution, in vitro and in vivo evaluation of fluorine-18-labeled isomers of 1-azabicyclo[2.2. 2] oct-3-ylα-(1-fluoropent-5-yl)α-hydroxy-α-phenylacetate (FQNPe) as new PET candidates for the imaging of muscarinic-cholinergic receptor. J. Label. Compd. Rad. **41**, 681–704 (1998)

M.R. Kilbourn, Fluorine-18 labeling of radiopharmceuticals, in *Nuclear Science Series* (National Academy Press, Washington DC, 1990) NAS-NS-3203

G. Angelini, M. Speranza, A.P. Wolf, C.-Y. Shiue, Nucleophilic aromatic substitution of activated cationic groups by 18-labeled fluoride. A useful route to no-carrier added (nca)18F-labeled aryl fluorides. J. Fluor. Chem. **27**, 177–191 (1985)

D.D. Nolting, M. Nickels, M.N. Tantawy, J.P. Xie, T.E. Peterson, B.C. Crews, L. Marnett, J.C. Gore, W. Pham, Convergent synthesis and evaluation of 18F-labeled azulenic COX2 probes for cancer imaging, Front. Oncology **2**, 1–8 (2013)

R. Ritawidya, B. Wenzel, R. Teodoro, M. Toussaint, M. Kranz, W. Deuther, S. Dukic, F.A. Ludwig, M. Scheunemann, P. Brust, Radiosynthesis and biological investigation of a novel fluorine-18 labeled benzoimidazotriazine-based radioligand for the imaging of phosphodiesterase 2A with positron emission tomography. Molecules **24**, 4149 (2019)

M.R. Kilbourn, M.S. Haka, Synthesis of [18F]GBR13119, a presynaptic dopamine uptake antagonist. Int. J. Rad. Appl. Instrum. A **39**, 279–282 (1988)

C. Lemaire, M. Guillaume, L. Christiaens, A.J. Palmer, R. Cantineau, A new route for the synthesis of [18F]fluoroaromatic substituted amino acids: no carrier added L-p-[18F]fluorophenylalanine. Int. J. Rad. Appl. Instrum. A **38**, 1033–1038 (1987)

M.S. Haka, M.R. Kilbourn, G.L. Watkins, S.A. Toorongian, Aryltrimethylammonium trifluoromethanesulfonates as precursors to aryl [18F]fluorides: improved synthesis of [18F]GBR-13119. J. Label. Compd. Radiopharm. **27**, 823–833 (1989)

J. Ballinger, B.M. Bowen, G. Firnau, E.S. Garnett, F.W. Tear, Int. J. Appl. Radiat. Isot. **35**, 1125–1128 (1984)

Y.Y. See, M.T. Morales-Colon, D.C. Bland, M.S. Sanford, Development of SNAr nucleophilic fluorination: a fruitful academia-industry collaboration. Acc. Chem. Res. **53**, 2372–2383 (2020)

V.W. Pike, Hypervalent aryliodine compounds as precursors for radiofluorination. J. Label. Compd. Radiopharm. **61**, 196–227 (2018)

M. Uyanik, T. Yasui, K. Ishihara, Enantioselective Kita oxidative spirolactonization catalyzed by in situ generated chiral hypervalent iodine(III) species. Angew. Chem. Int. Ed. Engl. **49**, 2175–2177 (2010)

S. Telu, J.H. Chun, F.G. Simeon, S. Lu, V.W. Pike, Syntheses of mGluR5 PET radioligands through the radiofluorination of diaryliodonium tosylates. Org. Biomol. Chem. **9**, 6629–6638 (2011)

V.V. Zhdankin, P.J. Stang, Chemistry of polyvalent iodine. Chem. Rev. **108**, 5299–5358 (2008)

M. Ochiai, M. Toyonari, T. Nagaoka, D.W. Chen, M. Kida, Tetrahedron. Lett. **38**, 6709–6712 (1997)

M. Bielawski, B. Olofsson, High-yielding one-pot synthesis of diaryliodonium triflates from arenes and iodine or aryl iodides. Chem. Commun. (Camb) 2521–2523 (2007)

M. Bielawski, D. Aili, B. Olofsson, Regiospecific one-pot synthesis of diaryliodonium tetrafluoroborates from arylboronic acids and aryl iodides. J. Org. Chem. **73**, 4602–4607 (2008)

E. Lindstedt, M. Reitti, B. Olofsson, One-pot synthesis of unsymmetric diaryliodonium salts from iodine and arenes. J. Org. Chem. **82**, 11909–11914 (2017)

N. Soldatova, P. Postnikov, O. Kukurina, V.V. Zhdankin, A. Yoshimura, T. Wirth, M.S. Yusubov, One-pot synthesis of diaryliodonium salts from arenes and aryl iodides with Oxone-sulfuric acid. Beilstein. J. Org. Chem. **14**, 849–855 (2018)

M. Ochiai, Y. Kitagawa, M. Toyonari, On the mechanism of α-phenylation of β-keto esters with diaryl-λ3-iodanes: evidence for a non-radical pathway. Archieve Org. Chem. **6**, 43–48 (2003)

J.H. Chun, S. Lu, Y.S. Lee, V.W. Pike, Fast and high-yield microreactor syntheses of ortho-substituted [(18)F]fluoroarenes from reactions of [(18)F]fluoride ion with diaryliodonium salts. J. Org. Chem. **75**, 3332–3338 (2010)

M. Ochiai, Y. takaoka, Y. Masaki, Y. Nagao, M. Shiro, J. Am. Chem. Soc, **112**, 5677 (1990)

T. Okuyama, T. Takino, T. Sueda, M. Ochiai, Solvolysis of cyclohexenyliodonium salt, a new precursor for the vinyl cation: remarkable nucleofugality of the phenyliodonio group and evidence for internal return from an intimate ion-molecule pair. J. Am. Chem. Soc. **117**, 3360–3367 (1995)

A. Shah, V.W. Pike, D.A. Widdowson, The synthesis of [18F]fluoroarenes from the reaction of cyclotron-produced [18F]fluoride ion with diaryliodonium salts. J. Chem. Soc. Perkin. Trans. **1**, 2043–2046 (1998)

V.V. Grushin, I.I. Demkina, T.P. Tolstaya, J. Chem. Soc. Perkin. Trans. **2**, 505 (1992)

M.S. Yusubov, D.Y. Svitich, M.S. Larkina, V.V. Zhdankin, Applications of iodonium salts and iodonium ylides as precursors for nucleophilic fluorination in positron emission tomography. Archieve Org. Chem. 384–395 (2013)

B.C. Lee, J.S. Kim, B.S. Kim, J.Y. Son, S.K. Hong, H.S. Park, B.S. Moon, J.H. Jung, J.M. Jeong, S.E. Kim, Aromatic radiofluorination and biological evaluation of 2-aryl-6-[18F]fluorobenzothiazoles as a potential positron emission tomography imaging probe for β-amyloid plaques. Bioorg. Med. Chem. **19**, 2980–2990 (2011)

B.S. Moon, H.S. Kil, J.H. Park, J.S. Kim, J. Park, D.Y. Chi, B.C. Lee, S.E. Kim, Facile aromatic radiofluorination of [18F]flumazenil from diaryliodonium salts with evaluation of their stability and selectivity. Org. Biomol. Chem. **9**, 8346–8355 (2011)

N. Satyamurthy, J.R. Barrio, No-carrier-added nucleophilic [F-18] fluorination of aromatic compounds, WO2010–117435 A2. 2010 (2010)

M.S. Yusubov, A. Yoshimura, V.V. Zhdankin, Iodonium ylides in organic synthesis. ARKIVOC-Revi. Acc. 342–374 (2016)

W. Kirmse, Carbene complexes of nonmetals. Eur. J. Org. Chem. 237–260 (2005)

S.R. Goudreau, D. Marcoux, A.B. Charette, General method for the synthesis of phenyliodonium ylides from malonate esters: easy access to 1,1-cyclopropane diesters. J. Org. Chem. **74**, 470–473 (2009)

C. Zhu, A. Yoshimura, P. Solntsev, L. Ji, Y. Wei, V.N. Nemykin, V.V. Zhdankin, New highly soluble dimedone-derived iodonium ylides: preparation, X-ray structure, and reaction with carbodiimide leading to oxazole derivatives. Chem. Commun. (camb) **48**, 10108–10110 (2012)

I.N. Petersen, J. Villadsen, H.D. Hansen, J. Madsen, A.A. Jensen, N. Gillings, S. Lehel, M.M. Herth, G.M. Knudsen, J.L. Kristensen, (18)F-Labelling of electron rich iodonium ylides: application to the radiosynthesis of potential 5-HT2A receptor PET ligands. Org. Biomol. Chem. **15**, 4351–4358 (2017)

N.A. Stephenson, J.P. Holland, A. Kassenbrock, D.L. Yokell, E. Livni, S.H. Liang, N. Vasdev, Iodonium ylide-mediated radiofluorination of 18F-FPEB and validation for human use. J. Nucl. Med. **56**, 489–492 (2015)

S.H. Liang, D.L. Yokell, R.N. Jackson, P.A. Rice, R. Callahan, K.A. Johnson, D. Alagille, G. Tamagnan, T.L. Collier, N. Vasdev, Microfluidic continuous-flow radiosynthesis of [(18)F]FPEB suitable for human PET imaging. Medchemcomm **5**, 432–435 (2014)

K. Gondo, T. Kitamura, Reaction of iodonium ylides of 1,3-dicarbonyl compounds with HF reagents. Molecules **17**, 6625–6632 (2012)

A.P. Wolf, C.S. Redvanly, Carbon-11 and radiopharmaceuticals. Int. J. Appl. Radiat. Isot. **28**, 29–48 (1977)

K. Dahl, C. Halldin, M. Schou, New methodologies for the preparation of carbon-11 labeled radiopharmaceuticals. Clin. Transl. Imaging **5**, 275–289 (2017)

A.C. Runkle, X. Shao, L.J. Tluczek, B.D. Henderson, B.G. Hockley, P.J. Scott, Automated production of [11C]acetate and [11C]palmitate using a modified GE Tracerlab FX(C-Pro). Appl. Radiat. Isot. **69**, 691–698 (2011)

F. Luzi, A.D. Gee, S. Bongarzone, Rapid one-pot radiosynthesis of [carbonyl-(11)C]formamides from primary amines and [(11)C]CO2. EJNMMI Radiopharm. Chem. **5**, 20 (2020)

C. Taddei, A.D. Gee, Recent progress in [(11) C]carbon dioxide ([(11) C]CO2) and [(11) C]carbon monoxide ([(11) C]CO) chemistry. J. Label. Comp. Radiopharm. **61**, 237–251 (2018)

J. Eriksson, G. Antoni, B. Langstrom, O. Itsenko, The development of (11)C-carbonylation chemistry: a systematic view. Nucl. Med. Biol. **92**, 115–137 (2021)

P.J. Riss, S. Lu, S. Telu, F.I. Aigbirhio, V.W. Pike, Cu(I)-catalyzed (11)C carboxylation of boronic acid esters: a rapid and convenient entry to (11)C-labeled carboxylic acids, esters, and amides. Angew. Chem. Int. Ed. Engl. **51**, 2698–2702 (2012)

A.A. Wilson, A. Garcia, S. Houle, N. Vasdev, Direct fixation of [(11)C]-CO(2) by amines: formation of [(11)C-carbonyl]-methylcarbamates. Org. Biomol. Chem. **8**, 428–432 (2010)

A.H. Dheere, S. Bongarzone, C. Taddei, R. Yan, A.D. Gee, Synthesis of 11C-labelled symmetrical ureas via the rapid incorporation of [11C]CO2 into aliphatic amines. Synlett **26**, 2257–2260 (2015)

M. Boudjemeline, R. Hopewell, P.L. Rochon, D. Jolly, I. Hammami, S. Villeneuve, A. Kostikov, Highly efficient solid phase supported radiosynthesis of [(11) C]PiB using tC18 cartridge as a "3-in-1" production entity. J. Label. Comp. Radiopharm. **60**, 632–638 (2017)

T.A. Singleton, M. Boudjemeline, R. Hopewell, D. Jolly, H. Bdair, A. Kostikov, Solid phase 11C-methylation, purification and formulation for the production of PET tracers. J. Vis. Exp. (2019)

R. Iwata, C. Pascali, A. Bogni, Y. Miyake, K. Yanai, T. Ido, A simple loop method for the automated preparation of (11C)raclopride from (11C)methyl triflate. Appl. Radiat. Isot. **55**, 17–22 (2001)

A.R. Studenov, S. Jivan, M.J. Adam, T.J. Ruth, K.R. Buckley, Studies of the mechanism of the in-loop synthesis of radiopharmaceuticals. Appl. Radiat. Isot. **61**, 1195–1201 (2004)

A.A. Wilson, A. Garcia, L. Jin, S. Houle, Radiotracer synthesis from [(11)C]-iodomethane: a remarkably simple captive solvent method. Nucl. Med. Biol. **27**, 529–532 (2000)

X. Shao, P.L. Schnau, M.V. Fawaz, P.J. Scott, Enhanced radiosyntheses of [(1)(1)C]raclopride and [(1)(1)C]DASB using ethanolic loop chemistry. Nucl. Med. Biol. **40**, 109–116 (2013)

L. Cai, R. Xu, X. Guo, V.W. Pike, Rapid room-temperature (11)C-methylation of arylamines with [(11)C]methyl iodide promoted by solid inorganic bases in DMF. Euro. J. Org. Chem. **2012**, 1303–1310 (2012)

C. Einhorn, J. Einhorn, J.L. Luche, Sonochemistry-the use of ultrasonic waves in synthetic organic chemistry. Synthesis 787–813 (1989)

Y. Andersson, A. Cheng, B. Langstrom, Palladium-promoted coupling reactions of [11C]methyl iodide with organotin and organoboron compounds. Acta. Chemica. Scandinvica. **49**, 683–688 (1995)

H. Doi, Pd-mediated rapid cross-couplings using [11C]methyl iodide: groundbreaking labeling methods in 11C radiochemistry. Label. Comp. Radiopharm. **58**, 73–85 (2015)

A.C. Frisch, M. Beller, Catalysts for cross-coupling reactions with non-activated alkyl halides. Angew. Chem. Int. Ed. Engl. **44**, 674–688 (2005)

V.L. Anderson, M.M. Herth, S. Lehel, G.M. Knudsen, J.L. Kristensen, Palladium-mediated conversion of para-aminoarylboronic esters intopara-aminoaryl-11C-methanes. Tetrahedron. Lett. **54**, 213–216 (2013)

L. Pizzorno, Nothing boring about boron. Integr. Med. **14**, 35–48 (2015)

E.D. Hostetler, S. Fallis, T.J. MCCarthy, M.J. Welch, J.A. Katzenellenbogen, Improved methods for the synthesis of [11C]palmitic acid. J. Org. Chem. **63**, 1348–1351 (1998)

K. Dahl, M. Schou, C. Halldin, Direct and efficient (Carbonyl)cobalt-mediated aryl acetylation using [11C]Methyl Iodide. Eur. J. Org. Chem. 2775–2777 (2016)

O. Rahman, T. Kihlberg, B. Langstrom, Eur. J. Org. Chem. 474–478 (2004)

F. Karimi, J. Barletta, B. Langstrom, Eur. J. Org. Chem. 2374–2378 (2005)

K. Dahl, M. Schou, N. Amini, C. Halldin, Eur. J. Org. Chem. 1228–1231 (2013)

T. Arai, M.R. Zhang, M. Ogawa, T. Fukumura, K. Kato, K. Suzuki, Efficient and reproducible synthesis of [1-11C]acetyl chloride using the loop method. Appl. Radiat. Isot. **67**, 296–300 (2009a)

T. Arai, K. Kato, M. Zhang, Tetrahedron Lett. **50**, 4788–4791 (2009b)

T. Arai, Nucl. Med. Biol. **39**, 702–708 (2012)

T. Kihlberg, B. Langstrom, Biologically active 11C-labeled amides using palladium-mediated reactions with aryl halides and [11C]carbon monoxide. J. Org. Chem. **64**, 9201–9205 (1999)

C.A. Tobias, J.H. Lawrence, F.J.W. Roughton, W.J. Rooth, M.I. Gregerson, Am. J. Physiol. **145**, 253–260 (1945)

A. Brennfuhrer, H. Neumann, M. Beller, Palladium-catalyzed carbonylation reactions of aryl halides and related compounds. Angew. Chem. Int. Ed. Engl. **48**, 4114–4133 (2009)

S. Kealey, A. Gee, P.W. Miller, Transition metal mediated [(11) C]carbonylation reactions: recent advances and applications. J. Label. Comp. Radiopharm. **57**, 195–201 (2014)

O. Rahman, [(11) C]carbon monoxide in labeling chemistry and positron emission tomography tracer development: scope and limitations. J. Label. Comp. Radiopharm. **58**, 86–98 (2015)

H. Doi, J. Barletta, M. Suzuki, R. Noyori, Y. Watanabe, B. Langstrom, Synthesis of 11C-labelled N, N'-diphenylurea and ethyl phenylcarbamate by a rhodium-promoted carbonylation via [11C]isocyanatobenzene using phenyl azide and [11C]carbon monoxide. Org. Biomol. Chem. **2**, 3063–3066 (2004)

I. Ryu, K. Kusano, A. Ogawa, N. Kambe, N. Sonoda, Free-radical carbonylation efficient trapping of carbon monoxide by carbon radicals. J. Am. Chem. Soc. **112**, 1295–1297 (1990)

O. Itsenko, B. Langstrom, Radical-mediated carboxylation of alkyl iodides with [11C]carbon monoxide in solvent mixtures. J. Org. Chem. **70**, 2244–2249 (2005a)

O. Itsenko, B. Langstrom, Photoinitiated free radical carbonylation enhanced by photosensitizers. Org. Lett. **7**, 4661–4664 (2005b)

P. Lidstrom, T. Kihlberg, B. Langstrom, J. Chem. Soc. Perkin. Trans. **1**, 2701–2706 (1997)

M.H. Al-Qahtani, V.W. Pike, J. Label. Compd. Radiopharm. **43**, 825–835 (2000)

H. Audrain, L. Martarello, A. Gee, D. Bender, Utilisation of [11C]-labelled boron carbonyl complexes in palladium carbonylation reaction. Chem. Commun. (Camb) **5**, 558–559 (2004)

S. Kealey, P.W. Miller, N.J. Long, C. Plisson, L. Martarello, A.D. Gee, Copper(I) scorpionate complexes and their application in palladium-mediated [(11)C]carbonylation reactions. Chem. Commun. (Camb) **25**, 3696–3698 (2009)

A.F. Scott, L.L. Wilkening, B. Rubin, Inorg. Chem. 2533 (1969)

M.I. Bruce, A.P.P. Ostazewski, J. Chem. Soc. Dlaton. Trans. 2433–2436 (1973)

J. Eriksson, J. van den Hoek, A.D. Windhorst, Transition metal mediated synthesis using [11C]CO at low pressure—a simplified method for 11C-carbonylation. J. Label. Compd. Radiopharm. **55**, 223–228 (2012)

J. Pardo, M.C. Lopez, J. Santafe, F.M. Royo, J.S. Urieta, Solubility of gases in butanols. I. Solubilities of nonpolar gases in 1-butanol from 263.15 to 303.15 K at 101.33 kPa partial pressure of gas. Fluid Phase Equilibria. **109**, 29–37 (1995)

F. Gibanel, M.C. Lopez, F.M. Royo, J. Santafe, J.S. Urieta, Solubility of nonpolar gases in tetrahydrofuran at 0 to 30C and 101.33 kPa partial pressure of gas. J. Solution. Chem. **22**, 211–217 (1993)

G. Kohler, C. Milstein, Continuous cultures of fused cells secreting antibody of predefined specificity. Nature **256**, 495–497 (1975)

K. Strebhardt, A. Ullrich, Paul ehrlich's magic bullet concept: 100 years of progress. Nat. Rev. Cancer. **8**, 473–480 (2008)

L. Shan, [(18)F]Fluorobenzoyl anti-HER2 Cys-diabody, molecular imaging and contrast agent database (MICAD), bethesda (MD) (2004)

T. Olafsen, A.W. Wu, Novel antibody vectors for imaging. Semin. Nucl. Med. **40**, 167–181 (2010)

G. Vaidyanathan, M.R. Zalutsky, Labeling proteins with fluorine-18 using N-succinimidyl 4-[18F]fluorobenzoate. Int. J. Rad. Appl. Instrum. B **19**, 275–281 (1992)

A.V. Maerle, M.A. Simonova, V.D. Pivovarov, D.V. Voronina, P.E. Drobyazina, D.Y. Trofimov, L.P. Alekseev, S.K. Zavriev, D.Y. Ryazantsev, Development of the covalent antibody-DNA conjugates technology for detection of IgE and IgM antibodies by immuno-PCR. PLoS ONE **14**, e0209860 (2019)

S. Madler, C. Bich, D. Touboul, R. Zenobi, Chemical cross-linking with NHS esters: a systematic study on amino acid reactivities. J. Mass. Spectrom. **44**, 694–706 (2009)

G. Mattson, E. Conklin, S. Desai, G. Nielander, M.D. Savage, S. Morgensen, A practical approach to crosslinking. Mol. Biol. Rep. **17**, 167–183 (1993)

L. Lang, W.C. Eckelman, One-step synthesis of 18F labeled [18F]-N-succinimidyl 4-(fluoromethyl)benzoate for protein labeling. Appl. Radiat. Isot. **45**, 1155–1163 (1994)

C.W. Choi, L. Lang, J.T. Lee, K.O. Webber, T.M. Yoo, H.K. Chang, N. Le, E. Jagoda, C.H. Paik, I. Pastan et al., Biodistribution of 18F- and 125I-labeled anti-Tac disulfide-stabilized Fv fragments in nude mice with interleukin 2 alpha receptor-positive tumor xenografts. Cancer. Res. **55**, 5323–5329 (1995)

T. Olafsen, S.J. Sirk, S. Olma, C.K. Shen, A.M. Wu, ImmunoPET using engineered antibody fragments: fluorine-18 labeled diabodies for same-day imaging. Tumour. Biol. **33**, 669–677 (2012)

C.J. Anderson, T.J. Wadas, E.H. Wong, G.R. Weisman, Cross-bridged macrocyclic chelators for stable complexation of copper radionuclides for PET imaging. Q. J. Nucl. Med. Mol. Imaging. **52**, 185–192 (2008)

T.K. Nayak, M.W. Brechbiel, Radioimmunoimaging with longer-lived positron-emitting radionuclides: potentials and challenges. Bioconjug. Chem. **20**, 825–841 (2009)

A. Obata, S. Kasamatsu, D.W. McCarthy, M.J. Welch, H. Saji, Y. Yonekura, Y. Fujibayashi, Production of therapeutic quantities of (64)Cu using a 12 MeV cyclotron. Nucl. Med. Biol. **30**, 535–539 (2003)

G.T. Hermanson, *Bioconjugate Techniques*, 2nd Edn, (2008), p. 499

W.C. Cole, S.J. DeNardo, C.F. Meares, M.J. McCall, G.L. DeNardo, A.L. Epstein, H.A. O'Brien, M.K. Moi, Serum stability of 67Cu chelates: comparison with 111In and 57Co. Int. J. Rad. Appl. Instrum. B **13**, 363–368 (1986)

G.J. Deblonde, M.P. Kelley, J. Su, E.R. Batista, P. Yang, C.H. Booth, R.J. Abergel, Spectroscopic and computational characterization of diethylenetriaminepentaacetic acid/transplutonium chelates: evidencing heterogeneity in the heavy actinide(III) series. Angew. Chem. Int. Ed. Engl. **57**, 4521–4526 (2018)

M. Gut, J.P. Holland, Synthesis and photochemical studies on gallium and indium complexes of DTPA-PEG3-ArN3 for radiolabeling antibodies. Inorg. Chem. **58**, 12302–12310 (2019)

M.L. Corman, S. Galandiuk, G.E. Block, E.D. Prager, G.J. Weiner, D. Kahn, H. Abdel-Nabi, E.P. Mitchell, V.L. Pascucci, A.N. Maroli et al., Immunoscintigraphy with 111In-satumomab pendetide in patients with colorectal adenocarcinoma: performance and impact on clinical management. Dis. Colon. Rectum. **37**, 129–137 (1994)

E. Boros, J.P. Holland, Chemical aspects of metal ion chelation in the synthesis and application antibody-based radiotracers. J. Label. Comp. Radiopharm. **61**, 652–671 (2018)

D.W. Margerum, G.R. Cayley, D.C. Weatherburn, G.K. Pagen-Kopf, Kinetics and mechanisms of complex formation and ligand exchange, in *Coordination Chemistry*, vol. 2, ed. by A.E. Martell (Chapter 1, American Chemical Society Monography 174, Washington, DC, 2 1978), p. 174

C.F. Meares, M.K. Moi, H. Diril, D.L. Kukis, M.J. McCall, S.V. Deshpande, S.J. DeNardo, D. Snook, A.A. Epenetos, Macrocyclic chelates of radiometals for diagnosis and therapy. Br. J. Cancer. Suppl. **10**, 21–26 (1990)

N. Wu, C.S. Kang, I. Sin, S. Ren, D. Liu, V.C. Ruthengael, M.R. Lewis, H.S. Chong, Promising bifunctional chelators for copper 64-PET imaging: practical (64)Cu radiolabeling and high in vitro and in vivo complex stability. J. Biol. Inorg. Chem. **21**, 177–184 (2016)

K.E. McCabe, B. Liu, J.D. Marks, J.S. Tomlinson, H. Wu, A.M. Wu, An engineered cysteine-modified diabody for imaging activated leukocyte cell adhesion molecule (ALCAM)-positive tumors. Mol. Imaging. Biol. **14**, 336–347 (2012)

J.R. Nedrow, A.G. White, J. Modi, K. Nguyen, A.J. Chang, C.J. Anderson, Positron emission tomographic imaging of copper 64- and gallium 68-labeled chelator conjugates of the somatostatin agonist tyr3-octreotate. Mol. Imaging **13** (2014)

T.J. Wadas, E.H. Wong, G.R. Weisman, C.J. Anderson, Coordinating radiometals of copper, gallium, indium, yttrium, and zirconium for PET and SPECT imaging of disease. Chem. Rev. **110**, 2858–2902 (2010)

R.D. Hancock, A.E. Martell, Ligand design for selective complexation of metal ions in aqueous solution. Chem. Rev. **89**, 1875–1914 (1989)

A.E. Martell, R.M. Smith, Critical stability constants (Plenum Press, New York, 1974, 1975, 1976, 1977, 1982, 1989), pp. 1–6

R.G. Pearson, J. Am. Chem. Soc. **85**, 3533 (1963)

A. Ramasubbu, K.P. Wainwright, Structurally reinforced cyclen: a rigidly trans-co-ordinating twelve-membered macrocycle. J. Chem. Soc. Commun. **5**, 277–278 (1982)

Z. Cai, C.J. Anderson, Chelators for copper readionuclides in positron emission tomography radiopharmaceuticals. J. Label. Compd. Radiopharm. **57**, 224–230 (2014)

C.A. Boswell, X. Sun, W. Niu, G.R. Weisman, E.H. Wong, A.L. Rheingold, C.J. Anderson, Comparative in vivo stability of copper-64-labeled cross-bridged and conventional tetraazamacrocyclic complexes. J. Med. Chem. **47**, 1465–1474 (2004)

J.S. Yoo, D. Pandya, EP Patent, 2,476,683 A472 (2012)

I. Kambali, F.A. Wibowo, Comparison of gallium-68 production yields from (p,2n), (α,2n) and (p,n) nuclear reactions applicable for cancer diagnosis. J. Phys.: Conf. Ser. **1198**, 022003 (2019)

Z. Talip, C. Favaretto, S. Geistlich, N.P.V. Meulen, A step-by-step guide for the novel radiometal production for medical applications: case studies with (68)Ga, (44)Sc, (177)Lu and (161)Tb. Molecules **25** (2020)

G.J. Meyer, H. Macke, J. Schuhmacher, W.H. Knapp, M. Hofmann, 68Ga-labelled DOTA-derivatised peptide ligands. Eur. J. Nucl. Med. Mol. Imaging. **31**, 1097–1104 (2004)

W.A. Breeman, M. de Jong, E. de Blois, B.F. Bernard, M. Konijnenberg, E.P. Krenning, Radiolablelling DOTA-peptide with 68Ga. Eur. J. Nucl. Med. **32**, 475–485 (2005)

K.P. Zhernosekov, D.V. Filosofov, R.P. Baum, P. Aschoff, H. Bihl, A.A. Razbash, M. Jahn, M. Jennewein, F. Rosch, Processing of generator-produced 68Ga for medical application. J. Nucl. Med. **48**, 1741–1748 (2007)

Chapter 4
Principles for the Design of MRI Probes

4.1 Introduction

Magnetic resonance imaging (MRI) is a medical imaging technique that fundamentally borrows the concept of magnetization of proton spins and relaxation from nuclear magnetic resonance (NMR) spectroscopy in chemistry for the identification and structural characterization of molecules. One of the inventors of MRI is chemist Paul Lauterbur, who has a deep interest in using NMR to detect organosilicon radicals. During this time, in 1971, Raymond Damadian from New York reported that normal tissue had shorter relaxation times than the tumor tissue in a rat model of cancer; subsequently, in 1974, Damadian obtained the first patent on the use of the NMR imaging technique for detecting cancer (Damadian 1974). In the same year, Paul Lauterbur and Peter Mansfield independently described the use of magnetic field gradients for spatial localization of NMR signals. These discoveries laid the foundation for a new field called MRI, which earned both of them a Nobel prize in medicine in 2003 (Edelman 2014). Since both NMR and MRI rely on the abundance of protons in the chemical samples and tissues, respectively, it is worthwhile to remind in this lecture about how proton spins generate a signal and are detected. Basically, the human body is comprised of over 70% of H_2O, which has two hydrogen atoms and one oxygen atom. Water is the largest source of protons; next is fat in the form of methylene groups (CH_2). For example, in a cubic millimeter of tissue, there are approximately 10^{19} hydrogen atoms (Edward Hendrick 2008). As mentioned in Chap. 1, each hydrogen atom has one proton (positive charge) and one electron (negative charge). Since each of them has an opposite charge, it is intuitive to rationale that the proton and electron are held together via electrostatic force. Both a proton and an electron have a mass, a charge, and a spin like every quantum particle, and thus, they possess magnetic dipole moments. However, the magnetic dipole moment of an electron is much stronger than that of a proton, and thus the electromagnetic waves needed to excite a proton occur in the right energy range (radio frequency) to penetrate tissue (Edward Hendrick 2008). In contrast, more energetic electromagnetic waves, such as microwaves, must be used to excite the electron. MRI takes advantage of this

W. Pham, *Principles of Molecular Probe Design and Applications*,
https://doi.org/10.1007/978-981-19-5739-0_4

abundant source of protons to produce anatomical-related images since each charged and fast-spinning hydrogen atom can produce a small but noticeable magnetic field, which is generally called a magnetic moment (Fig. 4.1). In a normal condition, the protons spin randomly, so there is no overall magnetic field. However, when there is an applied external magnetic source (B_0), most of the hydrogen protons align in the same direction as the external field B_0. Some align in the opposite direction. The larger the external B_0 field, the greater the excess number of protons aligned with the field. The sum of all aligned protons in the same direction of B_0 is called net magnetization. Now, the question is, how can MRI generate images out of this current? Basically, when a radiofrequency (RF) electromagnetic wave is applied, that will perturb the protons from the alignment with the magnetic field, and thus, a 90° flip angle is observed. When the RF pulse is turned off, the protons will process back to the original state, associated with two distinct phenomena called relaxation. During that process, they emit electromagnetic waves at characteristic or "resonant" frequency, falling into the radiofrequency (RF) signal (Yousaf et al. 2018) that is going to be detected by a scanner. The precession frequency of the spinning particle or nuclear magnetic moment (ω) is defined by the Larmor equation; it equals to the product of the gyromagnetic ratio (γ) and the strength of the externally applied field (B_0).

$$\omega = \gamma B_0$$

Initially, in phase, the excited protons begin to dephase, resulting in a decrease in transverse magnetization. This means an overall loss of a signal, called T2 relaxation (Fig. 4.2). The exponential decay of the T2 signal is different for different tissue, and it is the time the transverse magnetization decreases to 37% of its starting value (Fig. 4.2). In other words, each tissue has its own time constant, and this is one of the main methods to provide contrast in the images. In biological samples, T2 values

Fig. 4.1 Proton spin generates a magnetic field like a tiny magnet with a magnetic dipole moment

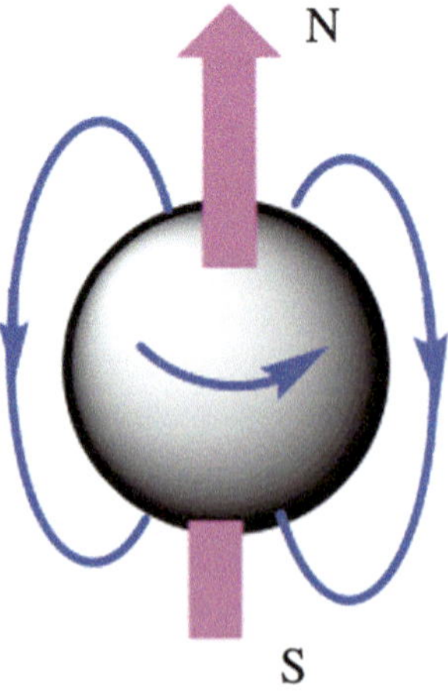

range from 30 to 150 ms. For example, the white matter has a short T2, resulting in dark pixels. In contrast, the cerebrospinal fluid (CSF) has a long T2 value and thus brighter signal, while the gray matter has intermediate T2, which produces signals in the middle range between the CSF and white matter (Fig. 4.3).

Eventually, the proton spin will return to align in parallel with the external magnetic field B_0, and this time constant is needed to grow back in the longitudinal magnetization (M_z) direction called T_1 relaxation or longitudinal relaxation (Fig. 4.2). In biological samples, T_1 values range from 300 to 2000 ms. T_1 is also a characteristic of tissue, and it is defined as the time that it takes the longitudinal magnetization to grow back to 63% of its final value (Pooley 2005). Because different tissue has its own T_1 constant, T_1-weighted contrast imaging concur a powerful means to distinguish different tissues within a biological structure. For example, the white matter has a short T_1 relaxation time; in contrast, the CSF has a long T_1 value, and the T_1 value of gray matter is somewhere in between. Then, in a T_1-weighted image, the white matter quickly realigns its longitudinal magnetization with B_0 and will appear bright. Conversely, the CSF has much slower longitudinal magnetization

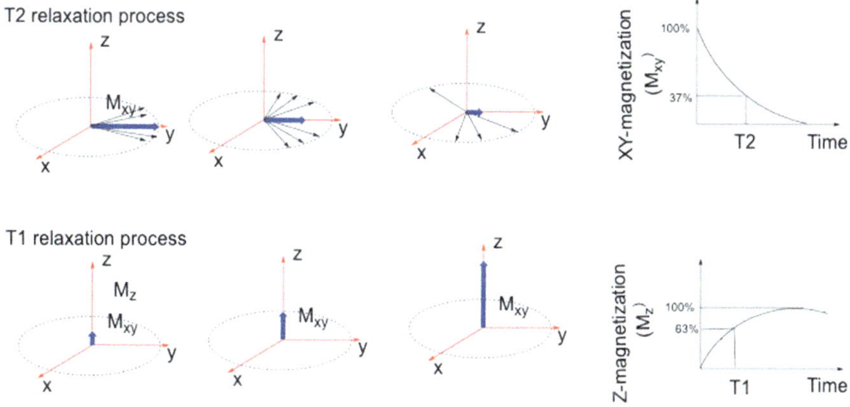

Fig. 4.2 Distinct T_1 and T_2 relaxation processes

Fig. 4.3 T_2 relaxation times of different tissues within a biological structure provide a distinct contrast to the T_2-weighted image. Data obtained from Pooley (2005) with permission from the author

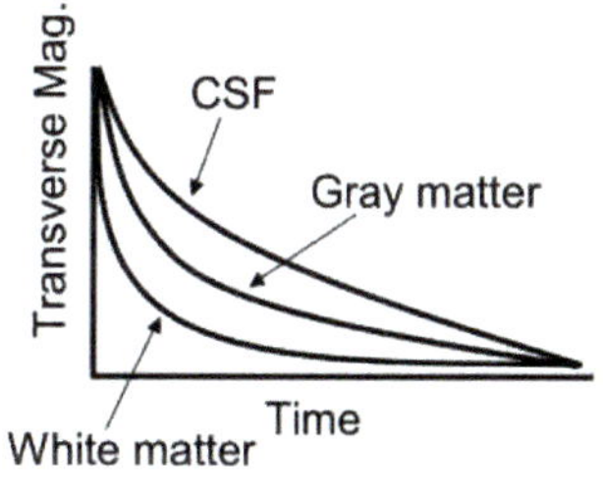

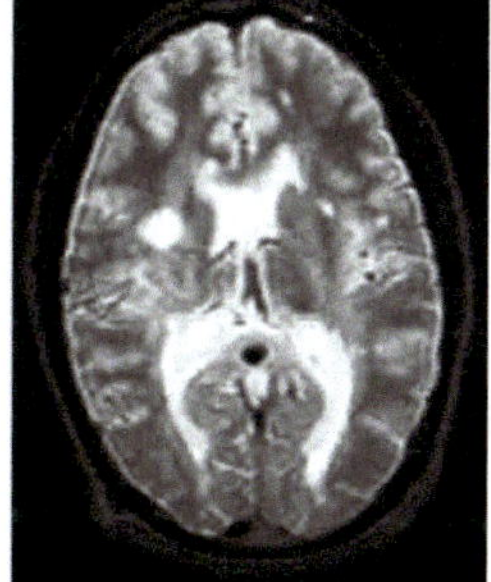

Fig. 4.4 T_1 relaxation
times of different tissues
provide a distinct contrast to
the T_1-weighted image. Data
is obtained from Pooley
(2005) with permission from
the author

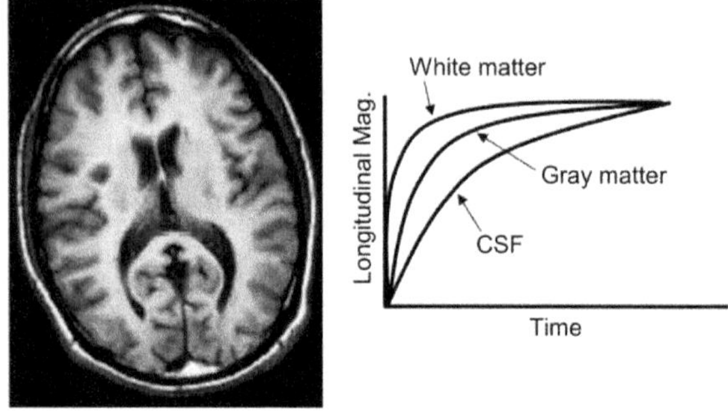

realignment, resulting in a low signal and appearing dark. While the gray matter
contributes to signals that have intermediate shades of gray (Fig. 4.4) (Pooley 2005).

Pulse sequences. In order to obtain specific MR images for a particular application,
it is necessary to optimize the RF pulses and signal acquisition in coordination
with magnetic field gradients, in a process called pulse sequences. The two most
commonly used pulse sequences are called gradient-echo and spin-echo sequences.
The former is produced by a single RF pulse, typically < 90°, in conjunction with
readout gradient reversal, while the latter is generated by pairs of RF pulses. Since
only one RF pulse is applied, the echo time (TE), which is the time between RF
excitation and signal acquisition, is generally shorter for gradient-echo sequences
than for spin-echo sequences. Shorter repetition time (TR), which implies the time
between subsequence RF excitations, can be achieved as well. Overall, gradient-echo
sequences are suitable for rapid imaging (Markl and Leupold 2012).

One of the reasons MRI became a golden diagnostic tool in clinical work compared
to other imaging modalities, such as CT, optical, PET, or X-ray, should be mentioned
related to the non-invasive, safe, and non-ionizing radiative features. One single MRI
scan can provide anatomical information with superior spatial resolution, regardless
of the depth of the tissues (Fig. 4.5) (James and Gambhir 2012). This unique feature
is only found in MRI and is unmatched by other imaging modalities. Particularly,
MRI is best suitable for neuroimaging to characterize neuroanatomy and functional
localization of the brain in the era where phrenology started to become an integral
part of the investigation. The brain is comprised of white matter, gray matter, and
CSF. White matter consists of the axons and appears bright due to the presence of
fatty materials of myelin. Gray matter consists primarily of neuronal cell bodies and
forms the cerebral cortex (Bridge and Clare 2006). In T_1-based MRI, water molecules
in free solution would appear dark on an image compared to water molecules in the
vicinity of a large fatty structure made up of myelin. Such imaging methods have been
exploited for the non-invasive detection of neuronal abnormalities for the diagnosis
of disorders of the white matter, such as multiple sclerosis (Bridge and Clare 2006).

In the era of precision neuroimaging, MRI is a notable modality, specifically
for the potential contribution to the detection of amyloid plaques. The process is
not only safe, but repeated scanning enables longitudinal assessment of the course

Modality	Temporal resolution	Spatial resolution	Tissue penetration	Sensitivity	Clinical use
CT	Minutes	50-200 µm	Limitless	ND	Yes
MRI	Minutes-hours	25-100 µm	Limitless	10^{-3} - 10^{-5} M	Yes
PET	Seconds-minutes	1-2 mm	Limitless	10^{-11} - 10^{-12} M	Yes
SPECT	Minutes	1-2 mm	Limitless	10^{-10} - 10^{-11} M	Yes
US	Seconds-minutes	0.01-0.1 mm	mm-cm	ND	Yes
OFI	Seconds-minutes	2-3 mm	<1cm	10^{-9} - 10^{-12} M	Limited
SERS	Minutes-days	mm	~5 mm	10^{-12} - 10^{-15} M	Limited
PAI	Seconds-minutes	10 µm-1 mm	6 mm-5 cm	ND	Not yet
IVM	Seconds-days	1-10 µm	~700 µm	10^{-15} - 10^{-17} M	ND

Fig. 4.5 Comparisons of major features of emerging imaging technologies. *ND* not determined; *CT* computed tomography; *MRI* magnetic resonance imaging; *PET* positron emission tomography; *SPECT* single photon emission tomography; *US* ultrasound; *OFI* optical fluorescence imaging; *SERS* surface-enhanced Raman scattering; *PAI* photoacoustic imaging; *IVM* intravital microscopy. Data modified from James and Gambhir (2012)

of development of plaque load to generate a biological mapping of the disease as amyloid plaques started the pathological aggregation from the hippocampus (Jacobs et al. 1995; Masur et al. 1994). In order to track the pathological amyloid plaques progression, high-resolution imaging such as MRI with submillimeter resolution capability is crucial because the hippocampus consists of a number of molecularly distinct subregions that are interconnected to form the hippocampal circuit (Amaral and Witter 1989). The functional structure of the hippocampus is highly organized in specialized modules, in which each submodule houses a distinct population of neurons expressing unique molecular profiles (Huddleston and Small 2005; Zhao et al. 2001). In that respect, if only global imaging of the whole hippocampus to be achieved, partly due to the lack of high-resolution capability, we will miss the chance to detect to role and sequence of each hippocampal module participating in the early plaque formation and the progress of the disease.

Aside from providing detailed anatomical information with excellent resolution, MRI is also used for imaging to acquire the body physiology and pathological information via a number of state-of-the-art MRI methods. It adds a new dimension to MRI examinations by adding functional information to the largely anatomical information gathered by the conventional sequences (Baliyan et al. 2016).

Diffusion-weighted imaging. The concept of diffusion-weighted imaging (DWI) is based on the random Brownian motion of water in the intracellular compartments versus their counterparts in extracellular environments. In general, water molecules in the extracellular matrix experience fairly free diffusion compared to the water molecules in the intracellular milieu, which have restricted diffusion. Different tissues in the human body have discerned cellular architectures and unique ratios of water molecules in these compartments, thus providing distinguish diffusion properties. Even for the same tissues, the proportion of water molecules and their diffusion can be altered in the face of pathological development, such as in the case of several neurological disorders or high-grade malignancies, or acutely infarcted tissues, thus revealing their microarchitecture. Since the original development of DWI in the early

1990s (Le Bihan 2014; Turner et al. 1990), DWI MRI has been considered an important modality for the diagnostic imaging of several clinical-related diseases, including brain tumors, brain ischemia, white matter-relate disorders, and oncological diseases.

Blood-oxygen-level-dependent MRI (BOLD-MRI). This technique forms the basis for functional MRI, particularly a very promising imaging technique for studying functional brain activity, whereby oxygen metabolism is notorious. The BOLD method directly measures the localized alterations of deoxyhemoglobin in the tissues. This measurement is possible thanks to the dynamic levels of oxy-/deoxyhemoglobins in a physiological context. Basically, oxyhemoglobin contains no unpaired electrons and is diamagnetic; when oxygen is released, the resulted deoxyhemoglobin inherits four unpaired electrons per heme and thus making the molecule strongly paramagnetic (Pauling and Coryell 1936). In deoxyhemoglobin blood, the water spins experience increase dephasing, which attenuates the T_2-weighted signal from tissue containing deoxyhemoglobin. This phenomenon can be exploited for imaging. A wide range of brain functions is ideal for study with BOLD-MRI. For example, when the brain actively performs a task, some related regions consume a large amount of oxygen, leading to a localized increase in deoxyhemoglobin. Consequently, the vascular systems in the vicinity start to dilate to compensate for the loss of oxygenated hemoglobin. This instant blood flow decreases deoxyhemoglobin concentration and leads to an increased BOLD signal (Gauthier and Fan 2019). BOLD-MRI has been used in clinical studies for the assessment of memory (Giesel et al. 2005), functional architecture of the brain (Lee et al. 2013), and tumors (Dunn et al. 2002).

CEST MRI. In the early 90s, the emergence of chemical exchange saturation transfer (CEST) MRI has served as a potentially versatile tool for clinical assessments of cerebral ischemia, neurological disorders, lymphedema, osteoarthritis, muscle physiology, and solid tumors (Jones et al. 2018). Different from the conventional approach that detects water in the tissues, CEST detects the presence of molecules to probe chemical compounds and metabolites associated with the body's physiological function and pathological conditions (Wu et al. 2016). The principle of CEST imaging is that a narrow bandwidth RF pulse is applied to a chemical molecule of interest that has its protons capable of exchanging with those of bulk water at its resonant frequency to reach a saturation state. The magnetic saturation is transferred to the bulk water protons via a chemical exchange, resulting in the attenuation of the water proton signal, which can be detected by MR imaging sequences. Overall, the method provides an indirect measurement of the concentration of the chemical species (Wu et al. 2016; Dula et al. 2013). Endogenous or exogenous CEST chemical molecules, including metabolic probes, contain key exchangeable moieties such as –NH_2 (amine), –$CONH_2$ (amide), and –OH (hydroxyl). Among these groups, CEST of amide protons, or sometimes also called amide proton transfer (APT), probably dominates in the literature, given the existence of amide bonds in peptides, proteins, and other active biological substances in the metabolic milieu. CEST of amine protons, using endogenous probes, including creatine and glutamate, is also commonly used to assess metabolic alternation related to neurodegenerative diseases (Haris et al.

2013) and strokes (Zaiss et al. 2014). Meanwhile, the CEST of the hydroxyl group is more challenging since the protons of alcohols are very close to those from water; thus, hydroxyl protons have a fast chemical exchange rate (Sherry and Woods 2008). Nevertheless, a number of metabolites with exchangeable hydroxyl protons have been incorporated in CEST imaging, including glycosaminoglycan (Gag), myo-inositol (MI), and glucose (Wu et al. 2016).

Hyperpolarized NMR

Despite as one of the most powerful spectroscopic methods, NMR spectroscopy has inherently poor sensitivity because the detected signal strength depends on the small population difference between spin states, even in high magnetic fields (Adams et al. 2009). In fact, the equilibrium nuclear spin determines the fraction of nuclear spins contributing to the detected signal; and this fraction is below 0.1% of all nuclear spins at currently available NMR spectrometer fields (Meier et al. 2014). This limitation is addressed by the merging of hyperpolarization methods, which are used to enhance nuclear spin polarization to improve NMR sensitivity by 4–8 orders of magnitude transiently at the time of measurements (Fig. 4.6) (Meier et al. 2014; Nikolaou et al. 2014). Regarding MRI probe chemistry, the development of hyperpolarized probes is one of the most active areas, including solids, gases, and liquid-state metabolites for applications in imaging lung disease and in vivo metabolism (Adamson et al. 2017). A number of nuclei can be directly hyperpolarized, including ^{1}H, ^{3}He, ^{7}Li, ^{13}C, ^{15}N, ^{19}F, ^{29}Si, ^{89}Y, ^{107}Ag, ^{109}Ag, ^{31}P, ^{83}Kr, and ^{129}Xe, among others (Barskiy et al. 2017). Specifically, carbon-13 (^{13}C) MRI is particularly attractive for metabolic imaging because carbon serves as the backbone of nearly all organic molecules, thus enabling the investigation of a wide range of biochemical processes relevant to human diseases (Wang et al. 2019). Hyperpolarized ^{13}C-substrate induces over the 10,000-fold enhancement of the ^{13}C MR spectroscopic signals, enabling the assessment of the substrate or its metabolic product in the biochemical pathways without background interference from surrounding tissues (Jennings and Bachmann 2008; Mansson et al. 2006). Thus, ^{13}C-hyperpolarized probes have been developed for the detection of biologically relevant metabolic pathways, such as ^{13}C-bicarbonate, ^{13}C-pyruvate, ^{13}C-fumarate, ^{13}C-lactate, ^{13}C-malate, and ^{13}C-carbon dioxide (Sharma et al. 2019). The next most popular probes are hyperpolarized gases, particularly ^{129}Xe gas, which has widespread application for functional lung imaging, metabolic brown fat imaging and more (Barskiy et al. 2017).

Among a number of different methods for hyperpolarization, dynamic nuclear polarization (DNP), parahydrogen-induced polarization (PHIP), and signal amplification by reversible exchange (SABRE) are mostly recognized in the field. In DNP, the substrate is doped with a polarizing agent, usually, a free radical molecule with unpaired electrons. As the mixture is placed inside a polarizer, immersed in liquid helium to achieve approximate temperature of about 1 K, at a high magnetic field, the unpaired electrons obtain a near-unity spin polarization. Then, microwave irradiation is applied to transfer the electron polarization from the polarizing agent to the substrate in a process DNP. An NMR coil surrounding the polarized substrate is used to monitor the polarization process (Barskiy et al. 2017; Yen et al. 2011). Each

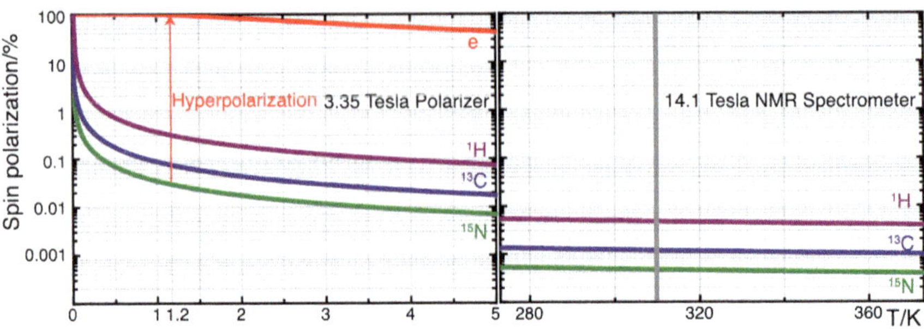

Fig. 4.6 Spin polarizations of ^{1}H, ^{13}C and ^{15}N at 3.35 T via dynamic nuclear polarization (DNP) compared to spin polarization at 14.1 T. DNP resulted in spin polarization that are several orders of magnitude enhanced compared to the thermal equilibrium polarization. Data obtained from Meier et al. (2014)

substrate has a different polarization time. For instance, it would take one hour to achieve 98% of the maximal level of solid-state polarization for [^{13}C]pyruvate. In the case of hyperpolarized gas, such as xenon, the super cold matrix-substrate mixture is rapidly warmed after microwave irradiation, sublimating the hyperpolarized xenon as a pure gas, hence the term "sublimation DNP" emanates from this phenomenon (Barskiy et al. 2017; Capozzi et al. 2015).

An alternative approach to hyperpolarization can be achieved using PHIP, which has the advantage of cost and simpler operations compared to DNP (Reineri et al. 2015). As a chemistry-based method, PHIP hyperpolarization requires the substrate molecule must have unsaturated bonds since the process involves the insertion of parahydrogen into the substrate via catalytic hydrogenation (Ardenkjar-Larsen et al. 2003). Therefore, unsaturated molecules, which can be hydrogenated to either naturally occurring metabolic products or similar analogs with ideal pharmacokinetics and safety profiles, are an excellent candidate for future contrast agents (Gloggler et al. 2013).

Another promising hyperpolarized technique suitable for developing biologically active substances is signal amplification by reversible exchange (SABRE). SABRE relies on a metallic catalyst to transfer polarization from parahydrogen to the substrate. The advantage of SABRE is that it can polarize a sample within a few seconds; however, the level of polarization is merely 10% compared to 30–40% polarization achieved after a few hours with DNP (Sinharay and Pagel 2016). The technique involves transferring spin order by hydrogenation reaction without changing the chemical structure of the substrate.

Despite providing excellent resolution and depth of penetration along with flexible tissue contrast by optimizing the relaxation times, MRI's intrinsic problem is the poor sensitivity compared to other imaging modalities. Since the 1970s, it has been proposed to use exogenous contrast agents to improve MRI's diagnostic utility (Lauterbur et al. 1978). Since then, a large number of contrast agents have been developed for T_1- and T_2-based imaging. The nature of the magnetic susceptivity of the probes suggests these agents should be made of metals, whether they are a single probe or clusters in form of nanoparticles. In the context of this lecture, the following sections will discuss the chemical development of MRI contrast agents.

4.2 Contrast Agents

4.2.1 Chelator-Free Metals

MR imaging contrast relies on the relaxation time differences between tissues, albeit this intrinsic signal is not always ideal for diagnostic purposes. So, the idea now is to administer contrast agents to increase water protons' relaxation times in tissues selectively. The question would come next as to what are the requirements of an MRI contrast agent? First and foremost, the agents must have strong magnetic suscepti-bility, the extent to which the material becomes magnetized when placed within an external magnetic field. The paramagnetic atoms or metals, such as Pt, O, W, Mn, Al, Eu, Ti, Te, Rh, have unpaired electrons, which have strong magnetic dipole moments because they can spin in either direction and attracted by a magnetic field. In contrast, some metal ions are not attracted, but on the other hand, repelled by the magnet. These are considered as diamagnetic materials, such as Hg, Ag, Cu, and C, Pb, which are unsuitable for use as MRI contrast agents. Another class of metal ions is classified as ferromagnetic materials because of their superparamagnetic characteristic. These metal ions, including Fe, Co, Ni, Gd, and Dy can be magnetized very easily, and they have a strong attraction to the magnetic field. Superparamagnetic materials are excel-lent MRI contrast agents because they can achieve a point of complete magnetization even when placed in a weak external magnetic field.

For an ideal contrast agent, the criteria would center on as low as possible the concentration of the paramagnetic materials sufficient enough to obtain maximal paramagnetic contributions to the relaxation rate (Mendonca-Dias et al. 1983). To achieve sufficient contrast capability, three factors should be considered during the design. First, the metal ions should have a great magnetic dipole moment, then concentration, and lastly, the physical proximity between the paramagnetic ion and the resonating nucleus, which is mostly the nuclear spin of the protons of the water molecules in the tissue. The relationship between the concentration and magnetic moment of the contrast agent in relation to T_1 can be rationale through this equation shown below (Bloembergen et al. 1948):

$$\Delta\left(\frac{1}{T_1}\right) = \frac{12\pi^2\gamma^2\varepsilon\mu^2 N}{5\kappa T}$$

where γ represents the gyromagnetic ratio for the hydrogen nucleus, ε represents the viscosity of the solvent, μ stands for the magnetic dipole moment, κ is the Boltzmann's constant, N is the number of ions per unit volume, and T is the abso-lute temperature. From this equation, it is apparent that increasing the metal ion concentration or using metal ions with large dipole moment will eventually result in a greater reduction in T_1. Meanwhile, the strength of the dipolar interaction is inversely proportional to the sixth power of distance ($1/r^6$), where r represents the mean distance between the paramagnetic center and the resonating nucleus.

In the earlier days, paramagnetic transition metals have been tested to influence the relaxation time T_1 and T_2 using a metal salt solution, like $Fe(NO_3)_3$ (Bloch et al. 1946), $MnCl_2$ or Mn-citrate (Dwek 1972; Eisinger et al. 1962), and they have been demonstrated to be effective relaxation agents. In the presence of paramagnetic ions, the relaxation rates R_1 and R_2 ($R = 1/T$) increase significantly as T_1 reduces and is directly proportional to the concentration. However, the effect varies depending on the ions. Mn^{2+}, Gd^{3+} and Fe^{3+} ions have the greatest effect upon T_1, followed by Cr^{3+} and Cu^{2+} have the least effect. Even these agents lack specificity, the solution containing these agents can distribute to tissues after intravenous injection in preclinical animal models for analysis. For example, a diluted manganese chloride in saline solution was injected in dogs at a dose of 0.1 mmol/kg of Mn^{2+} resulted in enhancement of the relaxation of water protons by Mn^{2+} in the distributed tissues (Fig. 4.7). Further, one of the first in vivo MR imaging of soft tissues was performed on live rabbits by intravenous injection of $MnCl_2$ solution at a concentration of 1–2 mg Mn^{2+}/kg body weight (Fig. 4.8) (Doyle et al. 1981). It is obvious from the in vivo cross-sectional data, tissues that have short T_1 appeared bright and vice versa. What is most noteworthy about this work is that it is the first to demonstrate that exogenous contrast agents such as Mn^{2+} can decrease the relaxation time leading to enhanced image contrast.

Aside from Mn^{2+}, other paramagnetic metal ions, such as Cr^{3+}, Mn^{2+}, Ni^{2+}, Fe^{3+}, and Gd^{3+}, have been demonstrated as potential contrast agents. Among them, Gd^{3+} with the greatest magnetic dipole moment (Fig. 4.9) serves as the most effective contrast agent with the shortest proton relaxation time as compared with other ions using the same concentration and magnetic strength (Morgan and Nolle 1959).

Fig. 4.7 Effects of Mn^{2+} on T_1 rates of water protons at 4 MHz after dogs were intravenously injected with 0.1 mmol/kg Mn^{2+}. Tissues were sampled 30 min post injection. Data obtained from Mendonca-Dias et al. (1983) with permission from Elsevier

	R_1 (s^{-1})	
Tissue	**Control**	**Mn^{2+} injected**
Myocardium (normal)	3.2	16.3
Myocardium (infarcted)	2.9	6.0
Lung	2.7	8.0
Liver	5.8	36.2
Spleen	3.4	7.2
Abdominal muscle	3.7	5.1
Large intestine	3.1	12.7
Small intestine		14.4
Kidney (medulla)	2.0	8.5
Kidney (cortex)	3.8	16.4

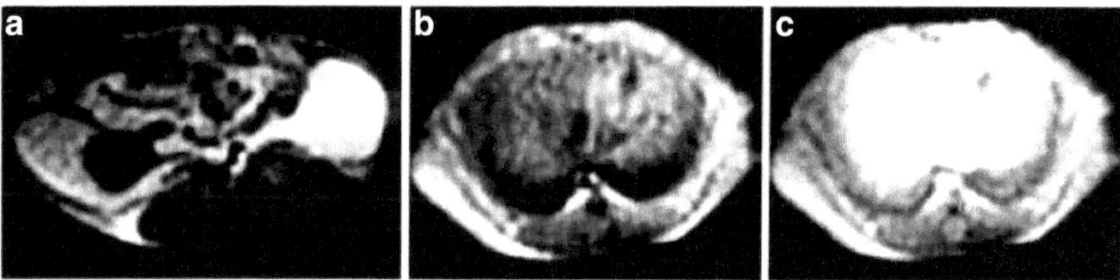

Fig. 4.8 Cross-sectional image of a rabbit abdomen, areas that appear bright have short T_1 and vice versa (**a**); cross-sectional image of a liver of an untreated rabbit (**b**); same cross section after intravenous injection of 1 ppm Mn^{2+}. The liver became brighter because the relaxation time has decreased by a factor of approximately 2 (**c**). Data obtained from Doyle et al. (1981) with permission from the author

Ion	Configuration	Unpair electron(s)	Magnetic Moment (Bohr magnetons)
Cr^{2+}	$3d^4$	4	4.8-4.9
Mn^{2+}	$3d^5$	5	5.9
Fe^{2+}	$3d^6$	4	5.3-5.5
Fe^{3+}	$3d^5$	5	5.9
Cu^{2+}	$3d^9$	1	1.7-1.8
Zn^{2+}	$3d^{10}$	0	0
Gd^{3+}	$4f^7$	7	7.9

Fig. 4.9 Magnetic moments of common paramagnetic ions

Based on the discussion above regarding the paramagnetism, it is not surprising that gadolinium (Gd^{3+}), which possesses seven unpaired electrons in the 4f-subshell, the most unpaired electrons known so far, generating a high magnetic moment that is effective at enhancing proton relaxation, and it is the most prevalent metal ion in MRI contrast agents (Caravan et al. 1999).

Acute and chronic toxicity is one of the major issues using free metal and lanthanide ions as MRI contrast agents. For example, it has been found that manganese causes cardiac toxicity in animal models. A dose of 0.1 mmol/kg of body weight, the free manganese ion can be lethal in animals (Mendonca-Dias et al. 1983). The same holds true for other metal ions, given their sizes are in the proximity of those of divalent cation calcium, and thus, they can inhibit calcium channels and compete for binding with calcium-binding membrane proteins and calcium-binding area on the endoplasmic reticulum CaATPase (Biagi and Enyeart 1990; Chevallier and Butow 1971). The interference with this secondary messenger implicates several important physiological cellular processes, including reproduction, muscle contraction, secretion, and wound healing (Baykara et al. 2019). In the early days, it has been known that chelating the metal ions with scavenging macromolecules or polycyclic compounds significantly attenuates the toxicity. The chelators not only help to prevent shielding the metal from direct interferences with the body's biochemical pathway, but the complex also alters and speeds up the washout of the contrast agent after they perform their task. For instance, the washout route of free manganese goes through the liver, but when associated with EDTA, the complex is excreted safely through the kidneys in the urine (Mendonca-Dias et al. 1983). Today, contrast agents are administered in about 25% of all MRI procedures, including brain and spine, for MR angiography, MRI examinations of the abdomen, breast, and heart (Lohrke et al. 2016).

4.2.2 Chelator-Associated Metals

In order to reduce paramagnetic metal ions toxicity, the concept of chelation of the metal ions with appropriate ligands was adopted, and metal chelates have been developed as MRI contrast agents (Moncelli et al. 1998). Although a number of scaffolded metal ion complexes using different types of paramagnetic ions, including Fe^{3+}, Mn^{2+} have been developed, such as Mn-(II)-EDTA and Mn-(II) citrate (de Haen 2001); however, the toxicity exerted by heavy metal salts, such as Mn^{2+} is notorious at the doses required for clinical imaging. The in vivo application of the bivalent Mn^{2+} has a number of recognized limitations (Weinmann et al. 1984). Among them, Mn^{2+} is easily oxidized, leading to a change or loss of paramagnetic properties. Furthermore, Mn^{2+} has extended retention time in the liver. Among the metal ions examined so far, gadolinium-based contrast agents dominate in clinical imaging due to their superior ability to shorten the T_1 relaxation times of hydrogen protons, resulting in remarkable contrast compared to other metallic ions. For a practical laboratory development of MRI contrast agents, the discussion of this subsection will focus solely on gadolinium-based contrast agents.

Complex stability would be one of the primary focuses for the design of safe and efficacious MRI pharmaceutical contrast agents. The stability of the coordination complex in vivo, or in other words, at the physiological pH, is governed by two factors: thermodynamic stability and kinetic inertness.

The thermodynamic stability of the metal/ligand complex refers to the Gibbs free energy associated with the coordination equilibrium. While it is not always true in vivo, this is a direct and initial assessment of the specific tendency of metal ions to bind to the ligand with a stability constant log K (log K_{GdL}). Based on the law of mass action, the K_{GdL} constant is defined by equation below (Idee et al. 2006):

$$\text{Metal} + \text{Ligand} \rightleftharpoons \text{Metal.Ligand complex}$$

$$K_{ML} = \frac{[\text{Metal.Ligand}]}{[\text{Metal}][\text{Ligand}]}$$

A number of assays can be used to assess log K by measuring the pH values or the magnitude of absorbance/emission spectra as a function of pH change or proton relaxometry. For instance, the metal–ligand association occurs when the ligand is fully deprotonated, which can only be available when the pH is high. The limitation

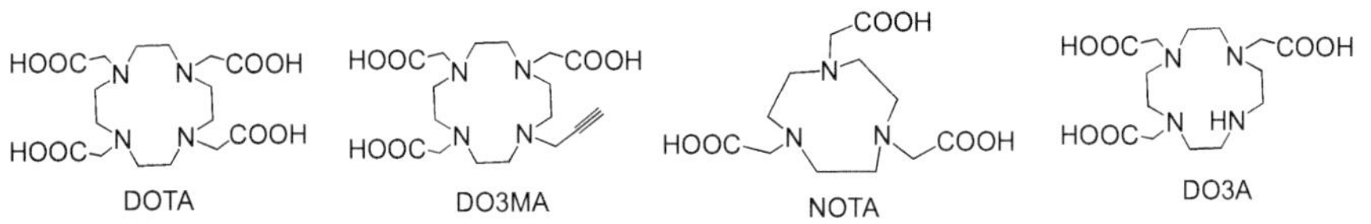

DOTA DO3MA NOTA DO3A

Fig. 4.10 Gadolinium chelators

of this measurement is that it may not reflect in vivo scenarios. A more realistic measurement called conditional thermodynamic stability constant (K_{cond}) or sometimes called K_{ML}', which measures the thermodynamic stability at physiological pH7.4 (Idee et al. 2006; Port et al. 2008):

$$K_{cond} = K_{ML}.[\text{Ligand}]/L_T$$

where L_T represents the total concentration of unchelated ligand, including ligand and its protonated forms of the free ligand species. Since the thermodynamic (K_{ML}) and conditional (K_{cond}) stability constants indicate the binding affinity of the metal for the chelators, the higher the values of these binding constants, the stronger the complex and vice versa. At the physiological conditions, proton competition is very significant, and each ligand behaves differently depending on its intrinsic basicity. For example, DOTA and DO3MA (Fig. 4.10) complexes of Gd^{3+} with equal stability; however, at pH7.4, the DO3MA complex is an order of magnitude less stable than the DOTA counterpart. Similarly, Gd-NOTA is 11.6 orders of magnitude less stable than Gd-DOTA, but at pH7.4, the NOTA complex is only 9 orders of magnitude less stable (Kumar et al. 1995).

As predicted from the chemical structure, the thermodynamic stability of the complex is dictated by the electrostatic interaction between the positive charge of Gd^{3+} and the chelate's carboxylates. It seems certain that the more basic these carboxylates, the stronger the interaction. This relationship is demonstrated by comparing the increasing basicity of the following ligands, DOTA > HP-DO3A > DO3MA > DO3A > NOTA; when chelating with Ca^{2+} the stability constants of the complex remained in the same order, $Ca(DOTA)^{2-}$ > $Ca(HP-DO3A)^{-}$ > $Ca(DO3MA)^{-}$ > $Ca(DO3A)^{-}$ > $Ca(NOTA)^{-}$ (Kumar et al. 1995).

Asides from that, increasing the number of the 5-membered structures, such as N-Gd-N and N-Gd-O rings generated from the Gd^{3+} with various donor moieties of the polyamino carboxylate complex also contributes to the stability. Lastly, the appropriate design of a chelator consisting of internal cavity size with oxygen and nitrogen binding sites that can wrap the metal ions is crucial to enhance the thermodynamic stability (Desreux and Barthelemy 1988).

A chelating complex might have excellent thermodynamic stability, but it does not mean the same holds true for kinetic stability. Given the abundance of metal ions in the serum and biological milieus, the release of Gd^{3+} from the ligand via protonation and/or transmetalation is high possible. For example, phosphorus-containing ligands (as shown in one representative structure of H_6L^1) generate thermodynamically stable

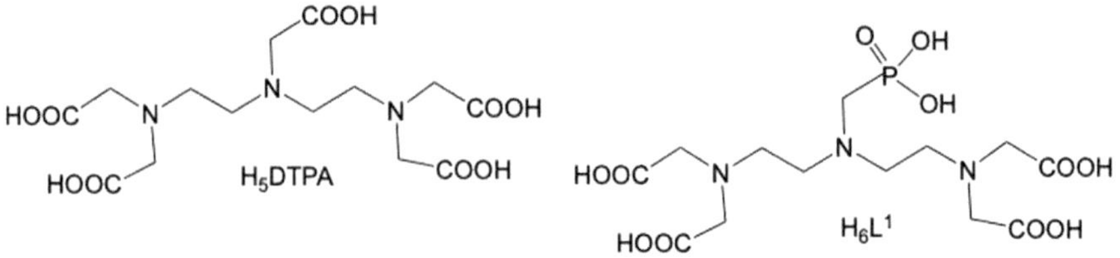

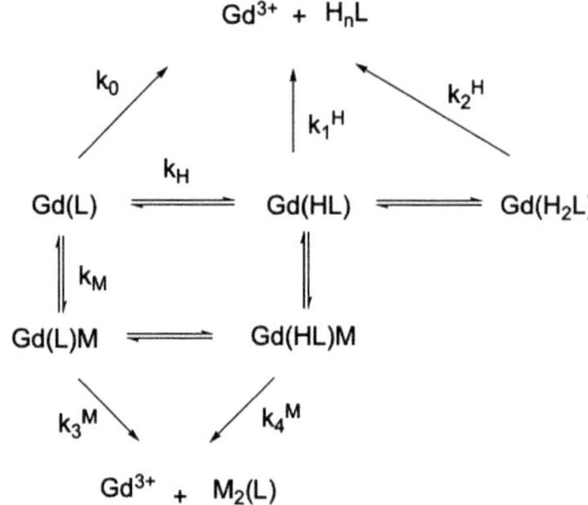

Fig. 4.11 Molecular structures of DTPA and modified ligand for use as gadolinium chelates

Fig. 4.12 Proposed dissociation of Gd^{3+} complex of H_5DTPA-like ligands. Data obtained from Kotek et al. (2006) with permission from the author

complexes with Gd^{3+} with stability constants comparable or even superior to that of the parent H_5DTPA (Fig. 4.11). However, the steric strain partly caused by the bulky phosphorus group resulted in poor kinetic stability of their Gd^{3+} complexes against acid- and transmetal-decomplexation (Kotek et al. 2006). Consequently, H_6L^1 and similar derivatives are unsuitable for medical applications. The rate of Gd-chelate dissociation is low at physiological pH, but accelerated at low pH conditions, ranging from 3.5 to 5.7. Since the large excess of physiological metals that participate in the transmetalation process, the mechanisms are considered a pseudo-first-order reaction. Through experimental analysis, it is found that the stability of the Gd^{3+} complexes with ligands is affected by proton- and metal-mediated dissociations (Sarka et al. 2000). Data presented in Fig. 4.12 provides a clear mechanism of Gd^{3+} dissociation from the ligand, where K_H and K_M are the protonation constant of the Gd^{3+}-ligand complex [Gd(HL)] and the stability constant depicting the transmetalation (such as Cu^{2+} or Eu^{3+}) intermediate Gd(L)M, respectively.

4.2.2.1 Acyclic Chelators

Gadolinium ion has excellent influence on the T_1 relaxation time; unfortunately, lanthanide metals have very high intrinsic toxicity (Moreau et al. 2004), thus prompting the design of methods of detoxification enabling in vivo translational application. Free gadolinium ions have an ionic radius with a similar size to that of calcium, thus interference with calcium-dependent pathways implicates muscular contraction and neurotransmission (Wolf and Fobben 1984). Releasing of gadolinium from its chelator within the liver may lead to cellular, hepatic, and systemic toxic effects (Bousquet et al. 1988). Like any rare-earth metal, gadolinium does not bind any other organic molecules through stable, covalent bonds (Weinmann et al. 1984). However, it does form strong chelation with ethylenediaminetetraacetic acid (EDTA) and diethylenetriaminepentaacetic acid (DTPA) (Fig. 4.13). For the design of safe MRI contrast agents, aside from the toxicity derived from the metal ions, the free chelators also implicate overall toxicity due to their potential scavenging of essential electrolytes in the body. Therefore, tight binding of the metal ions with the chelator during in vivo administration is crucial.

This phenomenon has been exploited and reported as one of the first MRI contrast agents for clinical imaging (Gries et al. 1981a). In a typical experiment, Gd_2O_3 was treated with DTPA in water, followed by heating to reflux for 48 h. Upon completion of the reaction, the undissolved materials were removed by filtration. The Gd-EDTA complex was further treated with N-methylglucamine to create water-soluble salts of the gadolinium chelates.

Gadolinium has approximately 8–10 coordination interactions (Reuben 1971; Dwek 1973). In aqueous conditions, $Gd(H_2O)_8^{3+}$ adapts a square antiprism conformation where the gadolinium ion is located between two planes, each containing 4 water molecules at the square apexes with the two planes staggered 45° with respect to each other (Sherry et al. 2009). The dynamic water exchange between the inner coordination site and the bulk water occurs promptly by the diffusion of additional

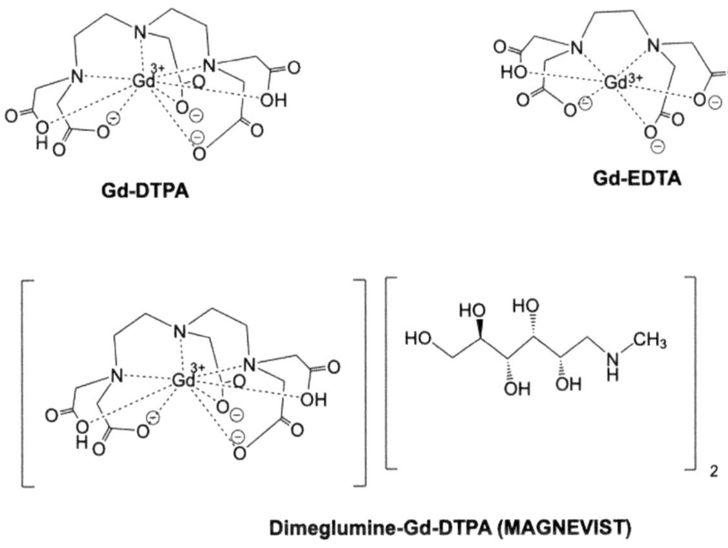

Gd-DTPA

Gd-EDTA

Dimeglumine-Gd-DTPA (MAGNEVIST)

Fig. 4.13 Linear scaffolds used to chelate gadolinium

water molecules to form an intermediate 9-coordination complex, then followed by rapid dissociation of one water molecule to reestablish the $Gd(H_2O)_8{}^{3+}$ conformation (Sherry et al. 2009). Upon chelation, the water molecules from the inner sphere of Gd(III) are sequestered out by the approaching chelating ligand, which contains nitrogen and carboxylate moieties, known for having a high affinity for Gd. Chelators usually have nine-coordinated sites, in which chelating ligands fill eight coordination sites at the metal center, thus leaving only one site to accommodate fast-exchange water protons approaching closely to the paramagnetic center to generate the MR signal-enhancing properties of the compound (Weinmann et al. 1984). By chelating the paramagnetic ions inside the caging, the number of coordination protons will decrease due to this. Furthermore, the distance between the metal ion the water protons increases, offsetting the rule $1/r^6$, as the electrostatic attraction between the atoms weakens. Altogether, it would be intuitive to the rationale that chelating Gd with these chelators reduces the paramagnetic property of nonchelated Gd. Thus, to achieve similar proton relaxation effects as those from free Gd, the concentrations of the Gd-chelator complex must be much higher than those from free Gd. In exchange, chelating Gd in EDTA or using related scaffolds reduces the potential toxicity significantly.

Gd-DTPA is more tolerant than Gd-EDTA or the free gadolinium analogs. In a systematic analysis of intravenous doses on the Wistar–Han–Schering rat model, it has been shown that for gadolinium chloride and Gd-EDTA, half of the tested animals died after receiving a dose of less than 1.0 mmol/kg, while the LD_{50} for Gd-DTPA was 10 mmol/kg (Weinmann et al. 1984). In vivo study of Gd-DTPA demonstrated remarkable tissue enhancement with doses ranging from 0.1 to 0.5 mmol/kg, which is barely 1–5% of the LD_{50} dose. The association with meglumine rendered the complex more hydrophilic, thus reducing long liver retention while increasing excretion through the renal-urea pathway and preventing Gd dissociation in vivo. Further study also suggested that Gd-DTPA does not permeate the cellular membranes, making it an exclusively extracellular contrast agent.

Using Gd-DTPA as a template, further modifications of the chelator to alter in vivo biodistribution aside from plasma-blood circulation for the development of a new family of MRI targeted contrast agent. Gd-BOPTA dimeglumine is a new octadentate chelate of Gd^{3+}, salified with meglumine for liver-specific imaging. In contrast to Gd-DTPA, which cannot traverse across the cellular membrane, the Gd-BOPTA, with the incorporation of the benzyloxymethyl group replacing a carboxylic group, serves as a steering force to permeate the cell membrane and homing in the cytoplasm of hepatocytes. While the presence of a chiral center in the Gd-BOPTA needs further analysis of their role, so far, no significant alteration in the ability of the compound to coordinate Gd^{3+}, and the probe is slightly less toxic than Gd-DTPA (Cavagna et al. 1990).

The synthesis of Gd-BOPTA started with treating diethylenetriamine with 2-chloro-3-benzyloxypropionic acid in the presence of sodium hydroxide solution (Fig. 4.14) (Gries et al. 1981b). The beauty of the benzyl group is that it provides an excellent means for convenient monitoring of the progress of the reaction. The primary and secondary amines in the subsequent intermediate were alkylated with

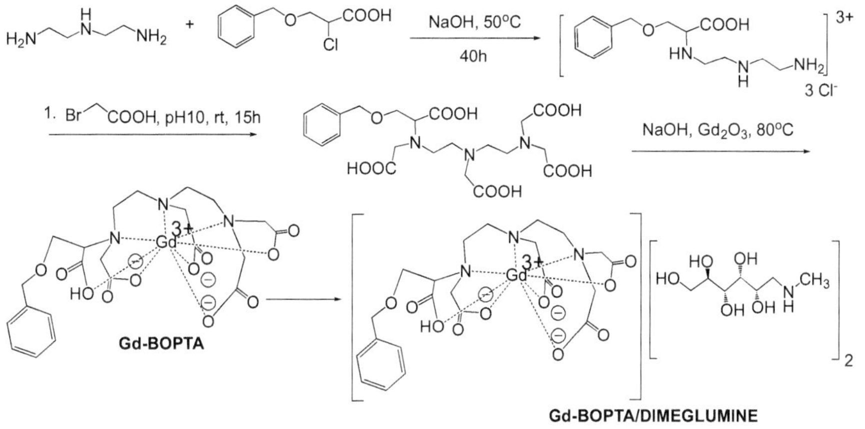

Fig. 4.14 Synthesis of Gd-BOPTA, deduced from the description of Cheng (2007)

bromoacetic acid at pH10 to afford the desired BOPTA. Finally, the Gd-BOPTA complex was achieved by sequestering the metal ions with BOPTA in aqueous conditions to provide the probe with a quantitative yield. The addition of N-methylglucamine to the Gd-BOPTA complex was intended for improving water solubility.

An alternative approach for the synthesis of BOPTA has been reported using commercially available starting materials, such as the protected serine (Patent CN102603550B). In this scheme (Fig. 4.15), diethylenetriamine was incorporated into the α-carbon of serine. But before that, the Sandmeyer reaction commenced the first step of the synthesis, in which the substitution of the primary amino group was performed via the diazonium salt intermediate, followed by subsequent displacement with a bromide. It seems reasonable to expect that the carboxylic group of the bromo intermediate was protected as a *tert*-butyl ester group before substitution reaction with diethylenetriamine. The *tert*-butyl ester group has two functions; first and foremost is to prevent an undesired reaction. But more than that, the organic reaction of amino acids in the presence of free carboxylic acid usually experiences poor solubility in organic solvents. Therefore, the other purpose of the protecting group aims to enhance solubility. The next step involves the alkylation of the primary and secondary amines with a protected version of bromoacetic acid. Finally, cleavage of the *tert*-butyl esters using TFA furnished the desired BOPTA chelator.

Aside from Gd-BOPTA, which is used in clinical imaging of the liver, a number of other Gd-based linear chelating probes have been developed using Gd-DTPA as a template, including Gadodiamide (Omniscan®), Gadoversetamide (Optimark™), Gadoxetate disodium (Primovist™) and Gadofosveset (Vasovist®). All of the Gd-based acyclic chelators distribute in the blood circulation, including the extravascular–extracellular space, and the probes are promptly washed out in the kidneys, except gadoxetic acid, BOPTA dimeglumine, and gadofosveset. These probes contain the lipophilic aromatic structures, which enhance binding to proteins, and thus, they

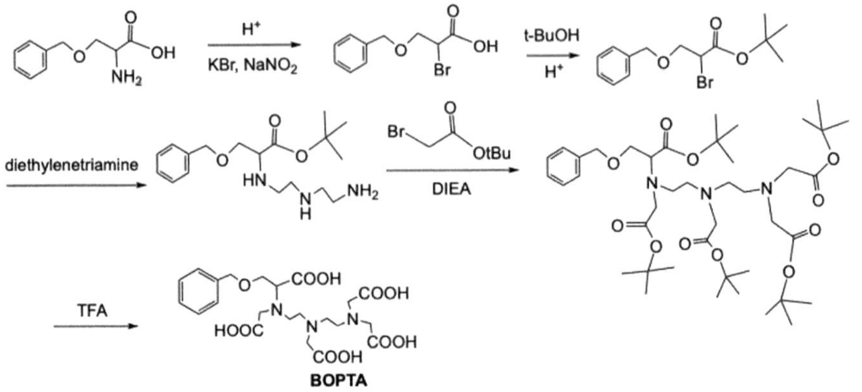

Fig. 4.15 Alternative approach for the synthesis of BOPTA. Data obtained from Patent CN102603550B

are taken up by hepatocytes and washed via the hepatobiliary route (Lohrke et al. 2016).

Aside from chelating Gd(III), several functionalized DTPA versions have been developed for conjugation to bioactive molecules, as well. Typically, the chemistry for synthesizing activated acyclic systems is straightforward. The propanamide starting material shown in Fig. 4.16 was developed through two steps. First, the *p*-nitrophenylalanine was esterified as a methyl ester derivative, followed by a reaction with ethylenediamine. The amide was reduced with BH_3.THF, then the pentetic acid compound was formed by a standard amine alkylation method using excess 2-bromoacetic acid in aqueous KOH (Brechbiel et al. 1986). The aromatic nitro group was converted into amine using hydrogenation in the presence of palladium on activated carbon. Finally, treating the aminated compound with thiophosgene to generate the desired DTPA with the amino-activated isothiocyanate with high yield. Aside from *p*-NCS-Bz-DTPA, a number of different acyclic chelators have been developed with activatable handles for labeling applications (Fig. 4.17).

4.2.2.2   Macrocyclic Chelators

The acyclic chelators are designed to hold the Gd^{3+} by wrapping around the metal ions via polydentate coordinations. The macrocyclic scaffolds have a similar function as the DTPA analogs, albeit with distinct chemical structures, and these differences are reflected in how these two classes of chelators dictate the complex stability. It is apparent that the Gd^{3+} complex must remain intact in vivo as a requirement of a safe MRI contrast agent. The advantage of acyclic chelators, like DTPA, is the fast labeling kinetics, which has unique applications when it comes to labeling unstable materials, like antibodies and antibody fragments due to their sensitivity toward elevated temperature (Liu and Edwards 2001). However, metal chelates of acyclic

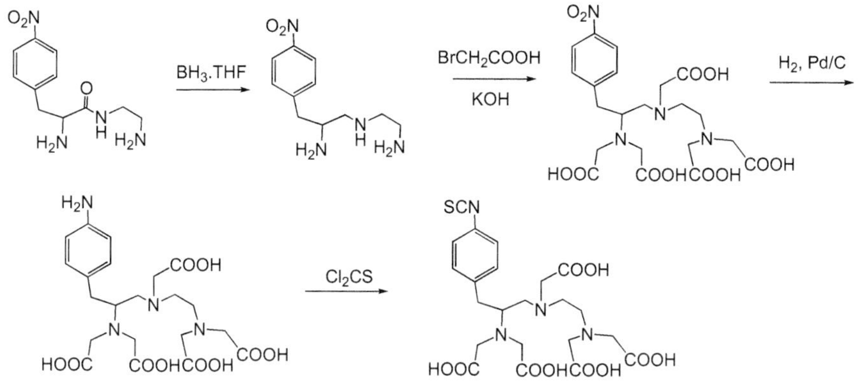

Fig. 4.16 Synthesis of functionalized DTPA, *p*-NCS-Bz-DTPA. Data obtained from Brechbiel et al. (1986) with permission from the American Chemical Society

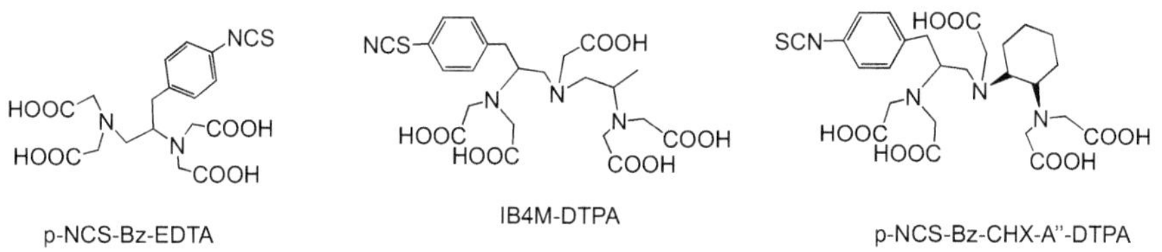

p-NCS-Bz-EDTA

IB4M-DTPA

p-NCS-Bz-CHX-A"-DTPA

Fig. 4.17 Activatable acyclic chelators

chelators are kinetically labile, contributing to the loss of metals, and leading to toxicity (Pippin et al. 1992).

The dissociation of metal ions from the chelator is undesirable, as both the free metal and the unchelated ligands are generally more toxic than the complex itself, particularly at the concentrations necessary for their use as contrasting agents (Caravan et al. 1999). As mentioned earlier, the LD_{50} of an intravenous dose of Gd-DTPA was found over 10 mmol/kg, while free Gd(III) in the form of $GdCl_3$ caused toxicity with LD_{50} with merely 0.4 mmol/kg. In contrast, LD_{50} of H_2DTPA (meglumine salt) was only 0.15 mmol/kg (Lauffer 1987). The toxicity of each chelating ligand is not truly well understood. In some less stable MRI contrast agents, such as in the formulation of Gd-DTPA-BMA and Gd-DTPA-BME, the compounding excipients include a large amount of excess chelating ligands, aiming to trap free Gd(III) may escape from the chelator (Idee et al. 2006; Port et al. 2008).

Since the first contrast agent Gd-DTPA (Magnevist) was first approved by the FDA for clinical use in 1988 (Niendorf 1988). Currently, a total of 11 GBCAs have been approved by FDA, with the five most popular agents used in clinical work, including $[Gd(OH_2)(BOPTA)]^{2-}$ (Multihance®), $[Gd(OH_2)(DTPA)]^{2-}$ (Magnevist®), $[Gd(OH_2)(DTPA\text{-}BMA)]$ (Omniscan®), $[Gd(OH_2)(DOTA)]^{-}$ (Dotarem®) and $[Gd(OH_2)(DO3A\text{-}butrol)]$ (Gadovist®). However, the notion of complex stability in regard to in vivo toxicity has been a complex and ongoing debate for many years. This is compounded by the administration of large doses of the contrast agents due to the low sensitivity of MRI.

The DTPA scaffolds are somehow relatively better than the EDTA versions in regard to complex stability but not strong enough to render them ideal. There are reports that indicate evidence of gadolinium dissociation from DTPA complexes (Duncan et al. 1994). The macrocyclic structure is recognized as more stable in terms of chelating metal ions, reducing metal dissociation, and thus less toxic compared to the open structure of DTPA. Designing metal-based MRI pharmaceutical contrast agents requires taking two complementary aspects to define the stability of the chelates, which are the thermodynamic and kinetics stability to prevent the metal from being released in vivo, which causes consequential toxicity. A number of reports showed that Gd(III) ions could compete with Ca(II) for the binding sites, and irreversibly bind to the skeletal tissue (Caravan et al. 1999; Cacheris et al. 1990). With an ionic radius of approximately 107.8 p.m., very close to the size of divalent Ca(II) (114 p.m.), this element competes with calcium for voltage-gated calcium channels at the nano to micromolar concentrations. Consequentially, it inhibits calcium-dependent physiological processes, such as contraction of smooth, skeletal, and cardiac muscle, the transmission of nerve impulses, blood coagulation, and more. Furthermore, gadolinium is also an inhibitor of Ca^{2+}-activated-Mg^{2+}-adenosine triphosphatase (ATPase) and dehydrogenases, kinases, and glutathione S-transferases (Idee et al. 2009). Gadolinium induces the overexpression of hepatic cytokines and several cytokine-regulated transcription factors such as c-JUN, C/EBP-β, and C/EBP-δ (Decker 1990; Rai et al. 1996). This explains why the presence of free Gd(III) in the biological system is utmostly concerned. So, the basic idea is that when gadolinium associates with a ligand (L) to form a Gd-L complex. This complex's stability requires that the equilibrium should drive in the way to ensure the intact complex maintains unaltered during in vivo study.

The Gd-DOTA complex (also called DOTAREM) exemplifies this remark. This molecule is very stable; it has five orders of magnitude stronger in regard to in vitro stability compared to Gd-DTPA (Bousquet et al. 1988). The combined higher relaxivity and stability of Gd-DOTA makes it an ideal alternative to Gd-DTPA for MRI (Nwe et al. 2010). It is reasonable to observe this since Gd(III) ion is a hard acid and thus prefers hard donor atoms. Further, the prominent basicity of DOTA compared to DTPA-BMA (Omniscan) also gives credence to form a thermodynamically stable complex with Gd(III) with log $K_{Gd.L} = 24.7$ versus log $K_{Gd.L} = 16.85$, respectively (Wahsner et al. 2019). This suggests the structure of the chelate partially dictates the kinetic lability of the metal ions and, apparently the rate how metals are released nonspecifically during in vivo application. The poorer thermodynamic stability of DTPA-BMA is the culprit behind the significant detection of free Gd(III) in tested subjects. By examining the concentration of Gd(III) using inductively coupled plasma mass spectroscopy (ICP-MS) on Gd-chelate-treated patients undergoing hip replacement, it is shown that Gd-DTPA-BMA left approximately 4 times more Gd(III) behind the bone than did Gd HP-DO3A (1.77 μg Gd/g bone vs. 0.477 77 μg Gd/g bone, respectively) (White et al. 2006). Further, the chemical structure of DTPA also reveals the weakness in regard to kinetic stability. It has been shown that the open-chain structure lacks conformational rigidity (Kumar 1997). Some other structures containing the amide moiety, such as DTPA-BMA ligand, which delocalized the electrons in the structure, rendering the carboxylate groups less basic and thus weak binding to Gd^{3+}, altogether, significantly reduce the kinetic stability.

DOTA represents many analogs that share a similar backbone, which is a twelve-membered tetraazamacrocyclic system. The chelating association occurs through the extended carboxylate arms and the cyclen amines. Gadolinium (III) associates with DTPA or DOTA through eight-coordinate complexes, and the ninth coordination site is reserved for the diffusing water molecule. So, the question is, why Gd-DOTA are better than Gd-DTPA in every aspect? When we discuss complex stability, a number of factors should be mentioned, such as the charge density, the nature of the medium, the size of the metal ion, and the rigidity of the ligand (Moreau et al. 2004). First, let us discuss the differences between the two in regard to the dissociation mechanism. There is a host of multiple competing equilibria known for their active roles contributing to the dissociation of metal ions in vivo. The full association of the metal ions to the chelator occurs when the carboxylic groups are deprotonated; a condition only exists at high pH environment. However, at very low pH, the protons in the aqueous solution would compete and sequestering the metal ions from the complex. Fortunately, not every tissue or pathological milieu has low pH, and the acid-catalyzed displacement significantly reduces in physiological pH. Meanwhile, the other dissociation mechanism to displace the metal ion comes from transmetallation. Given the vast abundance of endogenous competing metals and electrolytes available ubiquitously in every tissue, just mention a few, such as Ca^{2+}, Cu^{2+}, Zn^{2+} K^+, and Na^+ with concentration, sometimes nearly 100-fold over Gd^{3+}. In this respect, designing the chelating mechanism free of transmetallation probably represents one of the most daunting tasks in maintaining the safety of MRI pharmaceutical probes. Different scaffolds react to transmetallation differently; for example, DTPA is more susceptible to the transmetalation process than DOTA. Another noted characteristic of DOTA or macrocyclic ligands, in general, is the ability to contain the metal cations inside the molecular cavity, whose size can contract to accommodate the cation. There must be some sort of cavity selectivity for a particular metal ion with optimal association defined by electrostatic and van der Waals attraction and repulsion, nature and number of binding sites, ligand conformation changes, enthalpy/entropy contributions (Lehn and Sauvage 1975). Furthermore, DOTA is much more rigid than EDTA, enabling it to display binding selectivity. DOTA can discern the approaching metal ions, whether they are larger or smaller than their cavity. The small cavity of DOTA would experience some sorts of contraction or expansion of its cavity when it is in the proximity of an approaching metal ion. If we take all of the information regarding the thermodynamic, kinetic stability along with conformational strains, the rigidity of the ligand, and the pendant arm specificity into account, the information shown in Fig. 4.18 suggested that in all the tested conditions, the macrocyclic Gd(III) complexes are kinetically inert compared to the acyclic structures (Wahsner et al. 2019).

	Acid-catalyzed dissociation rate constant	Zn-catalyzed dissociation[a]	Dissociation in human plasma[b]
	$K_1/M^{-1}s^{-1}$	$t_{0.8}/min^c$	$t_{2\%}/h^d$
$[Gd(DTPA)(H_2O)]^{2-}$	0.58	260-280	8.73
$[Gd(DTPA\text{-}BMA)(H_2O)]$	12.7	50-60	0.51
$[Gd(DTPA\text{-}BMEA)(H_2O)]$	8.6		0.89
$[Gd(BOPTA)(H_2O)]^{2-}$	0.41	600	9.08
$[Gd(EOB\text{-}DTPA)(H_2O)]^{2-}$	0.16	1500	51.3
MS-325	2.9×10^{-2}	3800	34.4
$[Gd(DOTA)(H_2O)]^{-}$	8.4×10^{-6} 1.8×10^{-6}	>5000	>1000
$[Gd(HP\text{-}DO3A)(H_2O)]$	6.4×10^{-4} 2.6×10^{-4}	>5000	>1000
$[Gd(DO3A\text{-}butrol)(H_2O)]$	2.8×10^{-5}	>5000	>1000

[a] 1:1 eq. Zn (II) in the form of $ZnCl_2$:Gd(III)-based contrast agent in phosphate buffer (50 mM), pH 7.0 at 37°C; [b] 10 mM phosphate, 37°C; [c] $t_{0.8}$ = time for R_1 to reach 80% of its initial relaxivity; [d] $t_{2\%}$ = time when 2% of the Gd(III) ions are released from the complex.

Fig. 4.18 Kinetic constants for dissociation of Gd^{3+} from commercial ligands at different conditions. Data obtained from Wahsner et al. (2019) with permission from the American Chemical Society

Now let us focus the discussion on the design and chemical synthesis of the macrocyclic chelators. The earlier synthesis of the macrocyclic cyclams employed metal ion-assisted cyclization of the tetraamine macrocyclic systems. For example, the nickel (II) complex contributed to this process by serving as an excellent intermediate fostering the cyclization in aqueous conditions (Barefield 1972). The tetradentate ligands can be achieved selectively depending on the diamine starting materials. In a representative experiment, treating acrylonitrile with ethane-1,2-diamine resulted in the tetradentate product with a modest yield (Fig. 4.19). The tetradentate ligand was treated 1:1 ratio with $Ni(ClO_4)_2$ in water to provide the metal-amine solution with blue color. Then, the metal-amine solution was cooled to 5 °C, followed by the addition of 40% aqueous glyoxal solution, and the reaction was stirred overnight at room temperature. Next, hydrogenation was performed at 200–300 psi for 12 h using a catalytic amount of Raney nickel. Finally, nickel was removed by treating the solution with excess cyanide without isolating the metal complex leading to free cyclam.

This reaction is very robust. Replacement of ethane-1,2-diamine with similar and substituted versions will lead to different cyclams. Particularly, this approach enables a robust synthesis of versatile MRI pharmaceutical products, including the potential incorporation of handles for bioconjugation. Further, different nickel materials can be used; instead of using $Ni(ClO_4)_2.6H_2O$, other product like $NiCl_2.6H_2O$ also proved to be ideal for mediating the cyclization process. A number of useful tetradentates (Fig. 4.20) have been developed and reported using this approach during the early time (Barefield et al. 1976).

Simpler cyclization chemistry made its way to the field for the preparation of tetradentates or polyaza macrocyclic systems thanks to innovative chemistry for the

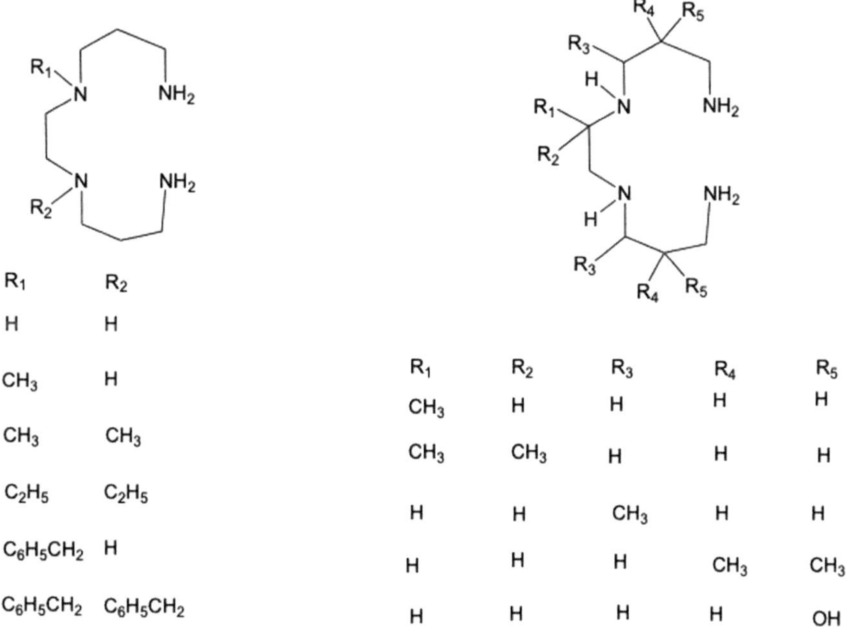

Fig. 4.19 Nickel ion-mediated cyclization of macrocyclic systems. Data obtained from Barefield (1972) with permission from the American Chemical Society

R₁	R₂
H	H
CH₃	H
CH₃	CH₃
C₂H₅	C₂H₅
C₆H₅CH₂	H
C₆H₅CH₂	C₆H₅CH₂

R₁	R₂	R₃	R₄	R₅
CH₃	H	H	H	H
CH₃	CH₃	H	H	H
H	H	CH₃	H	H
H	H	H	CH₃	CH₃
H	H	H	H	OH

Fig. 4.20 Different types of tetradentate can be developed using different versions of diamines. Data obtained from Barefield et al. (1976) with permission from the American Chemical Society

synthesis of the starting materials. Basically, the intermediate bis(α-chloroamide) can be prepared by using a diamine and chloroacetyl chloride (Fig. 4.21). The availability of two reactive alkyl chloride moieties from the crab-like bis(α-chloroamide) facilitates the cyclization with any di-primary amine to form the tetradentates. The cyclization occurs between two starting materials in the presence of anhydrous sodium carbonate in refluxing acetonitrile for 24–48 h. Two starting materials can be added simultaneously or slowly into the refluxing acetonitrile solution using a syringe pump. The second approach provided about 15% more yield than the other method. The cyclic polyamide can be reduced using borane-THF or borane-dimethyl sulfide to provide desired products. This versatile design generates a variety of macrocyclic di- and tetraamide ligands containing 9, 12, 14, 15, 17, and 18-ring systems (Krakowiak et al. 1990).

To generate fully functionalized cylams and cyclens in order to assess the complex formation with Gd(III), 1,4,7,10-tetraazacyclododecane (1), 1,4,7,10-tetraazacyclotridecane (2) and 1,4,8,11-tetraazacyclotetradecane (3) were treated with chloroacetic acid in an aqueous alkali solvent to afford the corresponding alkylated products, N,N′,N″,N‴-tetraacetic acid (Stetter and Frank 1976). The inorganic salts were isolated and purified using strongly basic ion-exchange chromatography (Fig. 4.22). The complex association strength was assessed via potentiometric titrations.

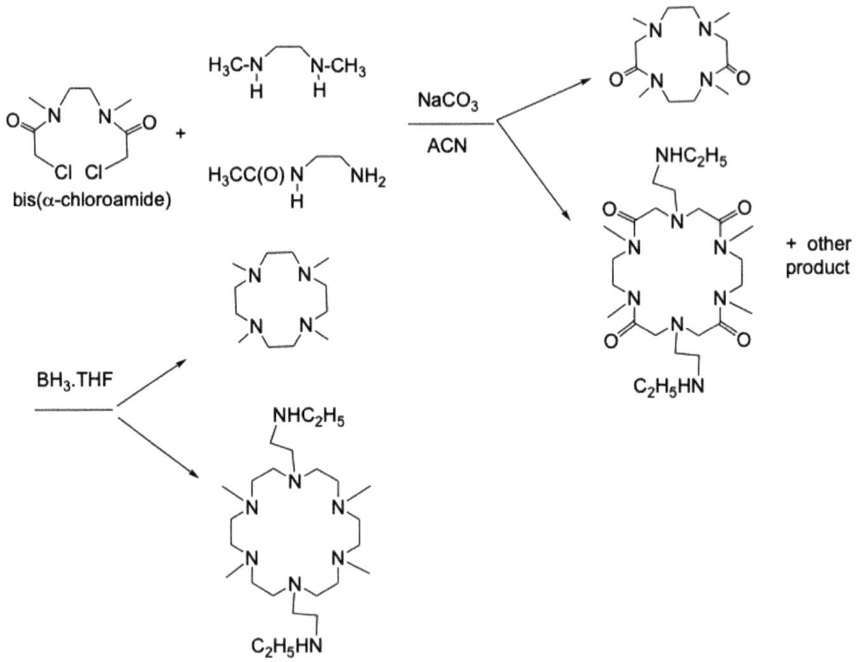

Fig. 4.21 Synthesis of tetra- and polydentates via the crab-like cyclization using Bradshaw method. Data deduced from Krakowiak et al. (1990)

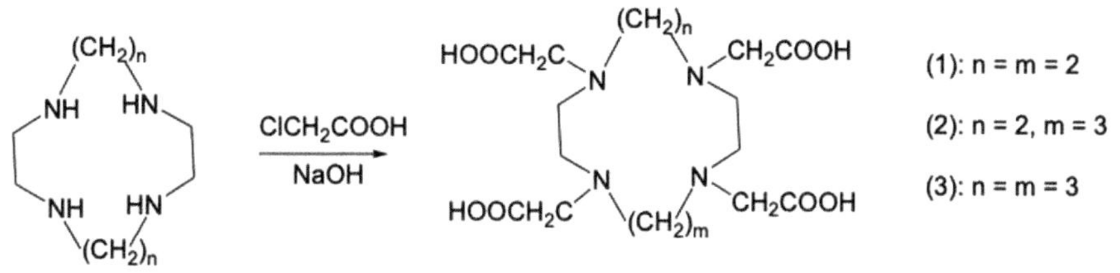

Fig. 4.22 First synthesis of fully functionalized macrocyclic chelator using Frank method. Data deduced from Stetter and Frank (1976)

Since then, several derivatives were developed through innovative chemistries to generate functionalized macrocyclic poly(aminocarboxylates) for bioconjugation to generate MRI pharmaceutical probes with improved pharmacokinetics and dynamic stability. To make Gd(III)-ligand complex more versatile for multiple applications in imaging, functionalized ligands were developed, particularly for the labeling of bioactive materials including peptides, peptidomimetics, and antibodies. Monoclonal antibody technology enables specific targeting of antigens overexpressed in pathological diseases. The most notable design in this aspect using a short chain of α-amino acids as the starting material, followed by intramolecular tosylamide ring closure (Moi et al. 1988).

The synthesis started with the treatment of the amino acids with borane-tetrahydrofuran complex (BH_3.THF) resulted in the chemical reduction of the C terminal carboxylic group and carbonyls to form the polyamino alcohol (Fig. 4.23). End-to-end-initiated cyclization via S_N2 reaction starting from N terminal amine can be achieved by generating a good leaving group at the other hydroxyl end. Treating an excess amount of p-toluenesulfonyl chloride in the presence of a strong organic base, such as triethylamine (TEA), produces a C terminal tosyl ester, an N terminal secondary tosylamide, and internal tertiary tosylamides. Cyclization occurred in the presence of a mild base, such as cesium carbonate in anhydrous N,N-dimethylformamide (DMF) at low temperature to provide the macrocyclic system with good yield. The side arms with affinity groups were attached to the ligand after the removal of the tosylate group in an acidic condition.

The aromatic nitro group can be reduced as an amine for serving as a handle for conjugation to antibodies and oligopeptide through a bifunctional linker using an orthogonal labeling approach. Further, the amino group can react directly to any amino-reactive intermediates, such as succinimide ester or isothiocyanate, available on bioactive materials or dyes.

Another synthetic approach for making C-alkylated macrocyclic ligands involves the aminolysis of malonates with the polyamine (Tabushi et al. 1977). This design has a significant advantage because of its simplicity and convenience, with reasonable reaction yield, albeit time-consuming. The underlying recognition of this work is that any substituent can be introduced into the carbon atom of the macrocyclic polyamine backbone during the condensation reaction using the corresponding

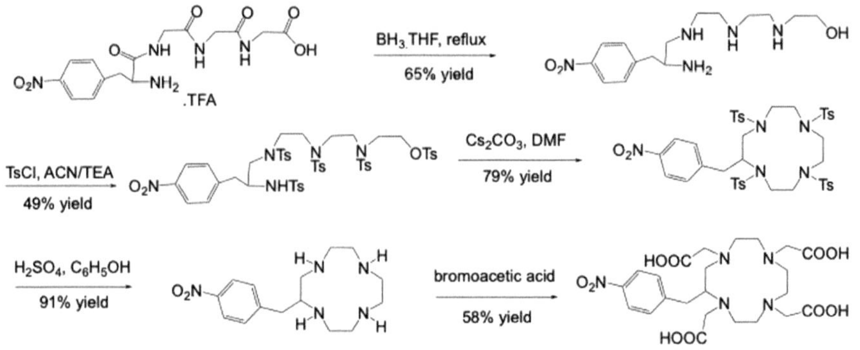

Fig. 4.23 Synthesis functionalized macrocyclic chelator from an amino acid backbone using Meares method. Data obtained from Moi et al. (1988) with permission from the American Chemical Society

substituted molanates (Fig. 4.24). This type of chemistry can be exploited to functionalize the ligand. But the disadvantage of this design starts with the condensation of the participating materials via refluxing for three days. Next, the final product was obtained after treating the cyclic diamide with a large excess of diborane in refluxing tetrahydrofuran for 24 h.

Another similar chemistry, worth mentioning regarding the synthesis of heteroatom-containing macrocyclic systems via the reaction of dimethylmalonyl dichloride with 2,2-dimethyl-1,3-propanediamine (Vellaccio et al. 1977). Aside from the 16-member ring of the tetraamine chelators, the 24-member ring of the hexamine and the 32-member ring of the octaamine systems were also generated along with a trace amount of larger cyclic homologs (Fig. 4.25). Typically, the reaction conditions involve the addition of acyl halide to an amine with three equivalents of triethylamine in acetonitrile at 0 °C. The tetramine product has the largest yield, approximately

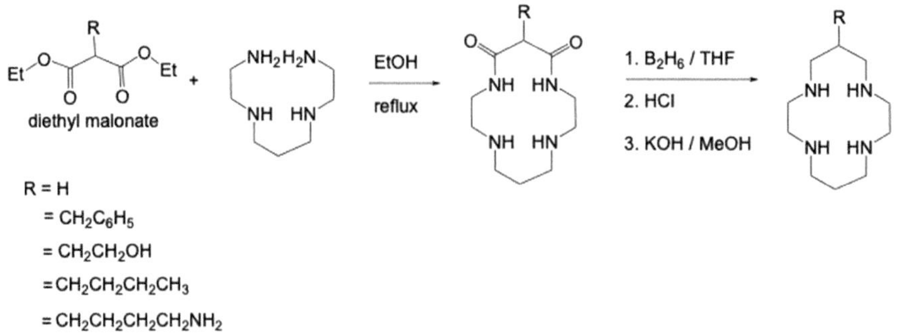

Fig. 4.24 Synthesis of functionalized macrocyclic ligands by condensation of malonate with polyamines using Tabushi method. Data was obtained from Tabushi et al. (1977) with permission from Elsevier

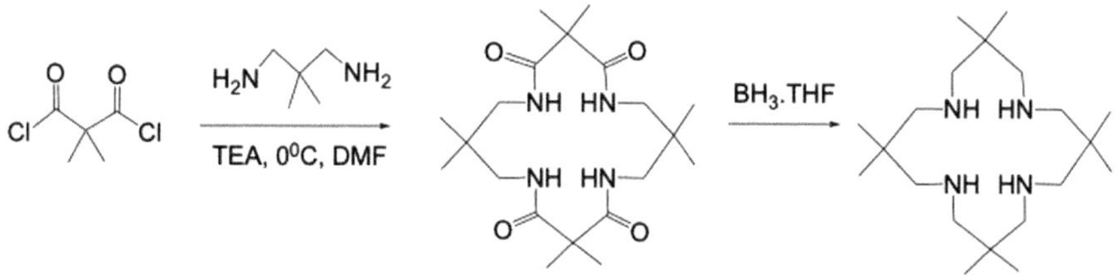

Fig. 4.25 Reaction of dialkylmalonyl dichlorides with diaminopropanes to generate tetraamine macrocyclic systems via Kemp method. Data obtained from Vellaccio et al. (1977) with permission from Elsevier

28% followed by the hexamine and octamine molecules with 7% and 3%, respectively. Reduction of the keto intermediate with excess borane in tetrahydrofuran under refluxing condition afforded the desired tetramine chelating agent in good yield.

Another development of an active macrocyclic system to foster metal chelation and biomimetic design involved the synthesis of C-functionalized cyclam with a hydroxymethyl substituent (Wagler and Burrows 1987). In this convergent approach, the (S)-2,4-diaminobutyric acid was first converted into the N,N'-ditosylamide, followed by reducing the carboxylic group to alcohol using borane in tetrahydrofuran (Fig. 4.26). The other portion of the macrocycle involved in the N-alkylation of the 1,3-N-tosylated propane amine with ethylene carbonate and conversion of the resultant diol into dimethanesulfonate. The macrocyclization of the alcohol and the mesylate starting materials using Cs_2CO_3 as described in Richman–Atkins reaction (macrocyclization of the disodium salt of tritosyldiethylenetriamine with N-tosyldiethanolamine ditosylate to provide tetratosylcyclen) to afford the desired product (Richman et al. 1979). However, a much better reaction yield occurred if using protected alcohols. For example, when the side chain alcohol is blocked with *t*-butyldimethylsilyl chloride (TBDMS-Cl), macrocyclization achieved 80% yield. Although additional steps involve during the synthesis, it is worthwhile pursuing this approach since the protection of alcohol via silylation and subsequent deprotection using TBAF (tetra-*n*-butylammonium fluoride) provide high yields.

Finally, tosyl protecting groups were removed using $Li-NH_3$ and precipitated as a tetrahydrochloride salt in aqueous ethanol. The free amine products were extracted in chloroform after washing with sodium hydroxide.

A milder cyclization for generating C-functionalized polyaza macrocyclic systems involved the condensation between primary amines and activated carboxylic acids. In a prototypical reaction, treating (4-nitrobenzyl)ethylenediamine with carbamate-protected amino disuccinimidyl esters in the presence of triethylamine afforded the cyclic diamides. Although typical succinimide ester coupling with primary amines can proceed smoothly at low temperature, that is not the case for the cyclization of macrocyclic systems. Each individual reaction needs optimization, some have lower barrier energy that can overcome easily at low temperature (Diederich et al. 1986), but in this reaction shown in Fig. 4.27, it is reported that lowering the temperature below 90 °C will adversely affect the yield of the macrocycles (McMurry et al.

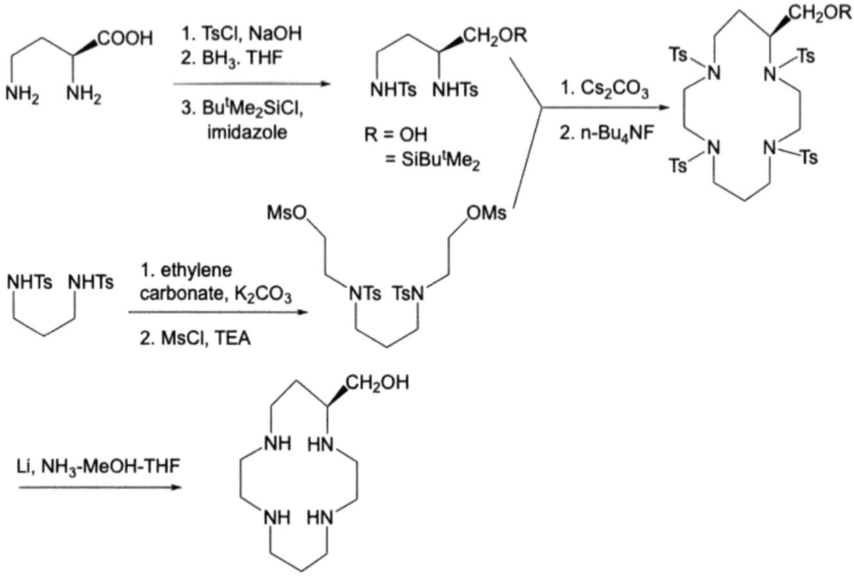

Fig. 4.26 Richman–Atkins method for the synthesis of C-functionalized cyclam. Data deduced from Wagler and Burrows (1987)

1992). After cyclization, the Boc protecting group was removed before the diamide amine hydrochloride was reduced using borane in tetrahydrofuran. To construct the chelating arms, alkylation of the macrocyclic amines with bromoacetic acid was performed, affording the polyamino carboxylates. As a precaution against undesired metal incorporation, mostly calcium, during the process of alkylation, the reaction should occur in a metal-free environment, which includes the use of acid-washed glassware of metal-free plasticware.

Through some conventional functional group maneuverings, the amino-activated handle was achieved to optimize the use of the macrocyclics, not only as a contrast agent, but it can conjugate with biomolecules for targeted imaging. Toward this path, the p-nitro group was easily converted to a reactive amino molecule (as shown in Fig. 4.16) using a neat hydrogenation reaction in the presence of catalytic amount of Pd on activated carbon, providing nearly quantitative yield. Treating this amino derivative with thiophosgene afforded the isothiocyanate. This activated group has prominent use in labeling monoclonal antibodies via the ε-amino moiety of lysine at moderate pH conditions. The covalent conjugation of the macrocyclics with antibodies was stable and suitable for in vivo applications (Reugg et al. 1990).

The versatile McMurry synthesis using readily available carbamate-protected amino diesters and the substituted diamine formed the foundation for successive development of different versions of C-functionalized tetraazamacrocycles. For instance, the 12-membered ring polyamine shown in Fig. 4.27 can be developed in the same manner regarding cyclization chemistry, albeit using different starting

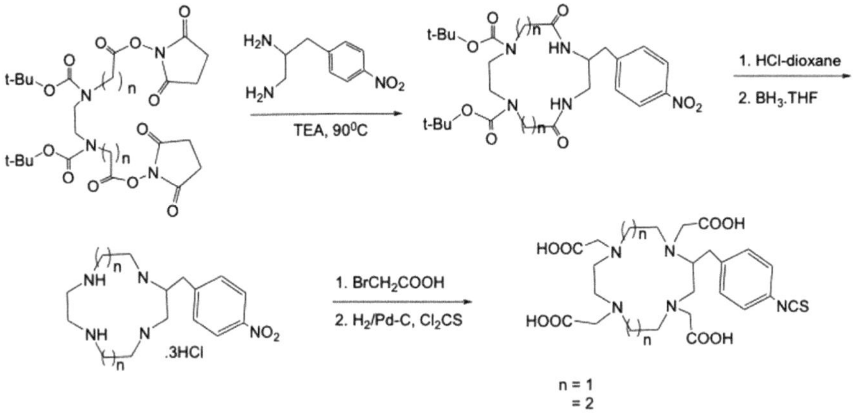

Fig. 4.27 McMurry reaction for the synthesis of *p*-SCN-Bz-DOTA. Data obtain from McMurry et al. (1992) with permission from the American Chemical Society

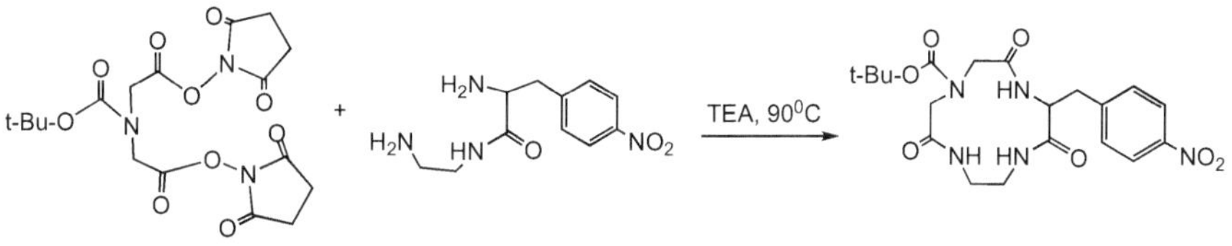

Fig. 4.28 McMurry synthesis. Data obtained from McMurry et al. (1992) with permission from the American Chemical Society

materials. Two practical advantages of this design are that the succinimidyl ester in Fig. 4.28 is more soluble in various solvents than the other counterpart shown in Fig. 4.27. Further, the triamide product precipitates as it forms, thus facilitating purification and improving the reaction yield.

A few years after the McMurry method was reported, a more simplified synthesis of similar functionalized macrocyclics was developed, using similar crab-like cyclization described in Bradshaw method. The process involves the reaction of a bisbromoamide with a *p*-nitrobenzyl ethylenediamine mediated by a base like Cs_2CO_3, followed by reduction of the cyclic diamide using excess BH_3.THF to afford the corresponding macrocyclic tetraamine. The macrocyclic tetraamine was then treated with bromoacetic acid in the presence of a strong base to afford the functionalized macrocyclic (Fig. 4.29).

The emergence of new synthetic methodologies is instrumental for the improvement of the synthesis of macrocyclic chelators. This enables the construction of more complex chelating structures, at the same time reducing strenuous reaction operations and cutting the product cost. For instance, the highly regioselective reductive cleavage of amidines with diisobutylaluminium hydride (DIBALH) during ring expansion of bicyclic amidines, in the case of 1,5-diazabicyclo[5.4.0]undec-5-ene

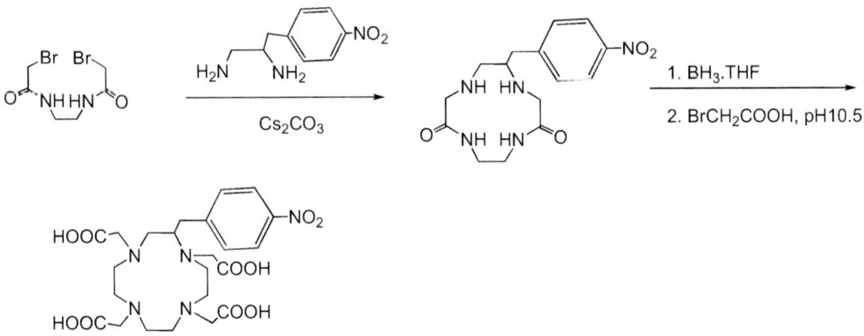

Fig. 4.29 Synthesis of functionalized macrocyclic using Mishra method. Data deduced from Mishra et al. (1996)

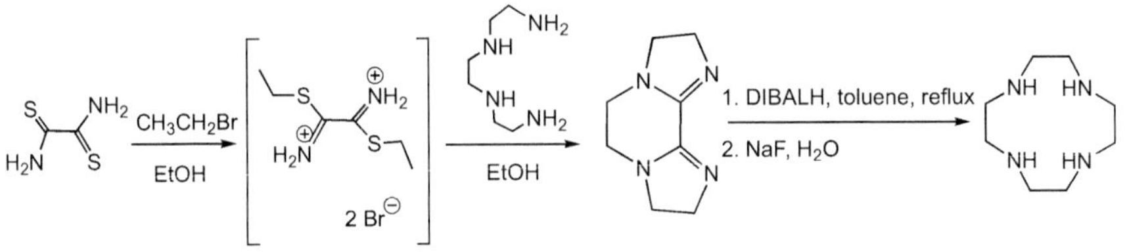

Fig. 4.30 Weisman method for the synthesis of cyclens via reduction of bis-amidines. Data obtained from Weisman and Reed (1997) with permission from the American Chemical Society

(DBU) to diazacycloalkanes (Yamamoto and Maruoka 1981) gave the impetus for the creative design of macrocyclic ring closure via reduction of oxalic acid-based bis-amidines using DIBALH (Fig. 4.30).

The first reaction in this 3-step reaction involves S-alkylation of dithiooxamide with an excess amount of bromothane in absolute ethanol. The generated bis-thioimido ester salt then reacted with triethylenetetramine to form the key intermediate tricyclic bis-amidine (Weisman and Reed 1997). The cyclization of tricyclic bis-amidine was performed with DIBALH in refluxing anhydrous toluene, followed by a typical workup for organo-aluminum reactions using NaF-H$_2$O.

So far, our discussion has focused on activable macrocyclic systems deployed with amine-reactive isothiocyanate moiety. The advantage of this chemistry justifies the reason why isothiocyanate macrocyclic dominant in the early days. Not only the conversion of primary amine to isothiocyanate can be achieved in a reproducible fashion using conventional chemistry, but it is also a selective process, and thus no protection of other functional groups necessary. In addition, isothiocyanate is very stable; no problem encounter during reaction workup. This strength, however, is also its shortcoming. Isothiocyanate has slow labeling kinetics. Long conjugation reaction time and high pH conditions are isothiocyanates' intrinsic parameters to achieve labeling efficiency. Considering all of these developments in the design, it should come as no surprise that other alternative approaches must have been derived to respond to the challenge, particularly when it comes to labeling unstable proteins, oligomers, peptides, and antibodies. One of the most recognized MRI contrast media, which has so much implications in MRI research, and a subject of intensive research is tris-carboxylated DOTA-NHS. The synthesis was started with mono-alkylation of the cyclen with benzyl bromoacetate, followed by alkylation of the rest of the available amines with t-butyl bromoacetate with an excess of K_2CO_3 in acetonitrile (Wangler et al. 2008). At this stage, it is intuitive to remove the benzyl ester selectively so that the carboxylic group would serve as a handle by the incorporating the N-hydroxysuccinimide ester. This activation reaction can be achieved using coupling agents, either dicyclohexylcarbodiimide (DCC) or, in this case, shown in Fig. 4.31 1H-benzotriazole tetramethyluronium hexafluorophosphate (HBTU). Finally, the t-butyl esters were cleaved off using TFA in methylene chloride at room temperature (Kovacs et al. 2005). The design seems to be simple as it shows; nevertheless, many laboratories have performed enormous efforts with continuous improvements over several years to deliver such an impeccable scheme. Particular attention is drawn to a few key steps, for example, maintaining low temperature during the reaction and purification of the first intermediate product (1,4,7,10-tetraaza-cyclododec-1-yl)-acetic acid benzyl ester (Fig. 4.31) is essential. Higher reaction temperatures and extended exposure with silica, used in chromatography would lead to intramolecular ring formations or fragmentations of the products (Wangler et al. 2008). The other steps involved the reaction workup of the final succinimidyl product. This ester is very susceptible to aqueous hydrolysis; the presence of a trace amount of water will result in hydrolysis of mono-NHS substituted DOTA. In this operation, TFA should be used as high-grade material. After the deprotection reaction finished, the solvent and TFA were removed under reduced pressure, and the suspension was washed with anhydrous diethyl ether and decanted to afford the solid off-white product (Kovacs et al. 2005).

With an increasing number of new biological products with diversified chemical functions with limitless applications, there is a need to develop MRI probes beyond the amine labeling domain. Thiol-reactive activated groups serve as an alternative option to expand the repertoire of biopharmaceutical MRI probes. A number of chemical substrates are suitable for reacting with thiol-active compounds, mostly found in proteins, peptides, and antibodies via the α, β-unsaturated carbonyl compounds in a thiol-Michael addition reaction. These substrates include maleimides, vinyl sulfones, acrylates, acrylamides, acrylonitriles, and methacrylates (Nair et al. 2013). The advantage of this type of reaction compared to amine-reactive molecules are many folds: first, the reaction is very spontaneous, like "click" chemistry, no catalysts are needed for the labeling operation. Second, no side products formed at the end of the reaction, simplifying the purification process. Third, the reaction condition is mild; the reaction kinetics is fast even at neutral pH, suitable for labeling unstable biomolecules. Among the substrates mentioned above, maleimidal molecules are mostly developed and proven versatile in this approach. Figure 4.32 shows an example of how to design a thiol-reactive DOTA chelator. DOTA-sulfonated NHS can be synthesized using water-soluble EDC as the coupling reagent (Lewis and Shively 1998). L-cysteine was introduced into the NHS to provide a handle for reacting with

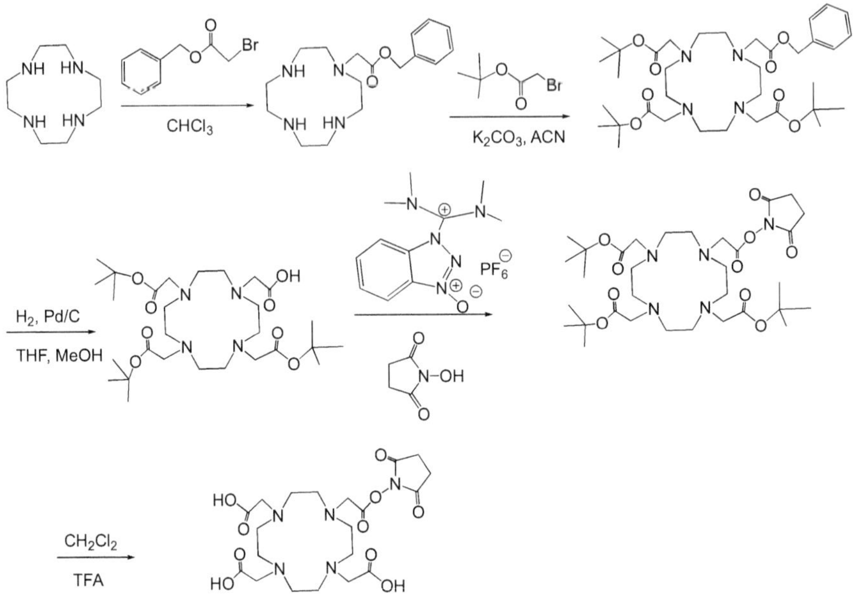

Fig. 4.31 Synthesis of tris-carboxylated DOTA-NHS. Data obtained from Wangler et al. (2008), Kovacs et al. (2005) with permission from Elsevier

the bis(maleimido)hexane. Although not discussed in the literature, it seems likely that the cysteine thiol group would displace the NHS first in a kinetic reaction; however, the stable thermodynamic amide would be a final product generated by the S-to-N transformation. After a mild reaction between cysteinyl-DOTA reacts with excess sulfhydryl-specific homo-bifunctional tag, the thiol moiety was oxidized as a sulfone derivative.

For labeling unstable bioactive compounds, it is necessary to develop functionalized macrocyclic chelators that facilitate bioconjugation in a mild reaction condition. The reaction should take place fast, robust, and neat, so no extensive purification is needed; otherwise, the compound will decompose. All of these criteria seem to be amenable to the Huisgen cycloaddition reaction, or sometimes called "click" chemistry. Aside from these unique characteristics offered by the "click" reaction, this copper-catalyzed 1,3-dipolar cycloaddition also provides remarkable product yield, in the absence of by-products. The reaction can occur in an aqueous medium and under physiological conditions. The simplest chemistry to achieve a "click" macrocyclic derivative is probably through the triacetic mono-alkyne ligand (DO3MA) (Vanasschen et al. 2011). The synthesis commenced with the alkylation of the free amino group on DO3A(t-Bu)$_3$ with propargyl bromide to afford the intermediate DO3MA(t-Bu)$_3$ with a high yield. Finally, TFA deprotects the triester to provide DO3MA after the mixture went through cationic and anionic ion-exchange chromatography (Fig. 4.33).

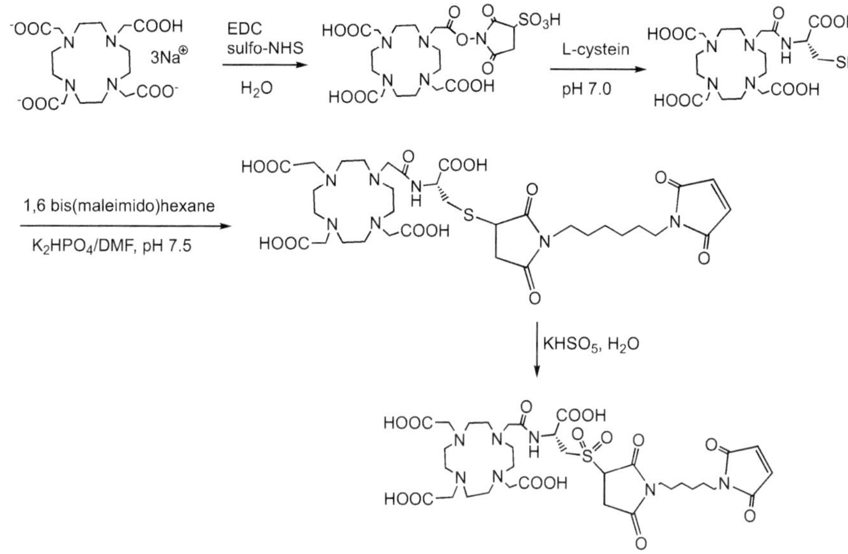

Fig. 4.32 Synthesis of maleimidyl-DOTA chelator. Data obtained from Lewis and Shively (1998) with permission from the American Chemical Society

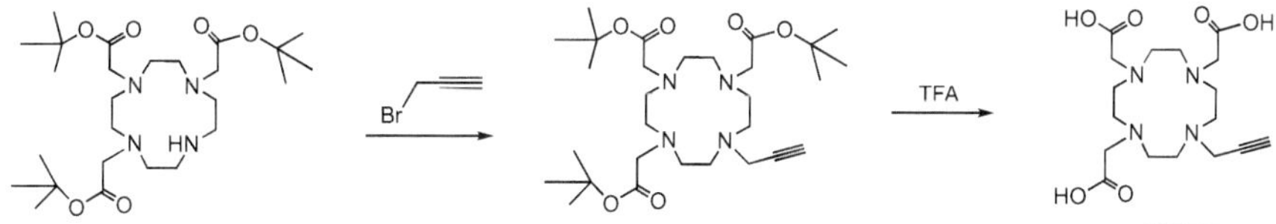

Fig. 4.33 Synthesis of DO3MA. Data obtained from Vanasschen et al. (2011) with permission from the American Chemical Society

To create a similar product with the alkyne moiety grafted directly on the tetraaza-cyclododecane ring requires a more complicated approach, in which reconstruction of the macrocyclic tetraaza unit is necessary. In a one-pot reaction starting from triethylenetetramine and butanedione to form the intermediate bisaminal, which was directly treated with 2,3-dibromopropan-1-ol to form the key bisaminal alcohol analog, which was treated in acidic condition to arrive the tetraazacyclic alcohol (Fig. 4.34) (Vanasschen et al. 2011). The free amines on the tetraazacyclic alcohol were alkylated with the bromo-*tert*-butyl esters in a basic condition, followed by a Williamson-type O-alkylation using propargyl bromide in the presence of cesium hydroxide in methylene chloride to afford the mono-alkyne product. It is noteworthy to mention that the presence of the bulky ester groups induced rigidity on the alkyne carbons, which differentiate the characteristics of R and S enantiomers as observed

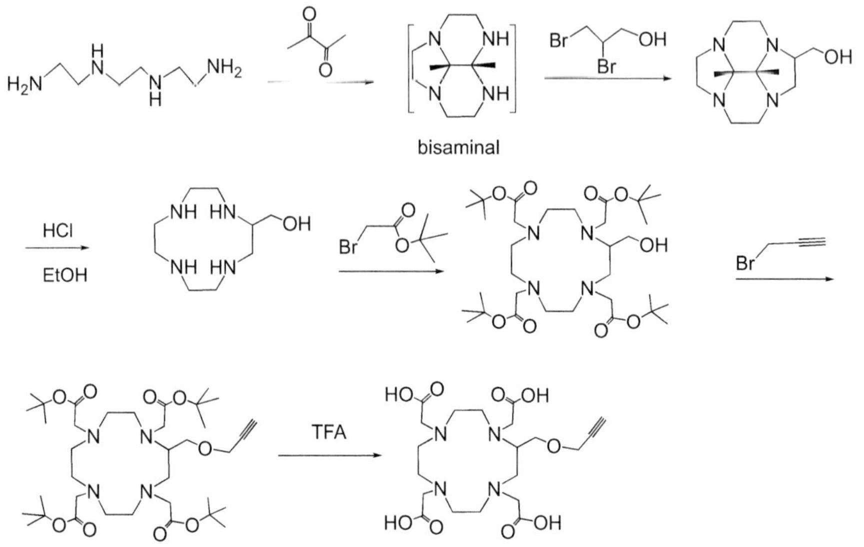

Fig. 4.34 Synthesis of DOTAMA analog. Data obtained from Vanasschen et al. (2011) with permission from the American Chemical Society

in the ^{13}C-NMR spectrum. After cleavage of the esters using TFA to afford pure DOTAMA.

4.2.3 Hyperpolarized Contrast Agents

4.2.3.1 [1-^{13}C]pyruvic Acid

Pyruvic acid, sometimes called 2-oxopropanoic acid, is one of the most popular ^{13}C pharmaceutical agents used in hyperpolarized MR imaging. Pyruvic acid plays an important role in several metabolic mechanisms. The molecules can penetrate the cellular membrane and home in the cytoplasm, with relatively long T_1 properties, making it a good contrast agent (Adamson et al. 2017). Pyruvate is a cascade product of the metabolism of glucose during glycolysis. One molecule of glucose breaks down into two molecules of pyruvate. Pyruvate is reduced into lactate under anaerobic conditions by the enzyme lactate dehydrogenase (LDH), or through oxidative (aerobic) decarboxylation to form acetyl-coA with production of [^{13}C]bicarbonate in the mitochondria. This biochemically induced altering of the pyruvate with altering chemical shifts that hyperpolarized MRI is capable of imaging with high temporal resolution (Kurhanewicz et al. 2019). Since tumor cells have altered metabolic phenotype characterized by increased glycolysis, exploiting hyperpolarized [^{13}C]pyruvate

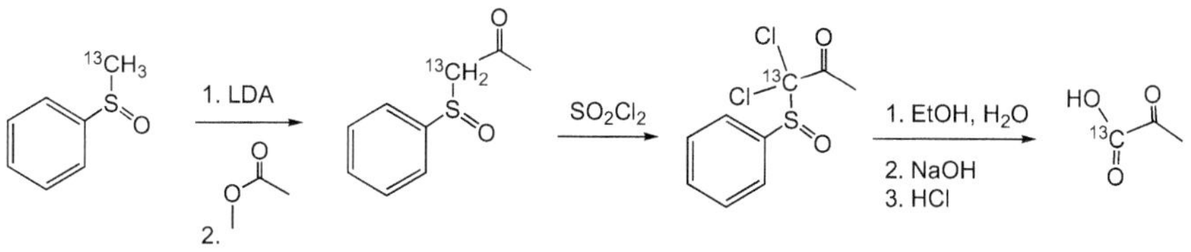

Fig. 4.35 Synthesis of [1-^{13}C] pyruvic acid. Data obtained from Martinez et al. (2000)

MR spectroscopic imaging has the potential to visualize glycolysis in de novo tumor formation and regression (Hu et al. 2011).

The synthesis of [^{13}C]pyruvate was previously reported (Fig. 4.35) (Martinez et al. 2000), where the reaction started with the dropwise addition of lithium diisopropylamide (LDA) to a solution of methyl-^{13}C-sulfinyl in tetrahydrofuran at -78 °C before adding methyl acetate neat to the mixture, and the reaction was stirred for 1 h followed by overnight stirring at room temperature. The reaction was quenched with 1 M HCl to afford α-(phenyl sulfinyl)[1-^{13}C]acetone, which was used without further purification. Next, sulfuryl chloride was treated to α-(phenyl sulfinyl)[1-^{13}C]acetone in methylene chloride provided the 1,1-dichloro-1-(phenylsulfinyl)[1-^{13}C]acetone in 1 h. The solvent was removed, and the reaction was reconstituted with a mixture of ethanol and water, followed by heating to reflux over the course of 12 h. After this period, there is a mixture of 2 products, namely [^{13}C]pyruvate and the unintended [^{13}C]pyruvate ethyl ester, with the latter, occupies up to 80% of the yield. Intuitively, the ester can be easily hydrolyzed in acidic conditions to provide a sole product of [^{13}C]pyruvate. The pH of the product was adjusted to 6.3 using sodium hydroxide to afford [^{13}C]pyruvate sodium salt.

4.2.3.2 [^{13}C]fumarate, [^{13}C]maleate and [^{13}C]succinate

Unsaturated fumarate is an ideal substrate for hydrogenative PHIP (parahydrogen-induced polarization). Hyperpolarized fumarate can be prepared by pairwise hydrogenation of the disodium acetylenedicarboxylate precursor in water (Ripka et al. 2018). This reaction is catalyzed by ruthenium for trans-hydrogenation of alkynes with stereoselectivity. The significance of this chemistry worth mentioning is the use of ruthenium for the selective homogenous (E)-hydrogenation of alkynes. Traditional hydrogenation usually yield Z-hydrogenation products; some provide (E)-hydrogenation products; however, the reactions frequently do not tolerate functional groups (Schleyer et al. 2001). Like every other hyperpolarized agent, it is necessary to label the carbon with some form; in this case, it is ^{13}C. Without this isotope, the protons in fumarate are chemically and magnetically equivalent; what it means is that the hyperpolarized proton singlet order, in this case, will be silent, and thus, enhanced NMR signals cannot be detected (Ripka et al. 2018). To generate para hydrogens, hydrogen gas must pass through a generator containing ion (III) oxide catalyst at 77 K.

Fumaric acid and maleic acid are isomers with the protons across the double bond structure arranged in trans- and cis-orientation, respectively. Thus, cognate preparation of [^{13}C]maleate can be performed as shown in Fig. 4.36, albeit different hydrogenation catalysts should be used to ensure Z-isomer. For example, [1,4-bis(diphenylphosphino)butane](1,5-cyclooctadiene)rhodium(I)tetrafluoroborate (dppb) (COD) is a common catalyst for this purpose (Eills et al. 2017).

In a similar approach, [1-^{13}C]succinate can be generated in a one-step reaction, directly from [1-^{13}C]fumarate or two steps starting from [1-^{13}C]acetylene dicarboxylate via [1-^{13}C]maleate intermediate (Fig. 4.37). From a practical point of view, the first approach is more favorable; not only short reaction operation but also fumaric acid is biologically safe (Ross et al. 2010).

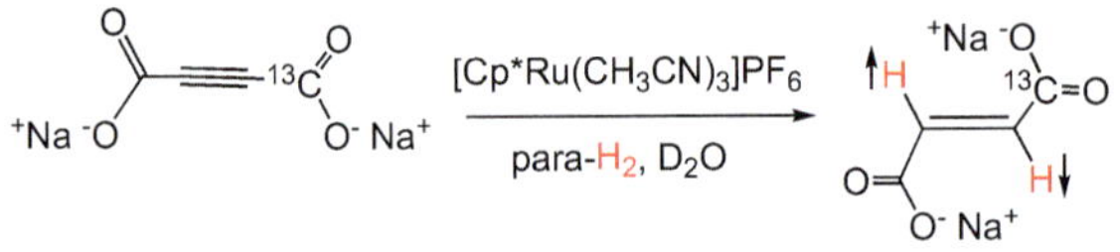

Fig. 4.36 Synthesis of [1-^{13}C]fumarate via parahydrogenation of acetylene [1-^{13}C]dicarboxylate catalyzed by [Cp*Ru(CH$_3$CN)$_3$]PF$_6$. The proton nuclear spin singlet state shown by opposing arrows. Data derived from Ripka et al. (2018) with permission from the author

Fig. 4.37 Different methods for the synthesis of [1-^{13}C]succinic acid. Data obtained from Ross et al. (2010) with permission from Williams & Wilkins publisher (American Society of Neuroradiology)

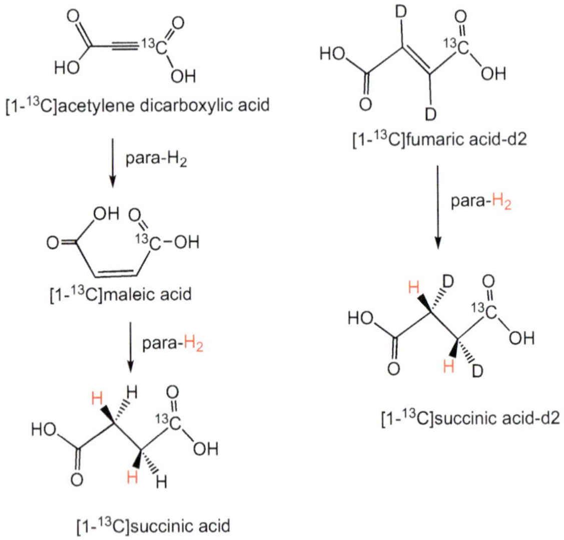

4.2.3.3 [$^{15}N_3$]metronidazole

Metronidazole is an FDA-approved antibiotic for oral or intravenous injection (Erickson et al. 1981). The nitroimidazole moiety of metronidazole serves as a substrate of up-regulated nitroreductases in tumors or ischemic tissues (Bolton and McClelland 1989; Kizaka-Kondoh and Konse-Nagasawa 2009). If this happens, there will be large ^{15}N chemical shifts of the ^{15}N sites on the imidazole, which can be detected by SABRE-SHEATH (Signal Amplification by the Reversible Exchange in SHield, Enables Alignment Transfer to Heteronuclei) technique (Shchepin et al. 2019). The innovative chemistry enabled ^{15}N-labeling of all three nitrogen sites of the substrate, achieved with minimal steps and straightforward spin-labeling synthetic chemistry (Fig. 4.38). First, the ^{15}N-labeled imidazole ring was constructed by a cyclization reaction using oxalaldehyde and acetaldehyde under acidic conditions in the presence of $^{15}NH_4Cl$, which is an economical source of ^{15}N. The ammonia moieties from ammonium chloride were added to the ends of oxalaldehyde through an acid-catalyzed reaction to form a bi-imine adduct. In this acidic environment, the acetaldehyde is converted into oxonium intermediate, which induces the attack from an imine leading to ring cyclization and rearrangement to form imidazole (Shchepin et al. 2017). Next, a conventional nitration reaction was performed using $Na^{15}NO_3$ catalyzed by sulfuric acid to afford the nitro-product with another nitrogen labeled as ^{15}N. This reaction also utilized inexpensive $Na^{15}NO_3$ as a source of ^{15}N. Finally, epoxy was added to the imidazole ring selectively to form the metronidazole-$^{15}N_3$. This prototypical synthetic experiment can serve as a template for introducing ^{15}N spins in other heterocyclic-based drugs, including but not limited to azomycin, benznidazole, secnidazole, ornidazole, nimorazole, evofosfamide, and more.

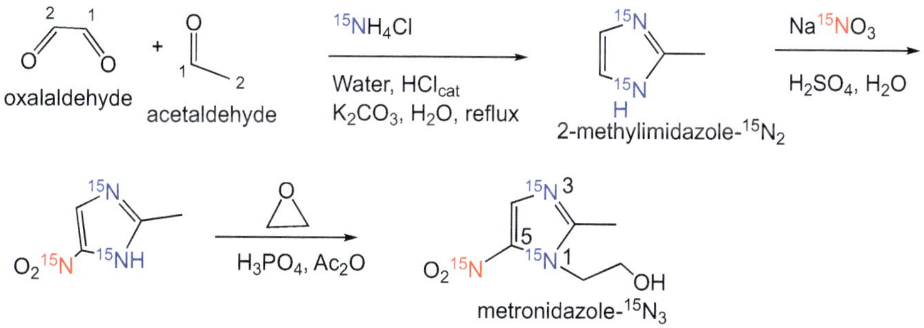

Fig. 4.38 Synthesis of metronidazole-$^{15}N_3$ as a hyperpolarized contrast agent for hypoxia imaging. Data obtained from Shchepin et al. (2019)

4.2.4 Superparamagnetic Nanoparticles

Molecular probes for MRI applications get more attention with the emergence of colloidal-based structures. So far, our discussion mentioned mostly Gd(III) as superparamagnetic materials; other elements such as Fe, Co, Ni, and Dy also belong to this category. Different from Gd(III), colloidal synthesis of Fe nanoparticles (magnetite, Fe_3O_4) in material chemistry by assembling elements with large magnetic domains roughly aligned in one direction, generating not only strong electromagnetic properties but also facilitates multimodal fabrication beyond traditional work and increasing blood half-life. Iron oxide nanoparticles dominate in this area because they are nontoxic, easy to achieve particle homogeneity, and enable surface fabrication. Many recognized applications of iron oxide nanoparticles have been reported for preclinical and clinical works, particularly in receptor imaging, cell trafficking, cancer, and biomarker imaging. These magnet nanoparticles also contribute significantly to neuroimaging, immunology, FRET-based imaging, and targeted drug delivery.

The superparamagnetic materials are prepared by reducing the overall size of a ferromagnetic or ferrimagnetic crystal until a single magnetic domain particle is formed. The benefit of superparamagnetic materials is that they achieve a point to complete magnetization even when placed in a weak external magnetic field. From a biological perspective, superparamagnetic particles are more attractive than paramagnetic counterparts since they can produce enhanced relaxation rates in specific organs at significantly lower doses and reduce if there is registered toxicity.

4.2.4.1   Methods of Synthesis

Nanoemulsion Synthesis

A number of different methods have been described for the synthesis of iron oxide nanoparticles. In nanoemulsion, the methods involve either water-in-oil or oil-in-water synthesis (Okoli et al. 2012). In the former case, the oil phase consists of *n*-octane mixed with a cationic surfactant (cetyltrimethylammonium bromide, CTAB) and a cosurfactant (1-butanol). The aqueous phase comprises a solution containing iron salt precursors with a mole ratio of 2:1 $FeCl_3$:$FeCl_2$. The addition of the aqueous solution into the mixture of oil phase/CTAB/1-butanol resulted in the formation of nanoemulsion. The formation of the magnetic nanoparticles is achieved with the addition of ammonia as a precipitating agent with vigorous stirring until the solution reaches pH11. The magnetic particle colloids are separated by centrifugation, dried, and resuspended in distilled water for storage at 4 °C.

In oil-to-water nanoemulsion synthesis, the oil phase contains iron precursor (iron(III) 2-ethylhexanoate) and non-ionic surfactant (ethyl oxide/propylene oxide block copolymer). The aqueous phase is purely distilled water. The nanoemulsion is formed by mixing these three components together with stirring and mild heating at 30 °C. The precipitation of the magnetic nanoparticles occurs by dropwise addition

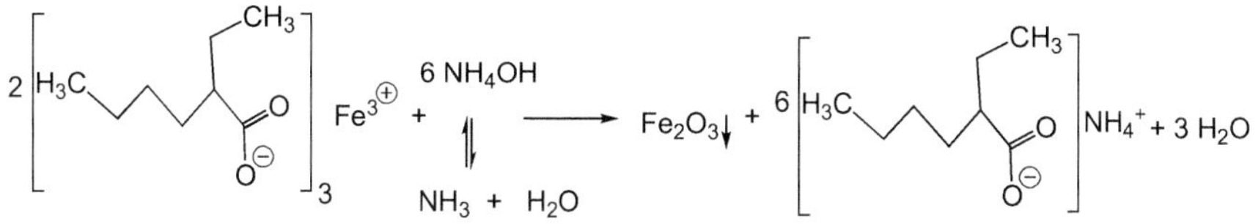

Fig. 4.39 Proposed reaction mechanism of organometallic precursor with ammonia to form iron oxide nanoparticle precipitates via oil-to-water microemulsion synthesis. Data obtained from Okoli et al. (2012) with permission from the American Chemical Society

of ammonia until achieving pH11. The suggested mechanism of magnetic nanoparticle precipitation is shown in Fig. 4.39. Purification and formulation are identical to the water-to-oil procedure described above. Nanoemulsion is useful for generating nanoparticles 4–8 nm.

It is worthwhile to mention the use of different types of surfactants in nanoemulsion synthesis. The choice of surfactants and cosurfactants depends on the overall design and desired outcome for each product, such as size and shape (Butter et al. 2005); and such anionic (sodium dodecylsulfate, dioctyl sodium sulfosuccinate, sodium lauryl ether sulfate, sodium dodecylbenzenesulfonate, perfluorobutanesulfonic acid), cationic (triethylamine HCl, octenidine.2HCl, adogen, cetylpyridinium chloride, benzethonium chloride, dimethyldioctadecylammonium chloride) or nonionic (triton X-100, Brij-97) surfactants can be used in the operation. For example, smaller particle sizes can be achieved when using ionic surfactants in the nanoemulsion process. It seems likely that the ionic surfactants' functional groups in the hydrated cores inhibit the crystallization of nanoparticles and usually result in aggregated magnetic nanoparticles (Kandori et al. 1991; Laurent et al. 2008).

Flow Injection Synthesis

In this method, the product will also have a size from 4 to 6 nm using a mixture of $FeCl_2$ and $FeCl_3$. The procedure involves continuous or periodic reproducible microinjection of reagents into a carrier stream followed by a chemical reaction that occurs as the reaction mixture travels through a capillary reactor (Salazar-Alvarez et al. 2006). In a flow injection setup, comprising of a propulsion unit, injection manifold, capillary reactor, and product collector (Fig. 4.40), a stream of acidic solution of 2:1 Fe^{3+}:Fe^{2+} was mixed with a stream of sodium hydroxide from another line. The colloidal synthesis occurs in the tiny capillary line in an inert environment with uniform temperature (80 °C). When the reaction completes, the product will be transferred to a separation vessel where a powerful magnet selectively collects the product. The reaction time of the nanoparticles is a function between the length of the manifold and the pumping rate. Under inert conditions at a high pH to form magnetite, as shown in Fig. 4.41.

The advantage of this technique is the ease of controlling the reaction conditions to tailor the particle size, morphology, behavior, and aggregation properties. In general,

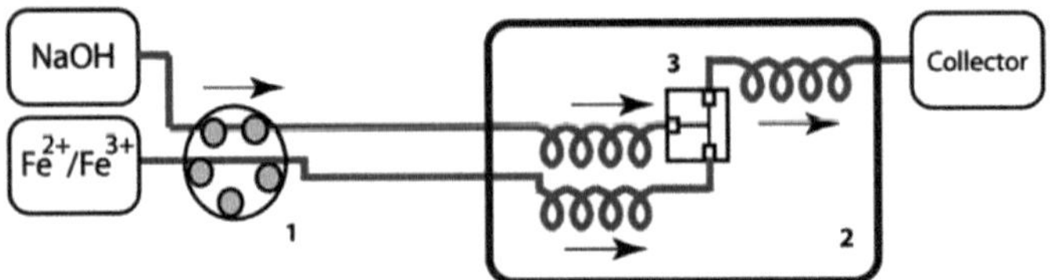

Fig. 4.40 Flow injection setup for the synthesis of magnetic iron oxide nanoparticles. Data obtained from Salazar-Alvarez et al. (2006) with permission from Elsevier

$$Fe^{2+} + 2Fe^{3+} + 8\,OH^- \longrightarrow Fe_3O_4 + 4\,H_2O$$

Fig. 4.41 Synthesis of iron nanoparticles using flow injection approach

the injection rate does not affect the size or homogeneity of the product. Nevertheless, it is crucial for maintaining the reproducibility of the system. Sodium hydroxide is more effective for making smaller sized particles by generating a large number of small nuclei, thus reducing the size of the product. Aggregation size also decreases with increasing the alkaline concentration.

Hydrothermal Synthesis

As the name suggests, the reaction occurs in either aqueous or an aqueous-organic environment, sometimes in the presence of ionic surfactants at high temperatures (150–250 °C) in a pressurized autoclave reactor. The main reagents used in this work include 1:2 Fe^{2+}:Fe^{3+} in mild alkaline conditions (Maurizi et al. 2011). It is found that the size and morphology of the magnetic nanoparticles are a function of pH, ionic surfactant, complexing reagents, temperature, and reaction time, as well as the source of alkaline materials. For example, in the absence of cationic surfactant cetyltrimethylammonium bromide (CTAB), at pH8, the particle size is about 100 nm. While in the presence of CTAB at pH4.6, the particle size is merely 20 nm (Giri et al. 2005).

Water-soluble magnetic nanoparticles can also be synthesized using hydrothermal synthesis. Specifically, only one source of iron, $FeCl_3.6H_2O$, was refluxed in 2-pyrrolidone. This solvent has a high boiling point, and it also works as a stabilizer since it can coordinate with metal ions. The reaction times dictate the particle sizes shape in this setup; 1 h, 10 h, and 24 h reactions correspond to 4, 12, and 60 nm, respectively. Prolonged reaction time and particles also change the original spherical shape to cubic particles (Li et al. 2005).

For larger magnetic nanoparticles, ranging from several hundred nanometers to micron size, hydrolysis, and thermolysis synthesis is the reaction of choice. The desired product can be achieved by the hydrolysis of iron tributoxide in octanol and acetonitrile solution.

Coprecipitation Synthesis

This is another popular technique for developing magnetite colloids in aqueous conditions using $FeCl_2$ and $FeCl_3$ salt. The desired physical property of iron oxide nanoparticles, such as size, can be fine-tuned readily by varying the pH, temperature, or the ratio of Fe^{2+}/Fe^{3+} salts. The particle size can be reduced significantly, as small as 4 nm, by calibrating the amount of bases, such as ammonia or sodium hydroxide, and the Fe^{2+}/Fe^{3+} ratio, the pH, and nature of the salts, such as perchlorates, chlorides, sulfates, or nitrates. Another way to reduce the particle size is to increase the concentration of citrate; this chelation of citrate on iron ions prevents nucleation and subsequently reducing size, as small as 3 nm. To maintain highly magnetic colloids, the level of oxygen exposure during synthesis must be well calibrated and controlled to reduce over oxidation of Fe^{2+} to Fe^{3+}, subsequently reducing the magnetic properties of the resulting nanoparticles. In general, the stock Fe^{3+} should be purged in an inert gas for at least 24 h before adding Fe^{2+} powder salt following neutralization of the reaction mixture and heating. During this whole process, inert gas should be bubbled in the solution.

4.2.4.2 Coating Iron Oxide Nanoparticles

A large number of iron oxide nanoparticles are fabricated with coating materials via electrostatic interaction with the polymer for a variety of applications, including citric acid (Hajdu et al. 2009), phosphonates (Basly et al. 2010), polyethylene glycol (Maurizi et al. 2009), DMSA (dimercaptosuccinic acid) (Bertorelle et al. 2006), or dextran (Josephson et al. 1999). In the context of this lecture, only dextran-coated iron oxide nanoparticles will be discussed; given dextran has been used extensively in clinical work due to its biocompatibility and biodegradability; these particles are already found their way to clinical work. Dextran was used to improve plasma circulation and prevent blood platelet aggregation (Dellacherie 1996). Different from other probe chemistry, nanotechnology is a multidimensional operation. Aside from size, shape, and the quantity of polymer ratios constituting the nanoparticles. Characterization of nanoparticles is challenging; the minimal factors that require the particles must have homogenous size and shape. Heterogenous nanoparticles possess many different physical properties, thus making characterization difficult. Other than that, it has been demonstrated in the past that heterogenous colloids have quick clearance by phagocytic cells of the reticuloendothelial system (RES). One of the significant applications for contrast enhancement using the T_2 signal of iron oxide nanoparticles relies on its long blood half-life for delineating blood pools in specific brain compartments. If the long blood half-life can be achieved; it means less colloidal concentration needed and less toxicity.

In a typical dextran-coating reaction, a highly concentrated aqueous and argon-purged solution of ferric chloride hexahydrate and dextran T-10 was stirred vigorously with a mechanical stirrer over 24 h (Palmacci and Josephson 1993). Then, ferrous salt was added to the ice-cooled reaction mixture as powder, and the acidic solution was neutralized slowly using ammonium hydroxide. Immediately after that, the solution was heated up to 85 °C for over an hour with constant stirring. For purification, the unreacted materials were eliminated from the product using 300 kD hollow fiber cartridges. The colloidal product was further concentrated using ultrafiltration cartridges to yield homogenous iron oxide nanoparticles with the size between 10 and 20 nm. The surface-covered dextran went further through the cross-linking using epichlorohydrin to stabilize the coating material and increase plasma half-life.

4.2.4.3 Activated Iron Oxide Nanoparticles

Modification of dextran. The holy grail of nanotechnology is the affordable multivalency that cannot find in other approaches. In conventional wet lab conjugation chemistry, probes can be labeled with targeted bioactive molecules on a 1:1 ratio. But that has been changed since the emergence of nanomaterials, which can harness hundreds if not thousands of molecules through surface fabrication. Toward that labeling approach, the dextran-coated iron oxide nanoparticles must be converted into active functional groups, such as amine or carboxylate groups. The simplest way involves the synthesis of aminated dextran, which then can be used for coating with the iron oxide nanoparticles during colloidal synthesis as described above. Dextran chemistry is unique in that the polymer is water soluble, so most of the reaction can occur in an aqueous condition, thus reducing potential toxicity concerns for the resulted product. Further, this exclusive solubility in one phase also enables simple purification, mostly via precipitation in a less polar solvent. With all of these characteristics, it seems likely pseudo-Schotten–Baumann reaction would fit this scheme (Fig. 4.42). To functionalize dextran with carboxylates, dextran 40 (MW 40 kg/mol) was treated with chloroacetic acid in strong basic condition (6 N NaOH), and the reaction was heated at 60 °C for nearly 1 h, before the polymer product was precipitated with the addition of methanol (Bouttemy 1960).

To generate aminated dextran polymer, the carboxymethyl dextran was coupled with excess ethylenediamine using 2-ethoxy-1-(ethoxycarbonyl)-1,2-dihydroquinoline (EEDQ) as a coupling reagent. Again, this coupling chemistry seems to be the most realistic option since the activation of the carboxylic group can be achieved in an aqueous condition. Further, in an industrial setting, large-scale reaction prefers economical reagents like EEDQ.

More concentrated loading of carboxylic groups can also be achieved by conjugation of DTPA to dextran. A number of different versions of activated dextran have been used, including DTPA bisanhydride or DTPA succinimide ester or DTPA mixed anhydride. Overall, DTPA was successfully interacting with all available amino groups to provide a modified dextran product with a high substitution of carboxylic acids.

Versatile linker(s) for modification of commercial dextran-coated magnet nanoparticles. So far, we have discussed the modification to develop activated dextran for coating magnetic nanoparticles during colloidal synthesis. A few reported procedures

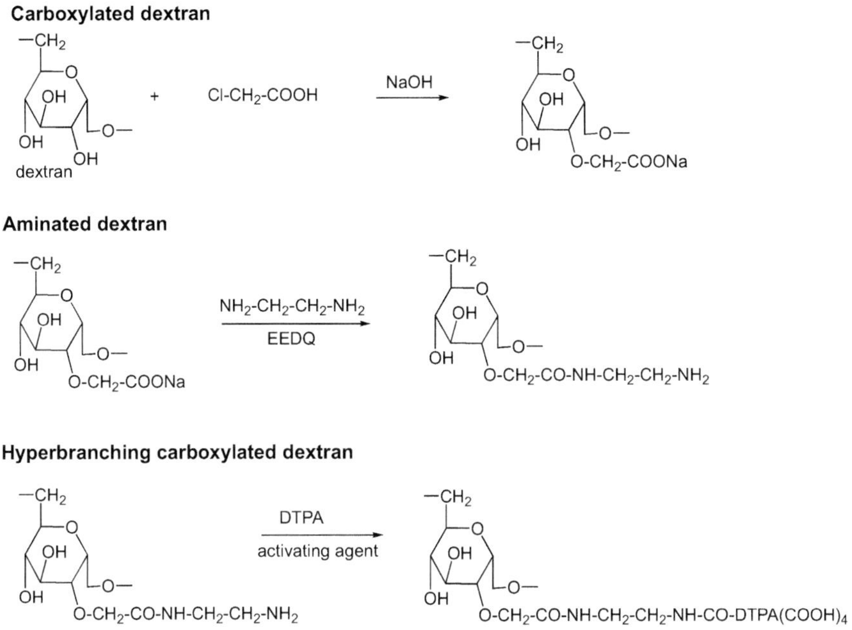

Carboxylated dextran

Aminated dextran

Hyperbranching carboxylated dextran

Fig. 4.42 Functionalization of dextran for coating magnetic nanoparticles. Data obtained from Rebizak et al. (1997) with permission from the American Chemical Society

manage to modify dextran-coated magnetic nanoparticles in a wet lab chemistry operation using highly specialized linkers.

The amine group appears to be an ideal chemical moiety for the functionalization of magnetic nanoparticles due to its strong nucleophilicity and compatibility with a wide range of available amine-activated biological materials. The idea of developing a versatile linker for surface modification of magnetic nanoparticles resonates very well with the need to tailor the development of commercial dextran-coated magnetic particles for biomedical applications. One of the requirements of the linkers is that they must be ready-to-use, water-soluble, and highly reactive so that the modification can occur in mild conditions without the use of a catalyst. Epoxide reactive agents are considered ideal for this application, given their stability and reactivity. As shown in Fig. 4.43, the synthesis of the epoxy-amine linker starts with Gabriel reaction for the conversion of 5-bromopent-1-ene into the corresponding alkenylamine. Protection of the newly formed amino group is necessary prior to the formation of an epoxide since amines can be oxidized faster than olefines once exposed to oxidizing agents. In addition, the final linker product must have the amino groups blocked to avoid intramolecular nucleophilic attack opening the epoxide ring. Theoretically, protection of amino groups in alkenylamines can be achieved using Fmoc, Boc or Cbz groups. However, removing these groups requires an extensive amount of work, contradicting the objective of this design. Instead, the amino group is protected as an alkenylammonium salt. The protonation of amine is efficient enough to prevent amine oxidation by a strong electrophilic O-transfer reagent. Another benefit to the design of the salt is that the compound is compatible with the aqueous condition, suitable for reaction with the magnet nanoparticles. But very surprising that olefin oxidation did not progress well with conventional reagents, such as mCPBA or hydrogen peroxide. While no mechanism has been studied to explain this, possible interference of the bulky counterion tosylate group could be the blame. To overcome this issue, dioxirane was developed, and it converts olefin into the desired epoxide linker as anticipated. Excess epoxide linker reacted with dextran hydroxyl groups in PBS buffer at physiological pH resulted in approximately 40–60 copies of amines on each magnetic nanoparticle.

Modification of commercial dextran-coated magnetic nanoparticles via organosilicon chemistry. Silicon chemistry is very similar to carbon chemistry, albeit with some unique features that make this chemistry so powerful that it covers almost every aspect of life; from cosmetics to household products, industry, such as heat transfer media, high/low temperature lubricants, solid-state catalysis, mix-phase catalysis, and health. There are thousands, if not millions, of silicon products in the market representing six major bonding groups, such as Si-H, Si-C, Si-Si, Si-halide, Si-O/SiO-X, Si-N/SiN-X. The silicon atom has four valencies like carbon, but it is

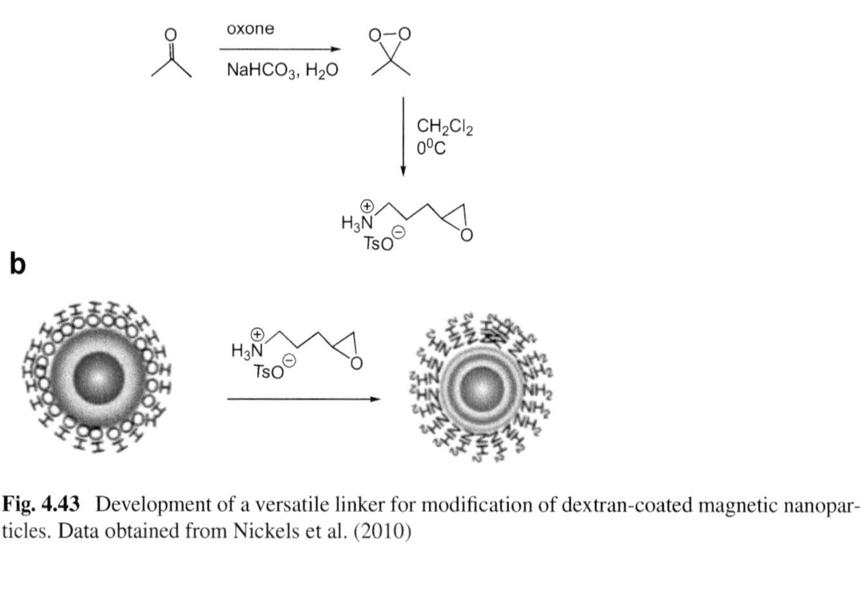

Fig. 4.43 Development of a versatile linker for modification of dextran-coated magnetic nanoparticles. Data obtained from Nickels et al. (2010)

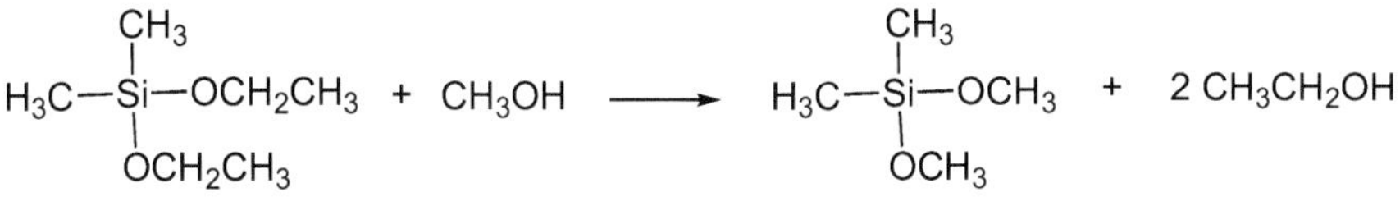

Fig. 4.44 Instant displacement of alkoxy groups by solvent

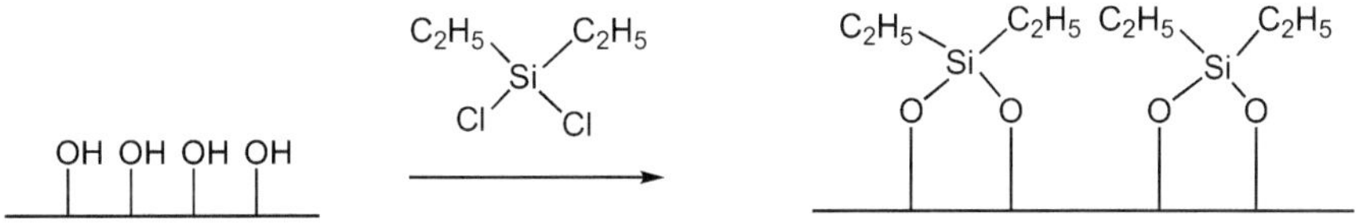

Fig. 4.45 Surface chemistry using chlorosilanes

more electropositive; as a result, the nucleophilic substitution at the silicon center is more robust than at the carbon center, and this is the underlying mechanism used in material chemistry with a variety of substrate that can be achieved with silanes, including surface fabrication, solid-phase synthesis, glass-bases surface synthesis.

Alkoxysilanes, sometimes called silane esters, react with alcohols instantly in aqueous or organic solutions without a catalyst. Using NMR spectroscopy, it is found the alkoxy group(s) are displaced by the hydroxyl-containing reagent in a manner analogous to the transesterification of carboxylic acid esters (Fig. 4.44).

Similar reactions also happen for chlorosilanes, and obviously, the byproduct is HCl instead of alcohols. Chlorosilanes are popular reagents for immobilizing materials in surface chemistry (Fig. 4.45). Several applications use this approach, such as high-throughput screening applications, extraction media, chromatography, and exchange resin applications.

These examples give a precedent of how organosilanes implicate the modification of hydroxyl-containing dextran. When it comes to designing experiments in aqueous conditions, alkoxysilanes are more favorable against amino- or acyloxy-substituted silanes since alkoxysilanes are more resistant to hydrolysis. Further, as the size of the alkoxy substituent increases, the silanes become increasingly resistant to hydrolysis. With this knowledge in mind, three commercially available organosilanes were tested for instant conversion the dextran hydroxyl groups on the surface of magnet nanoparticles into amines, including 3-aminopropyl-triethoxysilane (APTES), 3-aminopropyl-ethyl-diethoxysilane (APDES) and 3-aminopropyl-diethyl-ethoxysilane (APES) (Bini et al. 2012). The data demonstrated that alkoxysilanes are an excellent substrate for the modification of dextran-coated magnetic nanoparticles. As shown in Fig. 4.46, decreasing the number of alkoxide groups will reduce the concentration of the free amino groups to be derivatized on the surface of the nanoparticles, and thus, the order high density of amines for the three substrates is APTES > APDES > APES.

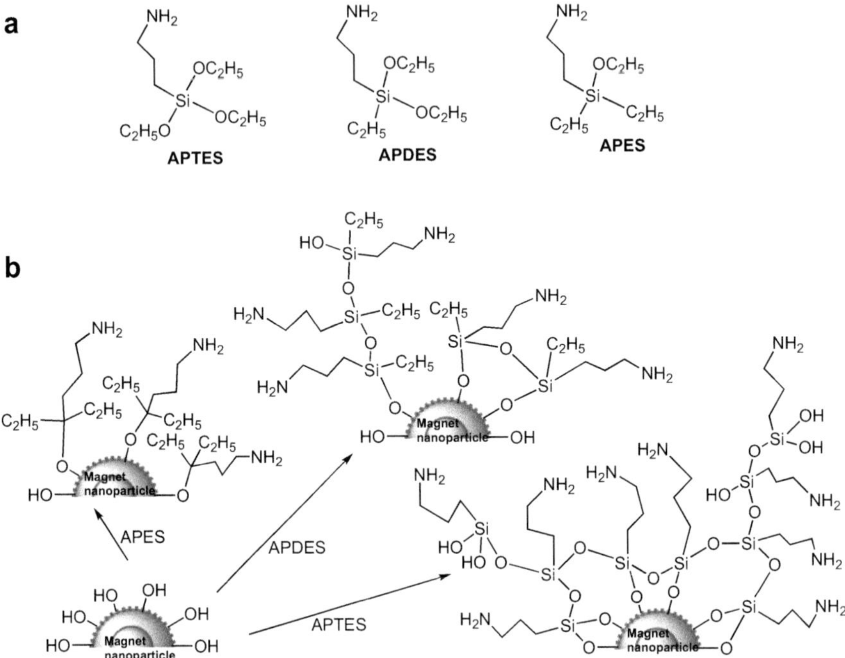

Fig. 4.46 Conversion of dextran hydroxyl group on the surface of magnet nanoparticles with amino groups using alkoxysilanes. Data obtained from Bini et al. (2012) with permission from Elsevier

4.3   Conclusion

The chemical development of MRI probes is still in the very nascent stage, and thus there are a lot of opportunities for new inventions. Despite having multiple advantages, MR imaging has a fundamental shortcoming when it comes to sensitivity. And thus, it is apparent that high doses of the contrast agents must be used to warrant signal detection. It seems reasonably certain that the next breakthrough in MRI probe design will focus on this issue. Now, we are moving to the next chapter, which will discuss solid-phase chemistry for molecular probes.

References

R.W. Adams, J.A. Aguilar, K.D. Atkinson, M.J. Cowley, P.I. Elliott, S.B. Duckett, G.G. Green, I.G. Khazal, J. Lopez-Serrano, D.C. Williamson, Reversible interactions with para-hydrogen enhance NMR sensitivity by polarization transfer. Science **323**, 1708–1711 (2009)

E.B. Adamson, K.D. Ludwig, D.G. Mummy, S.B. Fain, Magnetic resonance imaging with hyperpolarized agents: methods and applications. Phys. Med. Biol. **62**, R81–R123 (2017)

D.G. Amaral, M.P. Witter, The three-dimensional organization of the hippocampal formation: a review of anatomical data. Neuroscience **31**, 571–591 (1989)

J.H. Ardenkjar-Larsen, B. Fridlund, A. Gram, G. Hansson, L. Hansson, M.H. Lerche, R. Servin, M. Thaning, K. Golman, Increase in signal-to-noise ratio of >10,000 times in liquid-state NMR. Proc. Natl. Acad. Sci. U S A **100**, 10158–10163 (2003)

V. Baliyan, C.J. Das, R. Sharma, A.K. Gupta, Diffusion weighted imaging: technique and applications. World J. Radiol. **8**, 785–798 (2016)

E.K. Barefield, A new synthesis of 1,4,8,11-tetraazacyclotetradecane (cyclam) via the nickel(II) complex. Inorg. Chem. **11**, 2273–2274 (1972)

E.K. Barefield, F. Wagner, K.D. Hodges, Synthesis of macrocyclic tetramines by metal ion assisted cyclization reactions. Inorg. Chem. **15**, 1370–1377 (1976)

D.A. Barskiy, A.M. Coffey, P. Nikolaou, D.M. Mikhaylov, B.M. Goodson, R.T. Branca, G.J. Lu, M.G. Shapiro, V.V. Telkki, V.V. Zhivonitko, I.V. Koptyug, O.G. Salnikov, K.V. Kovtunov, V.I. Bukhtiyarov, M.S. Rosen, M.J. Barlow, S. Safavi, I.P. Hall, L. Schroder, E.Y. Chekmenev, NMR hyperpolarization techniques of gases. Chemistry **23**, 725–751 (2017)

B. Basly, D. Felder-Flesch, P. Perriat, C. Billotey, J. Taleb, G. Pourroy, S. Begin-Colin, Dendronized iron oxide nanoparticles as contrast agents for MRI. Chem. Commun. (Camb.) **46**, 985–987 (2010)

M. Baykara, M. Ozcan, M. Bilgen, H. Kelestimur, Effects of gadolinium and gadolinium chelates on intracellular calcium signaling in sensory neurons. Neurosci. Lett. **707**, 134295 (2019)

F. Bertorelle, C. Wilhelm, J. Roger, F. Gazeau, C. Menager, V. Cabuil, Fluorescence-modified superparamagnetic nanoparticles: intracellular uptake and use in cellular imaging. Langmuir **22**, 5385–5391 (2006)

B.A. Biagi, J.J. Enyeart, Gadolinium blocks low- and high-threshold calcium currents in pituitary cells. Am. J. Physiol. **259**, C515-520 (1990)

R.A. Bini, R.F.C. Marques, F.J. Santos, J.A. Chaker, M. Jafelicci Jr., Synthesis and functionalization of magnetite nanoparticles with different amino-functional alkoxysilanes. J. Magn. Magn. Mater. **324**, 534–539 (2012)

F. Bloch, W.W. Hansen, M. Packard, Physiol. Rev. **70**, 474 (1946)

N. Bloembergen, E.M. Purcell, R.V. Pound, Relaxation effects in nuclear magnetic resonance absorption. Phys. Rev. **73**, 679–710 (1948)

J.L. Bolton, R.A. McClelland, Kinetics and mechanism of the decomposition in aqueous solutions of 2-(hydroxyamino)imidazoles. J. Am. Chem. Soc. **111**, 8172–8181 (1989)

J.C. Bousquet, S. Saini, D.D. Stark, P.F. Hahn, M. Nigam, J. Wittenberg, J.T. Ferrucci Jr., Gd-DOTA: characterization of a new paramagnetic complex. Radiology **166**, 693–698 (1988)

M. Bouttemy, Contribution à l'étude dé carboxyméthylcelluloses. Préparation des carboxyméthyl-celluloes de hauts degrés de substitution. Bull. Soc. Chim. Fr. 1750 (1960)

M.W. Brechbiel, O.A. Gansow, R.W. Atcher, J. Schlom, J. Esteban, D.E. Simpson, D. Colcher, Synthesis of 1-(p-isothiocyanatobenzyl) derivatives of DTPA and EDTA. Antibody Labeling Tumor-Imaging Stud. Inorg. Chem. **25**, 2772–2781 (1986)

H. Bridge, S. Clare, High-resolution MRI: in vivo histology? Philos. Trans. R Soc. Lond. B Biol. Sci. **361**, 137–146 (2006)

K. Butter, K. Kassapidou, G.J. Vroege, A.P. Philipse, Preparation and properties of colloidal iron dispersions. J. Colloid Interface Sci. **287**, 485–495 (2005)

W.P. Cacheris, S.C. Quay, S.M. Rocklage, The relationship between thermodynamics and the toxicity of gadolinium complexes. Magn. Reson. Imaging **8**, 467–481 (1990)

A. Capozzi, C. Roussel, A. Comment, J.N. Hyacinthe, J. Phys. Chem. C **119**, 5020–5025 (2015)

P. Caravan, J.J. Ellison, T.J. McMurry, R.B. Lauffer, Gadolinium(III) chelates as MRI contrast agents: structure, dynamics, and applications. Chem. Rev. **99**, 2293–2352 (1999)

F. Cavagna, P. Tirone, E. Felder, C. de Haen, *Proceedings of the 2nd Special Topic Semina: New Development in Contrast Agent Research*, European Magnetic Resonance Forum, Blonay, Switzerland (1990), pp. 83–94

K.T. Cheng, Gadobenate, in *Molecular Imaging and Contrast Agent Database (MICAD)* (2007)

J. Chevallier, R.A. Butow, Calcium binding to the sarcoplasmic reticulum of rabbit skeletal muscle. Biochemistry **10**, 2733–2737 (1971)

R. Damadian, Apparatus and method for detecting cancer in tissue. Patent **3**(789), 832 (1974)

C. de Haen, Conception of the first magnetic resonance imaging contrast agents: a brief history. Top. Magn. Reson. Imaging **12**, 221–230 (2001)

K. Decker, Biologically active products of stimulated liver macrophages (KC). Eur. J. Biochem. **192**, 245–261 (1990)

E. Dellacherie, Polysaccharides in oxygen-carrier blood substitutes, in *Polysaccharides in Medicinal Applications*, ed. by S. Dumitriu (Marcel Dekker, New York, 1996), p. 525

J.F. Desreux, P.P. Barthelemy, Highly stable lanthanide macrocyclic complexes: in search of new contrast agents for NMR imaging. Int. J. Rad. Appl. Instrum. B **15**, 9–15 (1988)

F. Diederich, K. Dick, D. Griebel, Complexation of arenes by macrocyclic hosts in aqueous and organic solutions. J. Am. Chem. Soc. **108**, 2273–2286 (1986)

F.H. Doyle, J.C. Gore, J.M. Pennock, Relaxation rate enhancement observed in vivo by NMR imaging. J. Comput. Assist. Tomogr. **5**, 295 (1981)

A.N. Dula, S.A. Smith, J.C. Gore, Application of chemical exchange saturation transfer (CEST) MRI for endogenous contrast at 7 Tesla. J. Neuroimaging **23**, 526–532 (2013)

J.R. Duncan, F.N. Franano, W.B. Edwards, M.J. Welch, Evidence of gadolinium dissociation from protein-DTPA-gadolinium complexes. Invest. Radiol. **29**(Suppl 2), S58-61 (1994)

J.F. Dunn, J.A. O'Hara, Y. Zaim-Wadghiri, H. Lei, M.E. Meyerand, O.Y. Grinberg, H. Hou, P.J. Hoopes, E. Demidenko, H.M. Swartz, Changes in oxygenation of intracranial tumors with carbogen: a BOLD MRI and EPR oximetry study. J. Magn. Reson. Imaging **16**, 511–521 (2002)

R.A. Dwek, Proton relaxation enhancement probes: applications and limitations to systems containing macromolecules. Adv. Mol. Relax. Proc. **4**, 1 (1972)

R.A. Dwek, *Nuclear Magnetic Resonance (N.M.R.) in Biochemistry* (Clarendon, Oxford, 1973), pp. 174–284

R.R. Edelman, The history of MR imaging as seen through the pages of radiology. Radiology **273**, S181-200 (2014)

R. Edward Hendrick, *Breast MRI: Fundamentals and Technical Aspects* (Springer Science & Business Media, Berlin, 2008)

J. Eills, G. Stevanato, C. Bengs, S. Gloggler, S.J. Elliott, J. Alonso-Valdesueiro, G. Pileio, M.H. Levitt, Singlet order conversion and parahydrogen-induced hyperpolarization of (13)C nuclei in near-equivalent spin systems. J. Magn. Reson. **274**, 163–172 (2017)

J. Eisinger, R.G. Shulman, B.M. Szymanski, Transition metal binding in DNA solutions. J. Chem. Phys. **36**, 1721 (1962)

S.H. Erickson, G.L. Oppenheim, G.H. Smith, Metronidazole in breast milk. Obstet. Gynecol. **57**, 48–50 (1981)

C.J. Gauthier, A.P. Fan, BOLD signal physiology: models and applications. Neuroimage **187**, 116–127 (2019)

F.L. Giesel, N. Hohmann, U. Seidl, K.R. Kress, P. Schonknecht, H.U. Kauczor, J. Schroder, M. Essig, Working memory in healthy subjects and schizophrenics: studies using BOLD fMRT. Radiologe **45**, 144–152 (2005)

S. Giri, S. Samanta, S. Maji, S. Ganguli, A. Bhaumik, Magnetic properties of a-Fe_2O_3 nanoparticle synthesized by a new hydrothermal method. J. Magn. Magn. Mater. **285**, 296–302 (2005)

S. Gloggler, J. Colell, S. Appelt, Para-hydrogen perspectives in hyperpolarized NMR. J. Magn. Reson. **235**, 130–142 (2013)

H. Gries, D. Rosenberg, H.J. Weinmann, Patent Applications DE-OS, 3129906A3129901 (1981a)

H. Gries, D. Rosenberg, H.J. Weinmann, Patent Applications DE-OS, 3129906A1 (1981b)

A. Hajdu, E. Illes, E. Tombacz, I. Borbath, Colloids Surf. A **347**, 104–108 (2009)

M. Haris, K. Nath, K. Cai, A. Singh, R. Crescenzi, F. Kogan, G. Verma, S. Reddy, H. Hariharan, E.R. Melhem, R. Reddy, Imaging of glutamate neurotransmitter alterations in Alzheimer's disease. NMR Biomed. **26**, 386–391 (2013)

S. Hu, A. Balakrishnan, R.A. Bok, B. Anderton, P.E. Larson, S.J. Nelson, J. Kurhanewicz, D.B. Vigneron, A. Goga, 13C-pyruvate imaging reveals alterations in glycolysis that precede c-Myc-induced tumor formation and regression. Cell Metab. **14**, 131–142 (2011)

D.E. Huddleston, S.A. Small, Technology Insight: imaging amyloid plaques in the living brain with positron emission tomography and MRI. Nat. Clin. Pract. Neurol. **1**, 96–105 (2005)

J.M. Idee, M. Port, I. Raynal, M. Schaefer, S. Le Greneur, C. Corot, Clinical and biological consequences of transmetallation induced by contrast agents for magnetic resonance imaging: a review. Fundam. Clin. Pharmacol. **20**, 563–576 (2006)

J.M. Idee, M. Port, C. Robic, C. Medina, M. Sabatou, C. Corot, Role of thermodynamic and kinetic parameters in gadolinium chelate stability. J. Magn. Reson. Imaging **30**, 1249–1258 (2009)

D.M. Jacobs, M. Sano, G. Dooneief, K. Marder, K.L. Bell, Y. Stern, Neuropsychological detection and characterization of preclinical Alzheimer's disease. Neurology **45**, 957–962 (1995)

M.L. James, S.S. Gambhir, A molecular imaging primer: modalities, imaging agents, and applications. Physiol. Rev. **92**, 897–965 (2012)

G.T. Jennings, M.F. Bachmann, The coming of age of virus-like particle vaccines. Biol. Chem. **389**, 521–536 (2008)

K.M. Jones, A.C. Pollard, M.D. Pagel, Clinical applications of chemical exchange saturation transfer (CEST) MRI. J. Magn. Reson. Imaging **47**, 11–27 (2018)

L. Josephson, C.H. Tung, A. Moore, R. Weissleder, High-efficiency intracellular magnetic labeling with novel superparamagnetic-Tat peptide conjugates. Bioconjug. Chem. **10**, 186–191 (1999)

K. Kandori, M. Fukuoka, T. Ishikawa, Effects of citrate ions on the formation of ferric oxide hydroxide particles. J. Mater. Sci. **26**, 3313–3319 (1991)

S. Kizaka-Kondoh, H. Konse-Nagasawa, Significance of nitroimidazole compounds and hypoxia-inducible factor-1 for imaging tumor hypoxia. Cancer Sci. **100**, 1366–1373 (2009)

J. Kotek, F.K. Kalman, P. Hermann, E. Brucher, K. Binnemans, I. Lukes, Study of thermodynamic and kinetic stability of transition metal and lanthanide complexes of DTPA analogues with a phosphorus acid pendant arm. Eur. J. Inorg. Chem. 1976–1986 (2006)

Z. Kovacs, D.S. Mudigonda, A.D. Sherry, Process for making a chelating agent for labeling biomolecules using preformed active esters. US Patent, US 6,838,557 B1 (2005)

K.E. Krakowiak, J.S. Bradshaw, R.M. Izatt, Preparation of a variety of macrocyclic di-and tetraamides and their peraza-crown analogs using the crab-like cyclization reaction. J. Heterocycl. Chem. **27**, 1585–1589 (1990)

K. Kumar, Macrocyclic polyamino carboxylate complexes of Gd(III) as magnetic resonance imaging contrast agents. J. Alloys Compd. **249**, 163–172 (1997)

K. Kumar, M.F. Tweedle, M.F. Malley, J.Z. Gougoutas, Synthesis, stability, and crystal structure studies of some Ca^{2+}, Cu^{2+}, and Zn^{2+} complexes of macrocyclic polyamino carboxylates. Inorg. Chem. **34**, 6472–6480 (1995)

J. Kurhanewicz, D.B. Vigneron, J.H. Ardenkjaer-Larsen, J.A. Bankson, K. Brindle, C.H. Cunningham, F.A. Gallagher, K.R. Keshari, A. Kjaer, C. Laustsen, D.A. Mankoff, M.E. Merritt, S.J. Nelson, J.M. Pauly, P. Lee, S. Ronen, D.J. Tyler, S.S. Rajan, D.M. Spielman, L. Wald, X. Zhang, C.R. Malloy, R. Rizi, Hyperpolarized (13)C MRI: path to clinical translation in oncology. Neoplasia **21**, 1–16 (2019)

R.B. Lauffer, Paramagnetic metal complexes as water proton relaxation agents for NMR imaging: theory and design. Chem. Rev. **87**, 901–927 (1987)

S. Laurent, D. Forge, M. Port, A. Roch, C. Robic, L. Vander Elst, R.N. Muller, Magnetic iron oxide nanoparticles: synthesis, stabilization, vectorization, physicochemical characterizations, and biological applications. Chem. Rev. **108**, 2064–2110 (2008)

P.C. Lauterbur, M.H. Dias, A.M. Rudin, *Frontiers of Biological Energetics* (Academic Press, New York, 1978), pp.752–759

D. Le Bihan, Diffusion MRI: what water tells us about the brain. EMBO Mol. Med. **6**, 569–573 (2014)

M.H. Lee, C.D. Smyser, J.S. Shimony, Resting-state fMRI: a review of methods and clinical applications. AJNR Am. J. Neuroradiol. **34**, 1866–1872 (2013)

J.M. Lehn, J.P. Sauvage, [2]-Cryptates: stability and selectivity of alkali and alkaline-earth macrobicyclic complexes. J. Am. Chem. Soc. **97**, 6700–6707 (1975)

M.R. Lewis, J.E. Shively, Maleimidocysteineamido-DOTA derivatives: new reagents for radiometal chelate conjugation to antibody sulfhydryl groups undergo pH-dependent cleavage reactions. Bioconjug. Chem. **9**, 72–86 (1998)

Z. Li, Q. Sun, M. Gao, Preparation of water-soluble magnetite nanocrystals from hydrated ferric salts in 2-pyrrolidone: mechanism leading to Fe_3O_4. Angew. Chem. Int. Ed. **44**, 123–126 (2005)

S. Liu, D.S. Edwards, Bifunctional chelators for therapeutic lanthanide radiopharmaceuticals. Bioconjug. Chem. **12**, 7–34 (2001)

J. Lohrke, T. Frenzel, J. Endrikat, F.C. Alves, T.M. Grist, M. Law, J.M. Lee, T. Leiner, K.C. Li, K. Nikolaou, M.R. Prince, H.H. Schild, J.C. Weinreb, K. Yoshikawa, H. Pietsch, 25 years of contrast-enhanced MRI: developments, current challenges and future perspectives. Adv. Ther. **33**, 1–28 (2016)

S. Mansson, E. Johansson, P. Magnusson, C.M. Chai, G. Hansson, J.S. Petersson, F. Stahlberg, K. Golman, 13C imaging-a new diagnostic platform. Eur. Radiol. **16**, 57–67 (2006)

M. Markl, J. Leupold, Gradient echo imaging. J. Magn. Reson. Imaging **35**, 1274–1289 (2012)

R.A. Martinez, C.J. Unkefer, M.A. Alvarez, Synthesis of [1-13C]pyruvic acid, [2-13C]pyruvic acid, [3-13C]pyruvic acid and combinations thereof. US Patent, US 7,582,801 B2 (2000)

D.M. Masur, M. Sliwinski, R.B. Lipton, A.D. Blau, H.A. Crystal, Neuropsychological prediction of dementia and the absence of dementia in healthy elderly persons. Neurology **44**, 1427–1432 (1994)

L. Maurizi, H. Bisht, F. Bouyer, N. Millot, Easy route to functionalize iron oxide nanoparticles via long-term stable thiol groups. Langmuir **25**, 8857–8859 (2009)

L. Maurizi, F. Bouyer, J. Paris, F. Demoisson, L. Saviot, N. Millot, One step continuous hydrothermal synthesis of very fine stabilized superparamagnetic nanoparticles of magnetite. Chem. Commun. (Camb.) **47**, 11706–11708 (2011)

T.J. McMurry, M. Brechbiel, K. Kumar, O.A. Gansow, Convenient synthesis of bifunctional tetraaza macrocycles. Bioconjug. Chem. **3**, 108–117 (1992)

S. Meier, P.R. Jensen, M. Karlsson, M.H. Lerche, Hyperpolarized NMR probes for biological assays. Sensors (Basel) **14**, 1576–1597 (2014)

M.H. Mendonca-Dias, E. Gaggelli, P.C. Lauterbur, Paramagnetic contrast agents in nuclear magnetic resonance medical imaging. Semin. Nucl. Med. **13**, 364–376 (1983)

A.K. Mishra, J.F. Gestin, E. Benoist, A. Faivre-Chauvet, J.F. Chatal, Simplified synthesis of the bifunctional chelating agent 2-(4-aminobenzyl)-1,4,7,10-tetraazacyclododecane-N,N′,N″,N‴-tetraacetic acid. New J. Chem. **20**, 585–588 (1996)

M.K. Moi, C.F. Meares, S.J. Denardo, The peptide way to macrocyclic bifunctional chelating agents: synthesis of 2-(p-nitrobenzyl)-1,4,7,10-tetraazacyclododecane-N, N′, N″, N‴-tetraacetic acid and study of its yttrium(III) complex. J. Am. Chem. Soc. **110**, 6266–6267 (1988)

M.R. Moncelli, L. Becucci, C. de Haen, P. Tirone, Interactions between magnetic resonance imaging contrast agents and phospholipid monolayers. Bioelectrochem. Bioenerg. **46**, 205–213 (1998)

J. Moreau, E. Guillon, J.C. Pierrard, J. Rimbault, M. Port, M. Aplincourt, Complexing mechanism of the lanthanide cations Eu^{3+}, Gd^{3+}, and Tb^{3+} with 1,4,7,10-tetrakis(carboxymethyl)-1,4,7,10-tetraazacyclododecane (dota)-characterization of three successive complexing phases: study of the thermodynamic and structural properties of the complexes by potentiometry, luminescence spectroscopy, and EXAFS. Chemistry **10**, 5218–5232 (2004)

L.O. Morgan, A.W. Nolle, Proton spin relaxation in aqueous solutions of paramagnetic ions. II. Cr+++, Mn++, Ni++, Cu++, and Gd+++. J. Chem. Phys. **31**, 365 (1959)

D.P. Nair, M. Podgorski, S. Chatani, T. Gong, W. Xi, C.R. Fenoli, C.N. Bowman, The thiol-Michael addition click reaction: a powerful and widely used tool in materials chemistry. Chem. Mater. **26**, 724–744 (2013)

M. Nickels, J. Xie, J. Cobb, J.C. Gore, W. Pham, Functionalization of iron oxide nanoparticles with a versatile epoxy amine linker. J. Mater. Chem. **20**, 4776–4780 (2010)

H.P. Niendorf, Gadolinium-DTPA: a new contrast agent. Bristol Med. Chir. J. **103**, 34 (1988)

P. Nikolaou, A.M. Coffey, M.J. Barlow, M.S. Rosen, B.M. Goodson, E.Y. Chekmenev, Temperature-ramped (129)Xe spin-exchange optical pumping. Anal. Chem. **86**, 8206–8212 (2014)

K. Nwe, M. Bernardo, C.A. Regino, M. Williams, M.W. Brechbiel, Comparison of MRI properties between derivatized DTPA and DOTA gadolinium-dendrimer conjugates. Bioorg. Med. Chem. **18**, 5925–5931 (2010)

C. Okoli, M. Sanchez-Dominguez, M. Boutonnet, S. Jaras, C. Civera, C. Solans, G.R. Kuttuva, Comparison and functionalization study of microemulsion-prepared magnetic iron oxide nanoparticles. Langmuir **28**, 8479–8485 (2012)

S. Palmacci, L. Josephson, Synthesis of polysacharide covered superparamagnetic oxide colloids. US Patent 5,262,176 (1993)

L. Pauling, C.D. Coryell, The magnetic properties and structure of the hemochromogens and related substances. Proc. Natl. Acad. Sci. U S A **22**, 159–163 (1936)

C.G. Pippin, T.A. Parker, T.J. McMurry, M.W. Brechbiel, Spectrophotometric method for the determination of a bifunctional DTPA ligand in DTPA-monoclonal antibody conjugates. Bioconjug. Chem. **3**, 342–345 (1992)

R.A. Pooley, AAPM/RSNA physics tutorial for residents: fundamental physics of MR imaging. Radiographics **25**, 1087–1099 (2005)

M. Port, J.M. Idee, C. Medina, C. Robic, M. Sabatou, C. Corot, Efficiency, thermodynamic and kinetic stability of marketed gadolinium chelates and their possible clinical consequences: a critical review. Biometals **21**, 469–490 (2008)

Preparation method of intermediate BOPTA (4-hydroxy-5,8,11-tricarboxymethyl-1-phenyl-2-oxa-5,8,11-triazanaphthalenetridecane-13-acid). Patent, CN102603550B

R.M. Rai, S.Q. Yang, C. McClain, C.L. Karp, A.S. Klein, A.M. Diehl, Kupffer cell depletion by gadolinium chloride enhances liver regeneration after partial hepatectomy in rats. Am. J. Physiol. **270**, G909-918 (1996)

R. Rebizak, M. Schaefer, E. Dellacherie, Polymeric conjugates of Gd(3+)-diethylenetriaminepentaacetic acid and dextran. 1. Synthesis, characterization, and paramagnetic properties. Bioconjug. Chem. **8**, 605–610 (1997)

F. Reineri, T. Boi, S. Aime, ParaHydrogen induced polarization of 13C carboxylate resonance in acetate and pyruvate. Nat. Commun. **6**, 5858 (2015)

J. Reuben, Gadolinium (III) as a paramagnetic probe for protons relaxation studies of biochemical molecules. Biochemistry **10**, 2834–2838 (1971)

C.L. Reugg, W.T. Anderson-Berg, M.W. Brechbiel, S.M. Mirzadeh, O.A. Gansow, M. Strand, Improved in vivo stability and tumor targeting of bismuth-labeled antibody. Cancer Res. **50**, 4221–4226 (1990)

J.E. Richman, T.J. Atkins, W.F. Oettle, Org. Synth. 86 (1979)

B. Ripka, J. Eills, H. Kourilova, M. Leutzsch, M.H. Levitt, K. Munnemann, Hyperpolarized fumarate via parahydrogen. Chem. Commun. (Camb.) **54**, 12246–12249 (2018)

B.D. Ross, P. Bhattacharya, S. Wagner, T. Tran, N. Sailasuta, Hyperpolarized MR imaging: neurologic applications of hyperpolarized metabolism. AJNR Am. J. Neuroradiol. **31**, 24–33 (2010)

G. Salazar-Alvarez, M. Muhammed, A.A. Zagorodni, Novel flow injection synthesis of iron oxide nanoparticles with narrow size distribution. Chem. Eng. Sci. **61**, 4625–4633 (2006)

L. Sarka, L. Burai, E. Brucher, The rates of the exchange reactions between $[Gd(DTPA)]^{2-}$ and the endogenous ions Cu^{2+} and Zn^{2+}: a kinetic model for the prediction of the in vivo stability of $[Gd(DTPA)]^{2-}$, used as a contrast agent in magnetic resonance imaging. Chemistry **6**, 719–724 (2000)

D. Schleyer, H.G. Niessen, J. Bargon, In situ 1H-PHIP-NMR studies of the stereoselective hydrogenation of alkynes to (E)-alkenes catalyzed by a homogeneous $[Cp*Ru]^+$ catalyst. New J. Chem. **25**, 423–426 (2001)

G. Sharma, C.Y. Wu, R.M. Wynn, W. Gui, C.R. Malloy, A.D. Sherry, D.T. Chuang, C. Khemtong, Real-time hyperpolarized (13)C magnetic resonance detects increased pyruvate oxidation in pyruvate dehydrogenase kinase 2/4-double knockout mouse livers. Sci. Rep. **9**, 16480 (2019)

R.V. Shchepin, D.A. Barskiy, A.M. Coffey, M.A. Feldman, L.M. Kovtunova, V.I. Bukhtivarov, K.V. Kovtunov, B.M. Goodson, I.V. Koptyug, E.Y. Chekmenev, Robust imidazole-$^{15}N_2$ synthesis for high-resolution low-field (0.05 T) ^{15}N hyperpolarized NMR spectroscopy. ChemistrySelect **2**, 4478–4483 (2017)

R.V. Shchepin, J.R. Birchall, N.V. Chukanov, K.V. Kovtunov, I.V. Koptyug, T. Theis, W.S. Warren, J.G. Gelovani, B.M. Goodson, S. Shokouhi, M.S. Rosen, Y.F. Yen, W. Pham, E.Y. Chekmenev, Hyperpolarizing concentrated metronidazole (15) NO_2 group over six chemical bonds with more than 15% polarization and a 20 minute lifetime. Chemistry **25**, 8829–8836 (2019)

A.D. Sherry, M. Woods, Chemical exchange saturation transfer contrast agents for magnetic resonance imaging. Ann. Rev. Biomed. Eng. **10**, 391–411 (2008)

A.D. Sherry, P. Caravan, R.E. Lenkinski, Primer on gadolinium chemistry. J. Magn. Reson. Imaging **30**, 1240–1248 (2009)

S. Sinharay, M.D. Pagel, Advances in magnetic resonance imaging contrast agents for biomarker detection. Annu. Rev. Anal. Chem. (Palo Alto Calif) **9**, 95–115 (2016)

H. Stetter, W. Frank, Complex formation with tetraazacycloalkane-N'', N'''',-tetraacetic acids as a function of ring size. Angew. Chem. **88**, 686 (1976)

I. Tabushi, Y. Taniguchi, H. Kato, Preparation of C-alkylated macrocyclic polyamines. Tetrahedron Lett. **12**, 1049–1052 (1977)

R. Turner, D. Le Bihan, J. Maier, R. Vavrek, L.K. Hedges, J. Pekar, Echo-planar imaging of intravoxel incoherent motion. Radiology **177**, 407–414 (1990)

C. Vanasschen, N. Bouslimani, D. Thonon, J.F. Desreux, Gadolinium DOTA chelates featuring alkyne groups directly grafted on the tetraaza macrocyclic ring: synthesis, relaxation properties, "click" reaction, and high-relaxivity micelles. Inorg. Chem. **50**, 8946–8958 (2011)

F. Vellaccio Jr., R.V. Punzar, D.S. Kemp, The reaction of dialkylmalonyl dichlorides with 1,3-diaminopropanes, a new route to macrocyclic polyamides and polyamines. Tetrahedron Lett. **6**, 547–550 (1977)

T.R. Wagler, C.J. Burrows, Synthesis of an optically active C-functionalized cyclam: (S)-5-(hydroxymethyl)-1,4,8,1 I-tetra-azacyclotetradecane and its nickel(iI) complex. J. Chem. Soc. Chem. Commun. 277–278 (1987)

J. Wahsner, E.M. Gale, A. Rodriguez-Rodriguez, P. Caravan, Chemistry of MRI contrast agents: current challenges and new frontiers. Chem. Rev. **119**, 957–1057 (2019)

Z.J. Wang, M.A. Ohliger, P.E.Z. Larson, J.W. Gordon, R.A. Bok, J. Slater, J.E. Villanueva-Meyer, C.P. Hess, J. Kurhanewicz, D.B. Vigneron, Hyperpolarized (13)C MRI: state of the art and future directions. Radiology **291**, 273–284 (2019)

C. Wangler, B. Wangler, M. Eisenhut, U. Haberkorn, W. Mier, Improved syntheses and applicability of different DOTA building blocks for multiply derivatized scaffolds. Bioorg. Med. Chem. **16**, 2606–2616 (2008)

H.J. Weinmann, R.C. Brasch, W.R. Press, G.E. Wesbey, Characteristics of gadolinium-DTPA complex: a potential NMR contrast agent. AJR Am. J. Roentgenol. **142**, 619–624 (1984)

G.R. Weisman, D.P. Reed, A new synthesis of cyclen (1,4,7,10-tetraazacyclododecane). J. Org. Chem. **62**, 4548 (1997)

G.W. White, W.A. Gibby, M.F. Tweedle, Comparison of Gd(DTPA-BMA) (Omniscan) versus Gd(HP-DO3A) (ProHance) relative to gadolinium retention in human bone tissue by inductively coupled plasma mass spectroscopy. Invest. Radiol. **41**, 272–278 (2006)

G.L. Wolf, E.S. Fobben, The tissue proton T1 and T2 response to gadolinium-DTPA injection in rabbits: a potential renal contrast agent for NMR imaging. Invest. Radiol. **19**, 324–328 (1984)

B. Wu, G. Warnock, M. Zaiss, C. Lin, M. Chen, Z. Zhou, L. Mu, D. Nanz, R. Tuura, G. Delso, An overview of CEST MRI for non-MR physicists. EJNMMI Phys. **3**, 19 (2016)

H. Yamamoto, K. Maruoka, Regioselective carbonyl amination using diisobutylaluminum hydride. J. Am. Chem. Soc. **103**, 4186 (1981)

Y.F. Yen, K. Nagasawa, T. Nakada, Promising application of dynamic nuclear polarization for in vivo (13)C MR imaging. Magn. Reson. Med. Sci. **10**, 211–217 (2011)

T. Yousaf, G. Dervenoulas, M. Politis, Advances in MRI methodology. Int. Rev. Neurobiol. **141**, 31–76 (2018)

M. Zaiss, J. Xu, S. Goerke, I.S. Khan, R.J. Singer, J.C. Gore, D.F. Gochberg, P. Bachert, Inverse Z-spectrum analysis for spillover-, MT-, and T1-corrected steady-state pulsed CEST-MRI—application to pH-weighted MRI of acute stroke. NMR Biomed. **27**, 240–252 (2014)

X. Zhao, E.S. Lein, A. He, S.C. Smith, C. Aston, F.H. Gage, Transcriptional profiling reveals strict boundaries between hippocampal subregions. J. Comput. Neurol. **441**, 187–196 (2001)

Chapter 5
Solid-Phase Chemistry

5.1 Introduction

The chemical synthesis of imaging probes has crossed paths with solid-phase synthesis (SPS) in the recent past. Notably, in the era of proteomics and the emergence of recombinant DNA, there is a practical need for developing large compound libraries identified from these works. Never have so many people felt the need for a major change in how to synthesize these molecules chemically. In that regard, combinatorial SPS is a prime player. SPS involves the chemistry where all the reactants are immobilized onto the resin for interactions. At the end of the reaction, unreacted materials, chemical waste, and side products can be easily filtered and washed off. Thus, it simplifies the purification process. SPS is useful for high-throughput combinatorial synthesis, and it is impeccable if the synthetic operation requires repeated steps, or the reactions need multiple protection and deprotection sequences. This technique adds another level of exuberant research, not only in academia but also in industry. In most if not all, the reaction yields obtained from SPS are exceptionally high. The process eliminates unnecessary purifications and is amenable to automation, making it a promising modality from a commercial point of view since it saves time and effort and cuts the cost.

Peptide chemistry fits SPS because it requires repeated identical procedures. Peptides are the building blocks of proteins connected by amide bonds, which can be synthetically achieved by activating the carboxylic group of the prior amino acid and then coupling the next one. It sounds as simple as it is; in fact, the coupling reaction is not a concern, but the subsequent protection and deprotection of the amino acid side chains during the synthesis, sometimes more challenging or even it is impossible if performed in solution-phase chemistry. By incorporating this type of chemistry in SPS using the resin, the coupling of amino acids and deprotection of the side chains can proceed with no more than just a simple exposure of the excess mole equivalents of amino acid substrates along with coupling reagent and cleavage solution, respectively. This genius chemistry was first reported by Bruce Merrifield, who earned a Nobel prize in Chemistry in 1984. Since then, numerous chemical reactions and

W. Pham, *Principles of Molecular Probe Design and Applications*,
https://doi.org/10.1007/978-981-19-5739-0_5

analogs have been developed through SPS. Aside from making the amide bonds, SPS can do virtually every chemical conversion performed in solution-phase chemistry. Just mention a few, such as aromatic substitution, cycloaddition, condensation, reduction, oxidation, and heterocyclic reactions. SPS plays a major role in combinatorial chemistry, which is defined as the diversification of chemical structures of the targeted molecules deduced from a few simple and common monomers.

5.2 Solid Support Resins

If SPS is the soul in combinatorial chemistry, then resin is the heart in SPS. There are several different types of resins available for different reaction designs. In general, resins share similar characteristics, such as they must swell in the optimal solvent(s) to enhance the reactant's accessibility to the reaction site. As a rule of thumb, successful development of long peptides or molecules that require multi-step synthesis must all employ resins with good swelling capability. This holds true for long and branched peptides, and the extent to which a polystyrene resin swells is very important for a successful SPS (Pugh et al. 1992). Resin swelling is reversely proportional to the cross-linking. In almost organic solvents, it is found that the higher the percentage of cross-linking resins, the lower the swelling (Amadi-Kamalu et al. 2020). Thus, most of the resins used in solid-phase chemistry would have the divinylbenzene (DVB) cross-linking not more than 2%. For instance, most the resins, like Wang or Merrifield resins, are designed with 1–2% cross-linking to ensure good swelling across different organic solvents (Fig. 5.1).Further, other factors also contribute to successful syntheses, such as the bead size, uniformity of the beads, and the presence of functional groups on the resin (Pugh et al. 1992). When a complicated reaction needs optimization of the resin, it is crucial to ensure the number of beads should be way excess the number of the final products, usually at least ten times more beads than the product. And smaller bead size is more favorable due to its association with a fast reaction rate; generated by a higher surface area to volume ratio; somewhere from 37 to 75 μm (200–400 mesh) would suffice.

Different from solution-phase chemistry, aside from exposure to exothermic reaction conditions, in SPS, the resins are continuously shaken in each coupling cycle. Therefore, the mechanical stability of the resins is the utmost important criterion in the overall reaction. This phenomenon is called osmotic shock, which is responsible for morphological structure alteration (Al Musaimi et al. 2019). The use of bad solvents further exacerbates this problem. For instance, using scanning electron microscopy, it is found that CTC (2-chlorotrityl chloride) resins behave very differently from each solvent. The morphological structure of CTC resin was altered and appeared as punchers like dimples in diethyl ether (DEE) (Fig. 5.2). While the resins experienced severe fractures in *tert*-butyl methyl ether (MTBE), much surprisingly, 2-methyltetrahydrofuran (2-MeTHF) seems to be very compatible with CTC resin; it did cause damage to the resin. The morphological structure was altered, not serious as seen in the case of MTBE or DEE, but significant enough to concern. In contrast

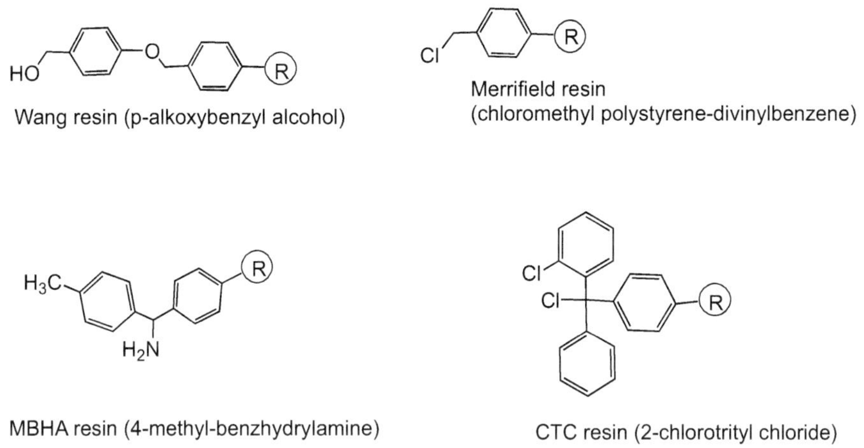

Wang resin (p-alkoxybenzyl alcohol)

Merrifield resin
(chloromethyl polystyrene-divinylbenzene)

MBHA resin (4-methyl-benzhydrylamine)

CTC resin (2-chlorotrityl chloride)

Fig. 5.1 Popular resins for SPS

to other ethers just mentioned, cyclopentyl methyl ether (CPME) stands out as the ideal solvent for CTC resin and probably the ether of choice in SPPS. Aside from providing excellent swelling properties for the resin, CPME is safer to use compared to other ethers, such as diethyl ether and *tert*-butyl methyl ether.

Osmotic shock apparently affects the reaction rates as the result of damaged resins. It turns out that cross-linking provides some sort of mechanical stability for the resin, preventing them from breakage. For instance, polystyrene resins used in most SPS are crosslinked with DVB to confer mechanical stability and consistency (Fig. 5.3). In lieu of the previous discussion about the reverse relationship between cross-linking and swell capability, it is tangible that there must be a good design to balance the percentage of cross-linking to maintain its dedicated role in SPS. Next, it is crucial to know the functional group available on the resin to anchor the chemical substrate before further chemical extension from there and the chemical conditions to dissociate the final product from the resin.

5.2.1 Merrifield Resin

Merrifield resin is compatible with Boc (t-butyloxycarbonyl)-based SPS, where the first substrate is introduced into the resin via a benzyl ester linkage between the resin and the carboxylic group of the first chemical entity. In the case of amino acids, it is the C terminal end. This mild esterification process aims to reduce racemization. Generally, the Boc-amino acids are treated with triethylamine to form the triethylammonium salt of the Boc-amino acid to form the benzyl ester linkage to the resin. Other organic bases can have the same effect, including tetramethylammonium hydroxide (Loffet 1971). Although this method proved to be effective and

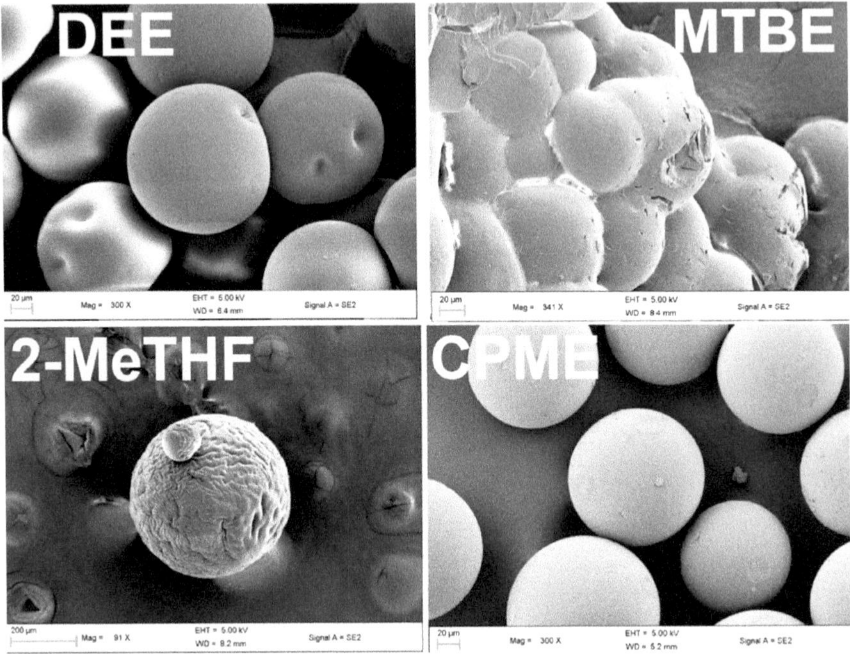

Fig. 5.2 Solvent effects on CTC resin as examined via scanning electron microscopy (SEM). Data obtained from Al Musaimi et al. (2019)

Fig. 5.3 Cross-linking of polystyrene with divinylbenzene (1–2%) copolymer

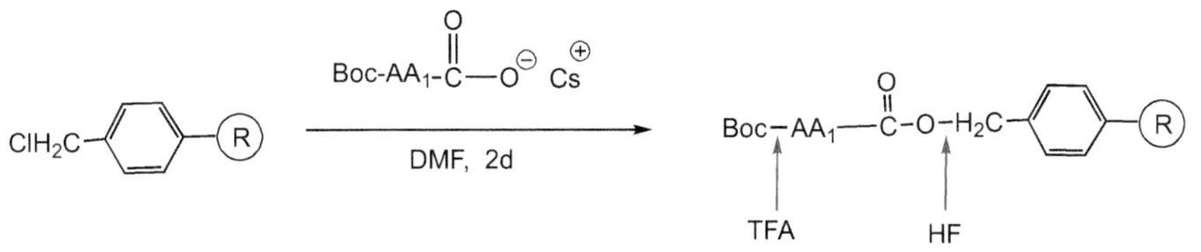

Fig. 5.4 — reaction scheme:

Boc-AA$_1$-C(=O)-O$^\ominus$ Cs$^\oplus$ / DMF, 2d

ClH$_2$C—(aryl)—R → Boc—AA$_1$—C(=O)-O-H$_2$C—(aryl)—R (TFA cleaves at Boc-AA$_1$; HF cleaves at benzyl ester)

Fig. 5.4 Loading first amino acid onto Merrifield resin. After each coupling cycle, TFA cleavage of the Boc group, and at the end of the synthesis, the resin can be removed by treating HF

resulted in good products. A few issues potentially implicate the chemistry, especially the ammonium moiety that can trap trifluoroacetic acid used in the deprotection step. The presence of TFA in the subsequent coupling step would interfere with the process and may lead to the termination of the peptide chains. To overcome this issue, a number of amino acid carboxylic salts have been explored using lithium, sodium, potassium, rubidium, or cesium. Among these, cesium salts of Boc-amino acids showed significantly enhanced loading efficiency (Fig. 5.4). Since then, cesium salt has become a recognized approach for loading the first substrate to Merrifield and related resins. Aside from preventing the aforementioned issues, this method is more efficient for incorporating the Boc-amino acids onto the Merrifield resin under anhydrous condition. The general procedure for the preparation of cesium salts involves adding aqueous cesium hydrogencarbonate into a solution of Boc-amino acid in ethanol: water (3:1) at pH 7. The solution can be removed completely using lyophilization, and the dried product can be used without further purification (Gisin 1973).

Merrifield resin is suitable for Boc/Bzl (benzyl) SPS strategy. This conventional chemistry indicates that the Boc group protects the N terminal residue of the amino acid, while the benzyl group protects the C terminal carboxyl moiety and the sidechains. Both Boc and benzyl are acidic labile molecules but at different levels. Boc can be deprotected with 30% TFA in methylene chloride, but benzyl group would survive well in those repeated cycles during the synthesis (Fig. 5.5). The cleavage of the benzyl group can be performed concomitant to the cleavage of the resin under a stronger acidic condition using HF.

5.2.2 PAM Resin

An extension design of Merrifield resin, again for Boc-amino acid coupling, is phenylacetamidomethyl (PAM)-resin, which is more stable in the presence of TFA compared to Merrifield resin. The key design involves the incorporation of the electron-withdrawing carboxamidomethyl group para to the peptide benzyl ester to enhance the acid stability of the PAM resin. This resin reduces the loss of peptides at each deprotection step, thus improving the overall product yield, particularly useful for developing larger peptides/structures. As shown in Fig. 5.6, PAM resin

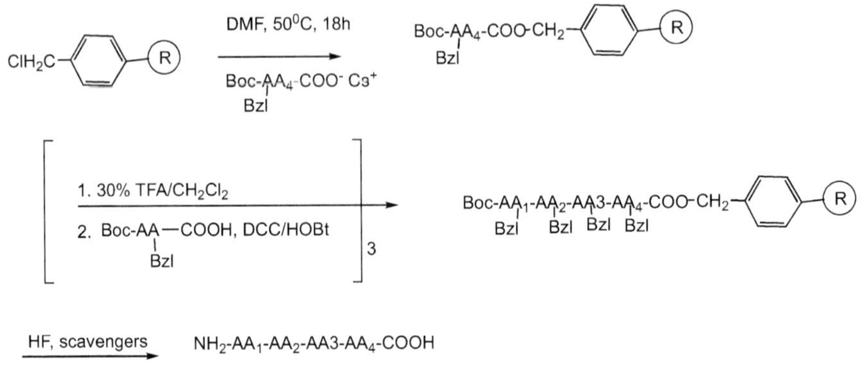

Fig. 5.5   An example of Merrifield resin in SPS

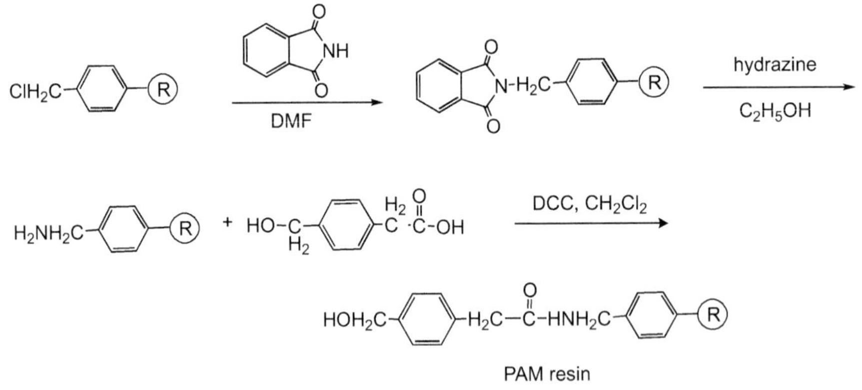

Fig. 5.6   Synthesis of PAM resin. Data obtained from Mitchell et al. (1976) with permission from the American Chemical Society

can be generated from Merrifield resin by the conversion of methylene chloride resin to an aminomethyl resin via Gabriel synthesis. Then, coupling of the free amine from the resin with 4-(hydroxymethyl)phenylacetic acid using dicyclohexylcarbodiimide (DCC) in methylene chloride provides the product. PAM resin is compatible with Boc-protected amino acids, and this resin has been found ideal for SPS since it simplifies chemistry. In addition, it prevents poor loading and reduces potential racemization.

5.2.3   *MBHA and Rink MBHA Resins*

From the chemistry of PAM resin, one would recognize the convenience and advantage of loading amino acids onto amine-resin. And that is the rationale for the synthesis of 4-methylbenzhydrylamine (MBHA) hydrochloride resin, which quickly became one of the most favorite resins in SPS because of its optimized conditions toward retaining stability under TFA while increasing lability toward HF. Another feature that distinguishes this resin from others that we discussed so far is that the carboxylic group of the final product is capped as an amide upon cleavage from the resin using acidic condition. As shown in Fig. 5.7, the reaction condition to generate functionalized resin can be achieved using a wide range of chemical transformations, albeit without the need to consider purification (Matsueda and Stewart 1981). First, the Friedel–Crafts acylation reaction started with the generation of a ketone on the resin's aromatic ring. Next, the ketone was converted to an amine through the reductive amination of the carbonyl group using the Leuckart reaction, mediated by in situ-generated ammonium formate. The coupling of the first amino acid was achieved by standard coupling procedure using DCC along with antiracemization reagents, such as HOBt (hydroxy benzotriazole) or HBTU (2-(1H-benzotriazol-1-yl)-1,1,3,3-tetramethyluronium hexafluorophosphate).

MBHA resin became popular in SPS for many reasons. The chemistry is compatible with Fmoc/tBu-amino acids. The advantages of Fmoc chemistry include less deleterious deprotection condition, using 20% piperidine instead of using TFA for the deprotection found in Boc chemistry. The reaction condition associated with Fmoc chemistry is mild and economical. Figure 5.8 demonstrated the use of MBHA resin for the synthesis of a peptidomimetic version of the diazepine derivatives using the Fmoc approach. The synthesis started with the reduction of an enamine, formed by the reaction between the N terminal amino group of MBHA-linked aspartic acid and an aldehyde by NaBH$_3$CN (Nefzi et al. 1997). The coupling of the secondary amine to the next Fmoc-amino acid mediated by HATU did not go to completion, and thus, doubling the coupling cycle is necessary to afford good yield. After the formation of the dipeptide, the Fmoc was removed in 20% piperidine in DMF, followed by the second reductive alkylation using the same condition. After deprotection of the *t*-Bu group in 60% TFA, the thermodynamically favorable cyclization occurred to form the stable 7-membered ring between the secondary amine and the

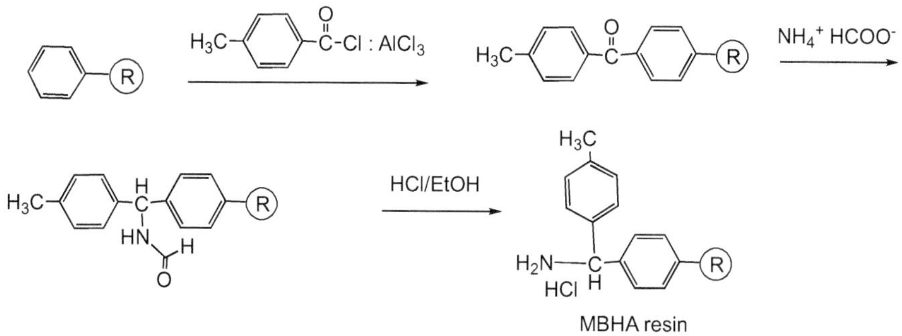

Fig. 5.7 Synthesis of MBHA resin. Data obtained from Matsueda and Stewart (1981) with permission from Elsevier

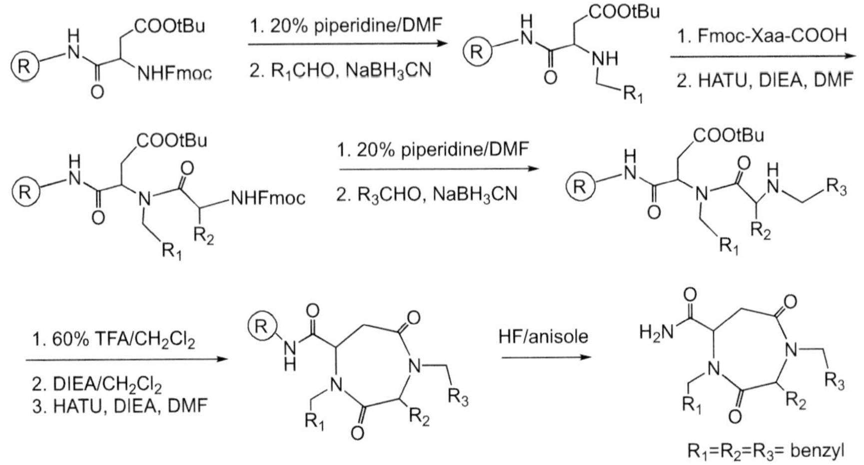

Fig. 5.8 Solid-phase synthesis using MBHA resin for the development of 1,3,4,7-tetrasubstituted perhydro-1,4-diazepine-2,5-diones using Fmoc-aspartic acid as a starting material. Data obtained from Nefzi et al. (1997) with permission from Elsevier

resulting carboxylic group, mediated by HATU. Finally, the peptidomimetic peptide was removed from the resin using HF in the presence of anisole as a scavenger. Aside from HF, the substrate product can be detached from the MBHA resin under a much milder condition, such as diluted HBr in TFA in the presence of pentamethylbenzene and thioanisole (Wang et al. 1992).

The Rink MBHA resin was developed based on the initial idea of the MBHA counterpart to overcome the issue of using strong acidic conditions for cleavage. Currently, two types of Rink resins are available for versatile application in SPS. The Rink acid resin can be removed from the peptide with merely 1–5% TFA in methylene chloride in 5–15 min, or it can also be removed in 10% acetic acid in methylene chloride over the course of 2 h. As the name suggested, the peptide product would have the carboxylic group. While Rink amide MBHA resin can be cleavage in 50% TFA in methylene chloride in 1 h. The C terminal end of the separated product is capped as an amide.

5.2.4 Wang Resin

Another resin typically used with Fmoc chemistry is p-alkoxybenzyl alcohol resin, Wang resin. An orthogonal protection strategy can be employed using Wang resin, where the Fmoc-amino acids, with the sidechains protected with t-Bu groups, which can be removed concomitantly under acidolytic cleavage of the resin. While the Fmoc is removed in each cycle using mild basic conditions. If considering the different acidic lability between *t*-Bu and 2-(4-biphenyl)isopropoxycarbonyl (Bpoc) (Fig. 5.9) in the design, there appears to have a number of different chemical maneuverings using Wang resin to design complicated structures. Different from the *t*-Bu group that requires moderate acidic conditions for the deprotection, Bpoc can be removed with as much as 0.2–0.5% TFA. This means the anchoring bonds are stable if Bpoc is used for amino protection during the synthesis. As a matter of fact, it has been demonstrated that the Wang resin anchoring bond remains unaffected after exposure to 0.5% TFA for 10 h.

What is most noteworthy about Wang resin is its good stability to many reaction conditions, and it can be easily removed in the final step. The cleavage step is simple, and it requires a short exposure to the mild acidic condition compared to its predecessors. In a typical synthesis, Wang resin can be removed in 50% TFA for 30 min, making it more universal to any peptide. Further, the anchoring chemistry enables to obtain peptides with greater purity. Tracing back to the original design, Wang resin was actually derived from Merrifield resin, by reacting the latter with methyl 4-hydroxybenzoate, followed by reduction using $LiAlH_4$ to provide the desired p-alkoxybenzyl alcohol resin (Fig. 5.10) (Wang 1973). Another synthetic possibility to develop Wang resin involved the treatment of Merrifield resin with 4-hydroxybenzyl alcohol in the presence of sodium methoxide.

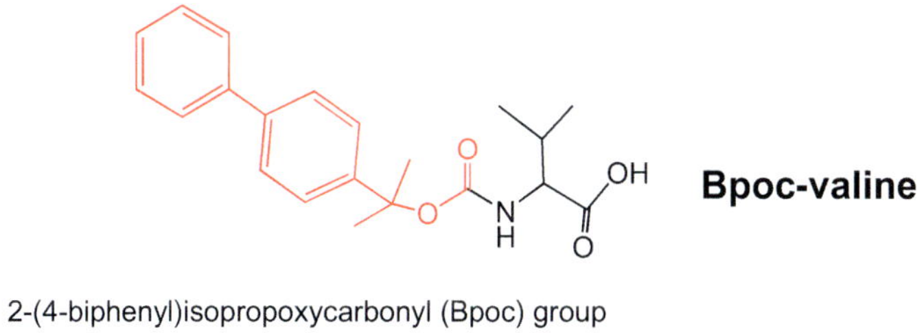

Fig. 5.9 Example of how Bpoc protects the amino ($^{\alpha}$N) group

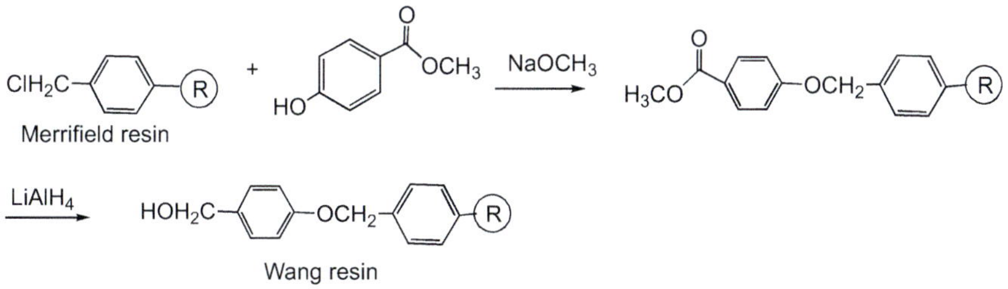

Fig. 5.10 Development of Wang resin. Data obtained from Wang (1973) with permission from the American Chemical Society

5.2.5 *2-Chlorotrityl Resin*

2-chlorotrityl chloride resin is another recognized resin suitable for use with Fmoc/t-Bu chemistry. This resin is similar to Merrifield resin in many aspects. Thus, the first amino acid loading can be achieved using the same protocol as discussed above with Merrifield resin. The loading of the first amino acid is accomplished with triethylammonium salt of the Fmoc-substrate/amino acid. This highly acidic labile resin is a perfect candidate for the synthesis of unstable amino acids or compounds that need gentle care, such as fluorescence dyes. The resin can be removed easily with 0.5–1% TFA in methylene chloride (Bollhagen et al. 1994).

The chemistry for developing 2-chlorotrityl chloride resin was improved after its introduction many years earlier (Barlos et al. 1991a, b). As shown in Fig. 5.11, in situ-generated phenyl lithium was transferred to a solution of 2-chlorobenzoyl polystyrene in tetrahydrofuran, followed by refluxing for 1 h. After the completion of the reaction, the anticipated 2′-chlorotriphenylcarbinol on the resin was washed with dimethylformamide and filtered. Next, the 2′-chlorotriphenylcarbinol resin was treated with trimethylsilyl chloride in methylene chloride and dimethyl sulfoxide with moderate stirring for 90 min to furnish the desired product. However, unpleasant odor occurs, probably due to the association with dimethyl sulfoxide. The wet resin was then treated with acetyl chloride in benzene, followed by refluxing for 30 min to afford the product (Orosz and Kiss 1998).

5.3 Resin Substitution

Assessment of resin substitution enables the determination of SPS batch size and other parameters, such as the number of amino acid equivalents for that particular resin. As the total number of active sites on a given resin is expressed in millimoles per gram of resin, the substitution would have the same units to reflect the inverse proportion of these two relationships. For example, when the substitution increases, there will be an increase in the molecular weight of the resin. In that case, the number of active sites in the resin will decrease. In general, the resin substitution can be calculated by differences in mass. In this approach, the resin should be genuinely dried carefully without interference from the solvent; particularly, the high-boiling-point

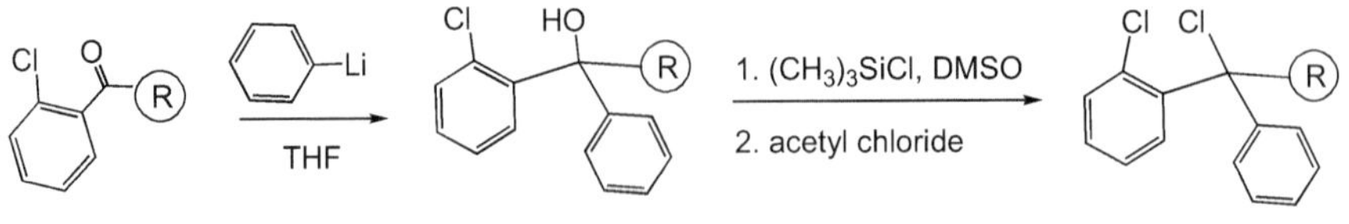

Fig. 5.11 Development of 2-chlorotrityl chloride resin. Data obtained from Orosz and Kiss (1998) with permission from Elsevier

solvent, such as DMF, must be removed completely. This can be done by stripping DMF from association with the resins several times with methylene chloride, followed by drying the resin under a high vacuum. The weight gain substitution (S) is expressed as

$$S = \frac{Wt_{gained} + Wt_{added}}{Wt_{total}} \times 1000$$

where Wt_{gained} and Wt_{total} are weight gained by resin and total weight of resin, respectively, expressed in grams. While Wt_{added} means gram/mol added to the resin. The value of 1000 is the conversion value from mol to mmol since the overall value of S is expressed in mmol per gram (Christensen 1998). This method of calculation is applicable to any loading amino acid.

Aside from this method, UV/V is spectroscopical analysis of resin substitution can be derived using Beer's law if the resin is associated with a Fmoc group. As suggested in Fig. 5.12, substitution calculation can be assessed by treating the Fmoc-protected starting resin with 20% piperidine in DMF to generate a dibenzofulvene-piperidine adduct. This chromophore has the absorbance lambda maxima and molar extinction coefficient at 301 nm and 7100–8100 mol^{-1} cm^{-1}, respectively (Eissler et al. 2017).

The absorption values are determined by either absorbance maxima, along with their respective molar extinction coefficient, to calculate the substitution of Fmoc-protected resins (Meienhofer et al. 1979). The Fmoc substitution can be quantified using modified Beer's law as shown below (Eissler et al. 2017):

$$S_{Fmoc} \ (mmol/g) = \frac{Abs_{289.8} \times V \times D \times 10^6 \ (mmol/mol) \ (mg/g)}{\varepsilon \times W_r \times l}$$

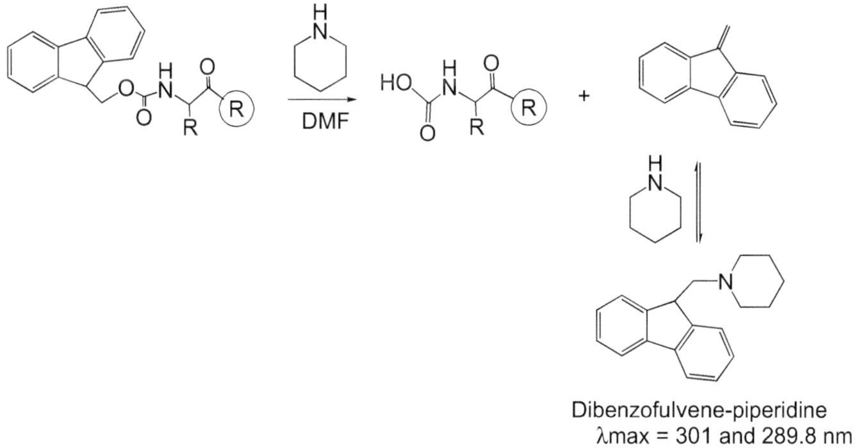

Fig. 5.12 Generation of dibenzofulvene-piperidine chromophore and quantitative analysis to determine resin loading. Data obtained from Eissler et al. (2017) with permission from John Wiley and Sons

where V represents sample volume, this is the volume of 20% (v/v) piperidine in DMF without dilution. D represents the dilution factor, and 10^6 is the conversion values from mol to millimole and gram to milligram. While epsilon (ε) represents the molar extinction coefficient, W_r is the sample weight of the resin expressed in milligrams. Finally, l is the optical path length of the cell in centimeters. The resin substitution for a number of polystyrene resins at 289.8 and 301 nm has been performed. The values vary about 14% between these two lambda maxima, and these substitution values are from 0.4 to 0.6 mmol per gram.

5.4 Solid-Phase Synthesis of Active Compounds and Dyes

After acquiring knowledge about the resins, how to load substrate molecules on them, and methods used for resin removal upon completion of the synthesis, the lecture now focuses on the applications of these resins. In this discussion, peptides and small organic molecules will be mentioned. The goal is to provide key experimental designs that can be applied in practical laboratory work, hoping that these examples will also serve as a catalyst for future breakthrough discoveries in probe development. The following model illustrates how SPS shapes the organic synthesis landscape for making small molecules and organic dyes. It is undoubtedly true that the organic synthesis of small molecules has been profoundly extended by the integration of conventional organic synthesis into solid-phase chemistry. Up to date, SPS can be performed by virtue of any reactions that are available in the solution-phase chemistry.

5.4.1 Cyclic Peptides

In this model experiment to demonstrate the versatile application of SPS, a cyclic dipeptide was synthesized via SPS using a number of creative operations more than just sequential attachment of the amide bonds. First, this diketopiperazine catalyst containing unnatural amino acid norarginine was found to display unique catalytic activity to facilitate the enantioselective Strecker synthesis of (S)-phenylglycine derivatives using N-substituted aldimines and hydrogen cyanide (Kowalski and Lipton 1996). The solution-phase development of (L)-norarginine has multiple issues related to extended reaction times, poor reaction yield, and solubility, along with significant side reactions. It is anticipated that by adopting the chemistry into SPS, these issues will be circumvented (Kowalski and Lipton 1996). The synthesis commences with Boc-L-phenylalanine immobilized on Merrifield resin (Fig. 5.13). The incorporation of protected glutamine was achieved without much effort using HBTU (2-(1H-benzotriazol-1-yl)-1,1,3,3-tetramethyluronium hexafluorophosphate) as a coupling reagent. The side chain amide moiety of glutamine was converted into the primary amine through Hofmann rearrangement using the oxidizing reagent of

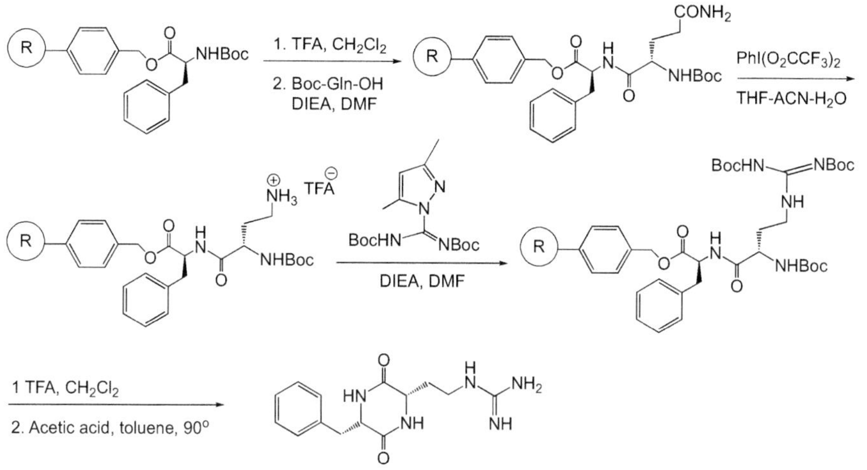

Fig. 5.13 Solid-phase synthesis of a diketopiperazine catalyst containing the unnatural amino acid (S)-norarginine. Data obtained from Kowalski and Lipton (1996) with permission from Elsevier

bis(trifluoroacetoxy)iodobenzene. In this condition, the reaction afforded the amino product with near-quantitative yield, as determined by the ninhydrin test. For the guanylation reaction, the guanidine was synthesized in the protected form to provide a consistent and high-yield product. Based on solubility and side reaction issues cited above, performing the SPS with acylated guanidine is more efficient than the synthesis of the parent species. Finally, the Boc groups were deprotected in TFA, and diketopiperazine cyclization was achieved after the resin was removed in acetic acid. First, it is worth mentioning that incorporating the guanidine synthesis in the solid-phase chemistry permits the use of organic-soluble guanylation reagents in the process, thus eliminating several issues. Second, no hassle related to purification of the product, thus eliminating many of the side products usually encountered in the original solution-phase synthesis.

Combinatorial synthesis is another area in which SPS has a unique contribution as a tool for discovering biologically active compounds. What was arguably one of most limitations in solution-phase chemistry when it comes to multi-component reactions came about the purification sagas, which can be just one among of many other issues. For example, bicyclic proline derivatives have been prepared in the solution-phase involving a three-component strategy (Aly et al. 1994). Theoretically, if the aldehyde, the maleimide, and the amine react equally, then the proline analog should be formed (Fig. 5.14). However, the presence of more than one component leads to side reactions between the various components and makes the process of purification and identification of the desired product more challenging. By derivatize one of the components onto the solid support, a mixture of resin-bound material could be employed to obtain a defined mixture of products, which can be analyzed and separated by LC–MS and HPLC, respectively (Hamper et al. 1996). As shown in one of such reactions (Fig. 5.14) to test this concept, the substituted benzaldehydes were anchored to the benzyl alcohol of Wang resin via the Mitsunobu reaction. The reaction proceeds with clean inversion and the resin substitution ranging from 0.24 to 0.89 meq/g. In the next step, the Mitsunobu product on the resin was treated with leucine methyl ester and N-phenylmaleimide in DMF in a sealed vial and heated to 100 °C. The desired product was removed from the resin by exposure to TFA and characterized by LC–MS without the need for purification. This reaction demonstrates that solid-phase chemistry can be performed in similar conditions as solution-phase chemistry, meaning that the immobilized material on resin can sustain under high pressure and temperature. Further, the reaction mixture can be stirred vigorously to assist with getting all the starting material into the solution as the reaction proceeds.

In a combinatorial synthesis feat, nine different aldehyde resins were treated with an equimolar amount of alanine methyl ester and N-phenylmaleimide to form a mixture of products, which were analyzed by the mass spectral identification of the molecular ions of the chromatographic peaks. Since each aldehyde resin reacts with single amino acid and maleimide, this multiplexed reaction proceeded neatly like a single-component reaction.

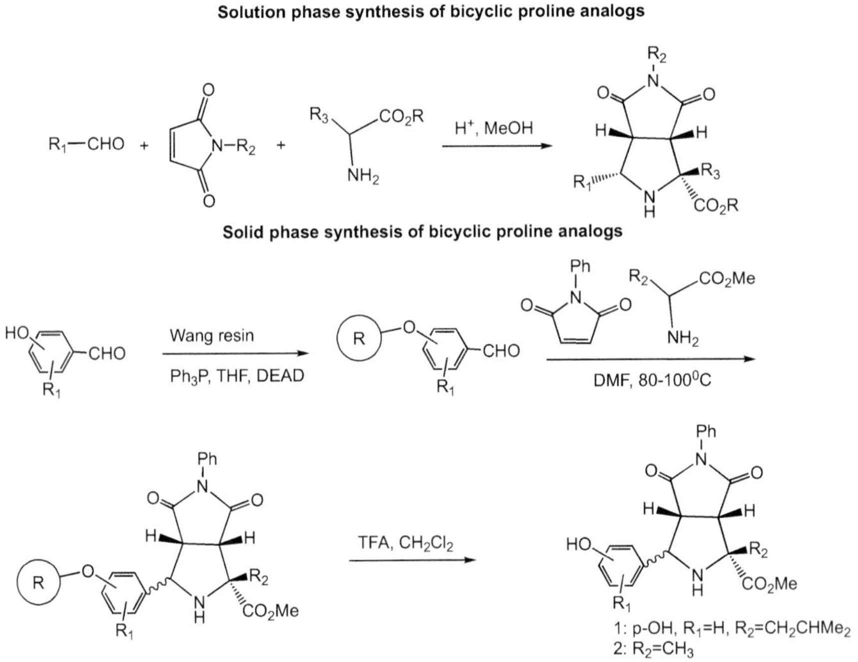

Fig. 5.14 Synthesis of proline analogs, a comparison between solution and solid-phase chemistry. Data obtained from Hamper et al. (1996) with permission from Elsevier

5.4.2 Indole Analogs

Substituted heterocyclic analogs are building blocks for the construction of more highly structural molecules with great diversity for use in therapeutics, diagnostic, and high-technological applications. Therefore, developing chemistry methods to incorporate the synthesis of these compounds, particularly the indoles, in a high-throughput manner via solid-phase chemistry attracts a great deal of interest. As mentioned in Chap. 2, to make indoles for practical dye synthesis, 2,3-substituted indole derivatives should be the target. This targeted chemistry started with the development of resin-anchored iodophenylsulfonamide intermediate (Fig. 5.15). The Fischer indole cyclization reaction via palladium-mediated heteroannulation of terminal alkynes transformed the resin-bound phenylsulfonamide into the key intermediate (Zhang et al. 1998). It is worthwhile to mention that the indole developed in this scheme is attributed to the presence of the sulfonamide moiety. The previous attempts to develop the same indoles using resin-bound aniline as a starting material were unsuccessful. It seems the organopalladium intermediate generated from aniline nitrogen is not reactive enough to cyclize into the indole. In contrast, when the amine is activated with a strong electron-withdrawing group, such as sulfonyl, the sp^2-sp coupling and indole cyclization can occur promptly under a mild reaction condition (Zhang et al. 2000).

Next, the sulfonamide was hydrolyzed easily by using 2% KOH. With commercially available indole 5- or 6-carboxylic acid coupled to Rink amide resin, the efforts for generating indole analogs, one of the key components for making cyanine dyes, seem never been easier. Further, using the robust Mannich reaction, a countless number of indoles with different substituents can be generated for pharmaceutical and dye research. Particularly, water-soluble indoles can be derived from this approach,

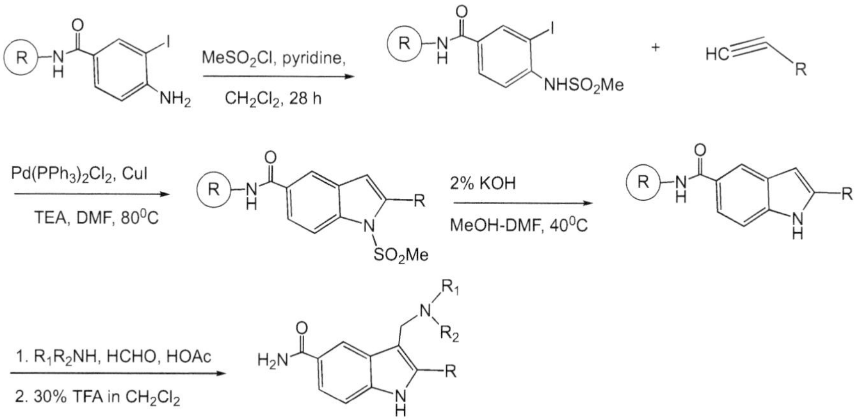

Fig. 5.15 SPS of 2,3-disubstituted indole analogs using Fischer indole synthesis, with the substitution facilitated by Mannich reaction. Data obtain from Zhang et al. (1998) with permission from Elsevier

given that SPS will overcome the issue of purification of water-soluble compounds. The reaction proceeded not only very smoothly but also furnished the product with excellent yields and purity.

Yet, this is just the beginning of what could be a very interesting process, and there are rooms for modification. For example, one of the recognized limitations of SPS of indoles is that the linkage to the resin via the amide, ester, or ether handles, which eventually becomes a remnant remaining in the final products after the compound is dissociated from the resin. Based on the interpretation of Fig. 5.15, how sulfonamide helps to induce the process of indole cyclization, and subsequent removal of the sulfone completely from the indole, the idea for a traceless sulfonyl linker was perceived. Basically, the position where the resin is attached in the indole structure is switched from the 6-membered ring to the nitrogen of the 5-membered ring, and thus, the sulfonyl linker would serve two purposes in this new construct. First, it will facilitate indole cyclization; and serve as a good substrate for cleavage under mild conditions (Zhang et al. 2000). In a proof-of-principle experiment, the resin-bound indole where the solid support attached to the 5-membered ring can be developed starting by loading 2-iodoaniline onto either polystyrene-based PS-TsCl resin or Merrifield resin-based sulfonyl chloride to provide resin-bound sulfonamide intermediate (Fig. 5.16). Next, this intermediate is treated with phenylacetylene and Pd(PPh$_3$)$_2$Cl$_2$ in a catalytic amount, along with CuI to furnish the resin-laden indole product. In this new construct, the resin is bound to a different position compared to the previous approach. The resin removal via the sulfonamide using the inorganic base such as KOH seemed to be sluggish. However, tetrabutylammonium fluoride (TBAF) cleaved the resin with an excellent yield to provide the desired indole.

This traceless linker concept was further expanded for the development of more complicated indole structures, which might contribute to imaging technology in manifold ways. As mentioned in Chap. 2, cyanine dyes are built based on indole rings, in which three functional groups are required to present in the indoles. These are necessary for the extending the polymethine bridge, increasing water solubility, and providing a handle for bioconjugation. These trisubstituted indole analogs play crucial roles in designing cyanine dyes with multimodal features contributing to in vivo imaging otherwise difficult to realize. As shown in Fig. 5.17, the 2,3,5-trisubstituted indoles have been synthesized in merely three steps starting from the resin-bound aniline (Wu et al. 2001) in this innovative design; it is undoubtedly to say that this work richly deserves all the attention from the imaging community. Like the process described above, 4-bromo-2-iodoaniline was loaded onto the PS-TsCl (polystyrene sulfonyl chloride) resin to provide the resin-bound sulfonamide key intermediate, followed by coupling to the terminal alkyne to initiate the intramolecular cyclization using a catalytic amount of palladium complex to form the resin-bound indole. Regioselective acylation at the C-3 position was successfully achieved via the Friedel–Crafts reaction. For solid-phase modification on the C-5 position, Suzuki coupling can be used to introduce the corresponding alkyl/aryl group. While alkyne-indole analogs can be obtained with more reproducible outcomes using the Sonogashira coupling reaction, particularly, this reaction is suitable for the insertion of terminal alkynes.

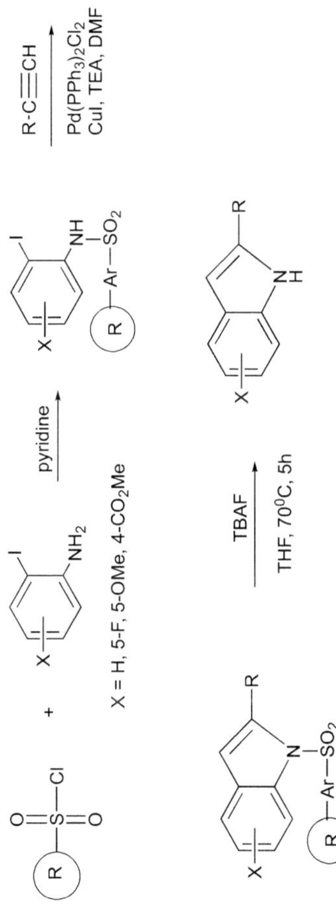

Fig. 5.16 SPS of indole derivatives using traceless and activating sulfonyl linker. Data obtained from Zhang et al. (2000) with permission from the American Chemical Society

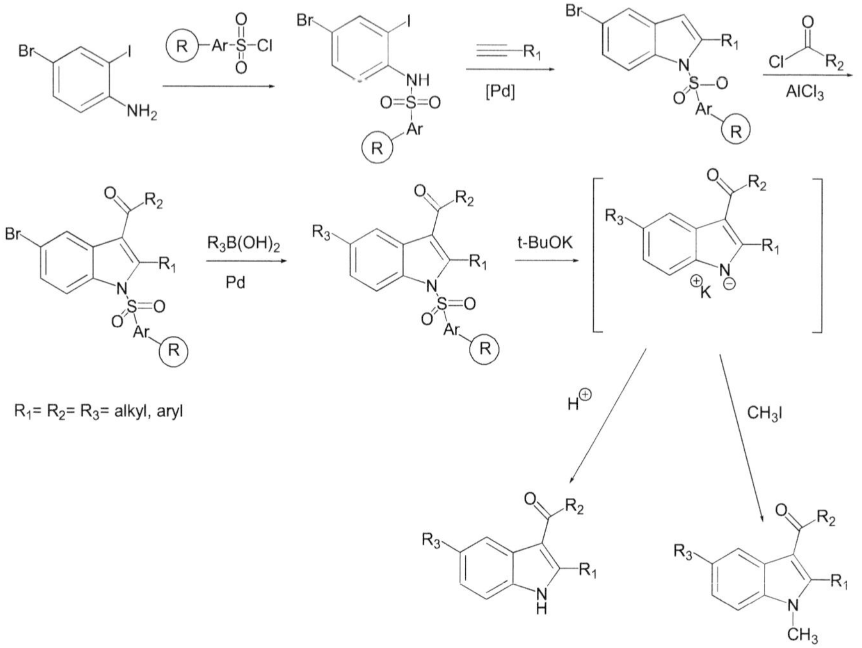

R₁= R₂= R₃= alkyl, aryl

Fig. 5.17 SPS of traceless linker-based development of 2,3,5-trisubstituted indoles. Data obtained from Wu et al. (2001) with permission from the American Chemical Society

5.4.3 Cyanine Dyes

As mentioned in Chap. 2, the asymmetrical synthesis of cyanine dyes suffered poor yield. And it is often more challenging in purification due to the presence of impurities, including the undesired symmetric dye products. A typical example of this notion is shown in Fig. 5.18. In a typical cyanine dye synthesis, the methylene base of the indole would react with a single-carbon synthon of the N,N′-diarylformamidine to afford the hemicyanine intermediate. The reaction does not stop there; instead, the hemicyanine can go (Fig. 5.18) (Mason and Balasubramanian 2002) all the way to react with excess amount of the indole to generate the undesired product of the trimethine symmetrical dye. In an alternative and preventive approach, the work focuses on capturing the hemicyanine intermediate by the resin-bound sulfonyl chloride. This creative design would selectively immobilize the hemicyanine intermediate, preventing the formation of the contaminating symmetrical dye. Further, when activating the hemicyanine carbon by N-sulfonylation, an electron-withdrawing group would facilitate toward nucleophilic attack from other heterocycles. This activation is more robust than the conventional approach, where activation of hemicyanine by N-acylation. It has been observed in the past that in N-sulfonylation, the less available the nitrogen lone pair is for delocalization, the more reactive the hemicyanine derivative is to nucleophiles (Mason and Balasubramanian 2002; Brooker

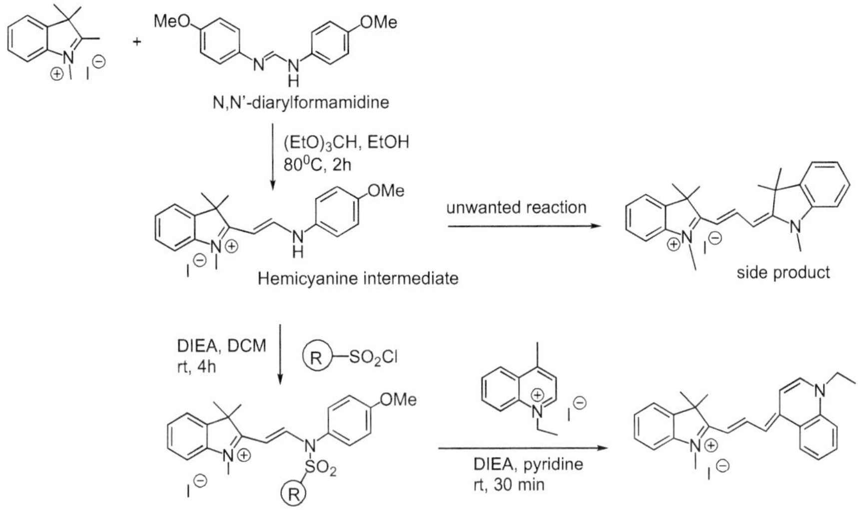

Fig. 5.18 Solid-phase synthesis of asymmetric cyanine dyes. Data obtained from Mason and Balasubramanian (2002) with permission from the American Chemical Society

et al. 1941). The process of hemicyanine capturing by sulfonyl chloride resin can be achieved under an optimal condition using DIEA in dichloromethane at room temperature, affording approximately 90% yield. At the same time, the dye formation with quinoline concurring with the resin decoupling step can be accomplished in DIEA/pyridine for 30 min at room temperature. This versatile chemistry is also suitable for a wide range of heterocyclic systems aside from quinoline.

Using the same resin-supported sulfonyl chloride "catch-and-release" strategy along with sulfonated indole analogs, water-soluble cyanine dyes can be obtained with high yield and no purification efforts. The purity obtained from this method ranges from 63 to 85%, with excellent chemical yields from 42 to 51%, several folds exceeding the conventional solution-phase approach (Mason et al. 2005). This is probably one of the signatory advantages of SPS. Given purification of water-soluble dyes is usually a tedious and time-consuming task. Most of cases, HPLC is only the tool of choice, and thus, cost-ineffectiveness only exacerbates the problems associated with the development of water-soluble dyes in a conventional approach.

Yet what makes this method so remarkable and earns some attention is not only the quality of the dyes that can be synthesized per se, but the highly versatile way in which it can be applied. If the resin-supported sulfonyl chloride is incorporated on the other end of the longer polymethine carbon chain, asymmetrically near-infrared fluorescent dyes will be formed as a result.

This synthesis has great implications in biomedical imaging, where dyes can be fine-tuned to improve tissue penetration and reduce autofluorescence. The successful synthesis of cyanine dyes with longer wavelengths relies on using several known malonaldehyde bisphenylimine salts to build the polymethine bridge. There are intermediates that are suitable for generating Cy5 or Cy7 dyes. The hemicyanine intermediate was immobilized by sulfonyl chloride resin in a similar manner as described in dimethine hemicyanines (Fig. 5.19). Nevertheless, coupling the second heterocycle to the hemicyanine with longer methine analogs is quite different from the dimethine counterparts. It is found that if the reaction occurred in 30 min with a reducing amount of organic base (from 10 to 1%), the pentamethine cyanine (Cy5) dye could be obtained with over 95% yield. In contrast, extended reaction time resulted in poor yield due to the hydrolysis of the resin-bound hemicyanine to form a side product.

Cationic hemicyanine dyes are another class of attractive probes used to detect anionic nucleotides, assuming the electrostatic interaction represents the key mechanism for this union. Nucleotides play an essential role in metabolism and cell signaling at all levels of cellular activities. The process includes extracellular signaling via the cell surface receptors to sense the input information regarding the environment for the well-being of cells and receiving messages conveyed from other cells. In this regard, it is worthwhile to discuss G-protein coupled receptors (GPCRs) as a model. Indeed, GPCRs play a major role in signaling in response to diverse inputs and regulate a plethora of cellular processes (Wootten et al. 2018). These receptors are implicated in several diseases ranging from whooping cough to cancer and neurodegenerative diseases. Thus, it is not surprising that GPCR-related drugs occupy, in an estimate, between one-third and one-half of all marketed drugs. The superfamily of

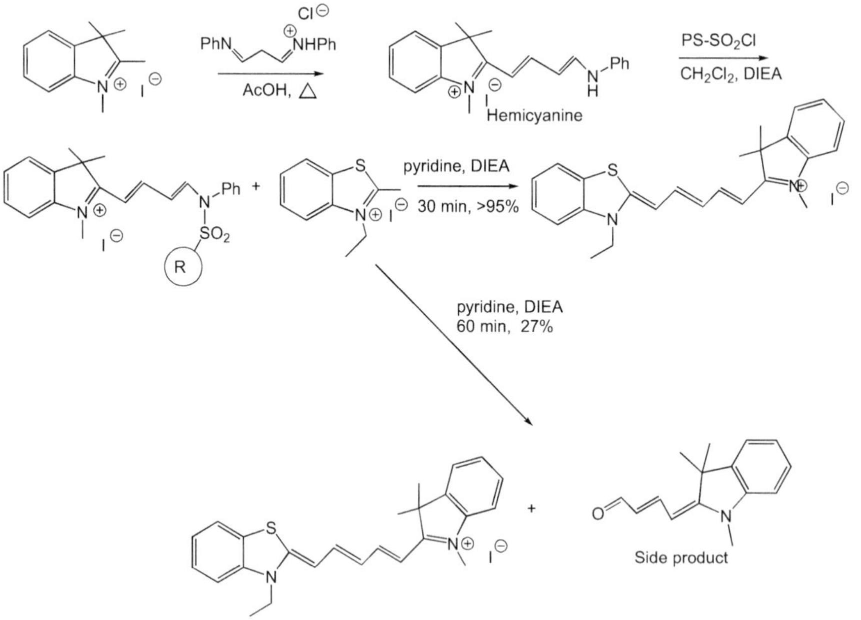

Fig. 5.19 Solid-phase catch-and-release synthesis of pentamethine (Cy5) dye. Data obtained from Mason et al. (2005) with permission from the American Chemical Society

GPCRs includes approximately 800 receptors; however, the endogenous ligands of more than 140 GPCRs remain unknown, and these are often called orphan receptors (Tang et al. 2012). Therefore, the development of molecular sensors for the detection of the dynamic interactions between regulatory proteins with nucleotides in on/off modes is of paramount importance for drug discovery projects. For example, in the inactivated form heterotrimeric G-protein, which is comprised of three subunits $G\alpha,\beta/\gamma$ subunits, stay together with the $G\alpha$ binds to GDP. Upon activation by a GPCR, GTP physically displaces GDP bound to the $G\alpha$ subunits and separates it from the $G\beta/\gamma$ dimers, and each can activate downstream signaling. Any interference with this signaling will result in severe consequences. Thus, the steadily increasing demand for these dyes for monitoring nucleotide binding is the reason behind the development of a solid-phase approach ensuring the combinatorial synthesis of a large library of compounds can be achieved not only in a high-throughput manner but also guarantees high purity and cost-effectiveness. One of the approaches shown in Fig. 5.20 uses benzimidazolium analog, fixed on the resin for diversification of the other half structure of the dye spanned over by a single methine bridge. Through a simple condensation of resin-bound and substituted benzimidazolium with a variable number of aromatic aldehydes provides a large library of nucleotide fluorescent sensors. The first task in this design involves the alkylation of benzimidazole with any desired alkylating reagents. The purpose is multifold, it will convert the molecule into benzimidazolium salts, with the goal of making cationic hemicyanine dyes for detecting of anionic nucleotides. Further, the extended alkylated moiety provides a handle helping to immobilize the salts onto the resin. The condensation of the resin-bound benzimidazolium intermediate with an aromatic aldehyde proceeded smoothly in N-methyl-2-pyrrolidinone (NMP) and pyrrolidine as an organic base. Finally, resin cleavage in mild acidic conditions to provide the dye. Although this is considered a Cy2 dye, whose emission is in the blue range, the wavelength can be fine-tuned with a bathochromic shift in the fluorescence by manipulating the electronic effect. In one example, introducing two chloro moieties to the benzimidazolium structure resulted in the emission in the green–red range (Wang and Chang 2006).

5.4.4 Rhodamine Dyes

As mentioned in Chap. 2, cyanine and rhodamine dyes are the backbone and driving force behind most of the development of optical probes for biomedical imaging work. It is not surprising to see several great SPS of rhodamine dyes about the same time as seen in cyanine. Rhodamine dyes are compatible with SPS because of their recognized stability. One of the first reports of SPS for rhodamine dyes involved the use of solid-phase materials for improving dye purification. In this work, SPS was employed to solve the long-standing issues with the isolation of rhodamine B after modification. This type of dye has a penchant for forming a spirolactam ring via cyclization in alkaline pH. In this "closed" structure, the dyes quench the fluorescence signal and significantly decrease absorbance, thus preventing the dyes from serving

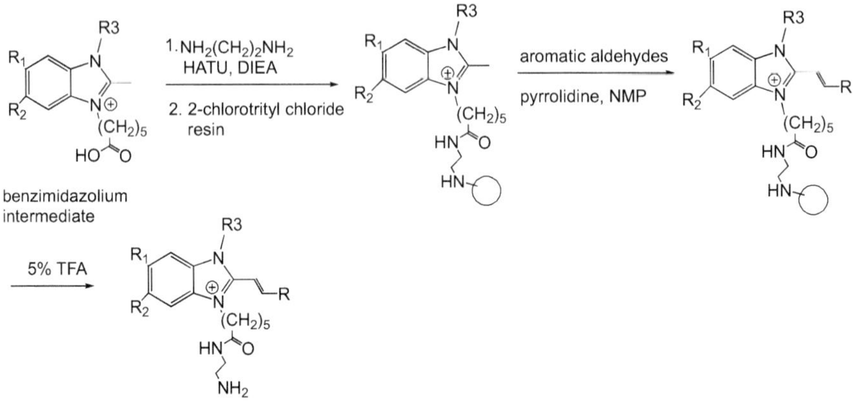

Fig. 5.20 Solid-phase synthesis of cationic hemicyanine dyes for the detection of nucleotide. Data obtained from Wang and Chang (2006) with permission from the American Chemical Society

as a contrast agent. One of the approaches to prevent lactamization is to block it as an N, N-disubstituted amide functional group (Bossi et al. 2006). In the earlier days, piperazine was considered as an effective structure for creating the tertiary amide, inhibiting lactamization in rhodamine dyes by conversion of the esters to amides (Basha et al. 1977). This approach is ideal for the large-scale synthesis of active dyes utilizing the economical and commercially available rhodamines. However, it is challenging to isolate the amidated product from the starting dye materials. In addition, the highly polar character of the amidated rhodamine products made conventional purification using silica gel difficult. To overcome this issue, Wang resin was functionalized with 4-benzyloxybenzyl alcohol to capture the piperazine moiety on the dye (Fig. 5.21). After washing to remove the starting materials and other reagents, the desired chromophore was released from the resin while exposed to a slightly acidic condition in methylene chloride (Nguyen and Francis 2003).

5.4.5 Fluorescein Dyes

Like rhodamine dyes, the fluorescein counterparts are also great candidates for incorporation into the solid-phase platform. Given their compact structure, robust labeling of peptides using these dyes provided an essential tool for biochemical and biomedical research. For instance, in fluorescence polarization assays, protein–protein interactions can be detected and quantified when the interaction is deduced to protein-peptide interactions. A fluorescence-labeled analog of a peptide of interest and its chemically mutated sequence can be used as a probe to screen for binding specificity or inhibitors using fluorescence polarization competition binding assays. Fluorescein was selected in this design for not only its compact size but also because the fluorescence lifetime of fluorescein and its related analogs are easily reproducible regardless

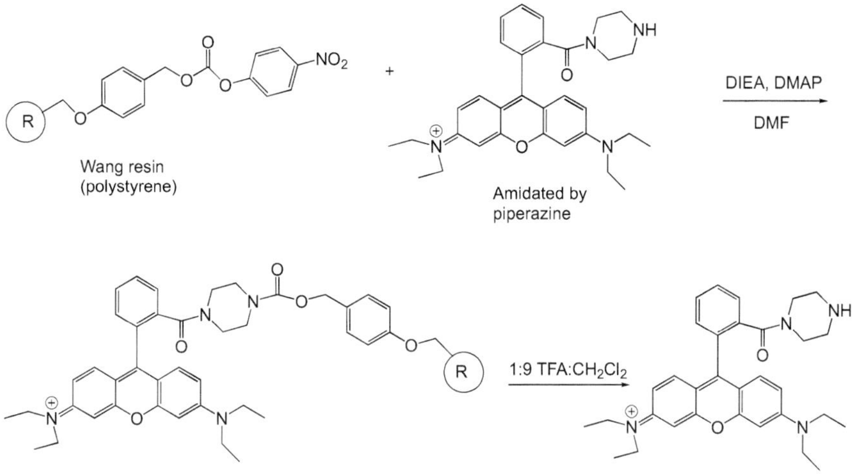

Fig. 5.21 Trapping the amidated rhodamine dyes among the undesired materials via Wang resin to improve purification. Data obtained from Nguyen and Francis (2003) with permission from the American Chemical Society

of the change in the emission wavelength and the dye concentration. The fluorescence of fluorescein dyes is also very consistent; thus, they are usually preferred for several applications (Zhang et al. 2014). For instance, the aminofluorescein dye used in this work has been shown that, if acylated, the compound exerts a long emission wavelength (519 nm), far away from the autofluorescence generated by the screening molecules, and thus reducing the concern of potential false positives. The other advantage of using aminofluorescein in solid-phase chemistry is that it can be treated in the same manner as a building block as of amino acid, meaning that the 2-chlorotrityl chloride resin can be incorporated in one of the hydroxyl group while the aniline nitrogen is protected with a Fmoc group (Fig. 5.22). Obviously, from the scheme that without protecting the aniline amine, direct coupling of the resin would go to the amino group exclusively. Further, before initiating amino acid coupling, the other hydroxyl group on the dye should be blocked. As the compound is now immobilized on the resin, the protection step using MEM-chloride (2-methoxyethoxymethyl) can be achieved with ease. After finishing the anticipated dye-labeled peptide sequence using diisopropylcarbodiimide (DIC) with HOBt as an antiracemization reagent, the Fmoc group was removed. Subsequently, the N terminal of the peptide was capped with an acetylated reagent. Finally, the MEM protecting group and the resin were removed simultaneously with TFA with scavenger reagents.

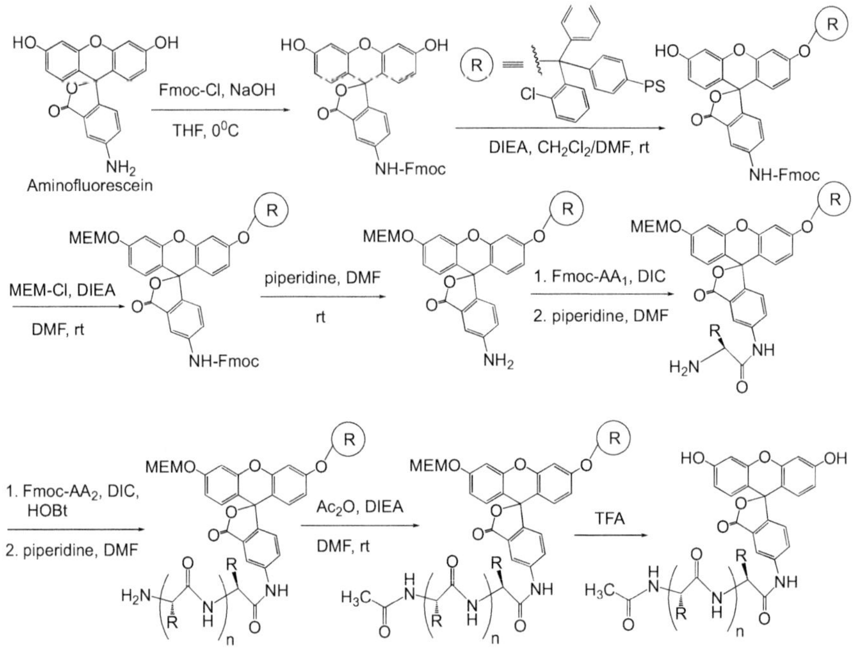

Fig. 5.22 Incorporation of an aminofluorescein dye into SPPS as a building block. Data adapted from Uryga-Polowy et al. (2008)

5.4.6 Coumarin Dyes

Coumarin dyes are another popular fluorochromes worth mentioning, given their huge impact on biomedical imaging research. Different versions of coumarins have been used for cancer detection (Kumagai et al. 2013; Sakuma et al. 2015), as well as therapeutic agents (O'Kennedy and Thornes 1997) and other optical sensing technology (Hemmila 1989). Coumarin, probably one of the most compact dyes that can exert long emission wavelengths, is suitable for molecular imaging. There is also a historical event related to this dye; in 1898, Knoevenagel described the first synthesis of coumarin analogs using the condensation reaction between malonic acid and ortho-hydroxyarylaldehyde; later on, this type of reaction carries his name (Loncaric et al. 2020). Substituted coumarin can be achieved using the same condensation reaction on Wang resin in solid-phase chemistry. It has been reported in the past that the Knoevenagel condensation between malonate-Wang resin and substituted arylalde-hydes resulted in a mixture of E:Z isomers (Hamper et al. 1998). The same holds true in this case, as seen in Fig. 5.23, E:Z isomers were formed after Knoevenagel conden-sation between malonate-Wang resin and the corresponding 2-hydroxybenzaldehyde. During the process of ring closure of the Z-isomer, the resin was cleaved off, and its subsequent coumarin ethyl ester was found in the solution phase. Meanwhile, the E-isomer proceeded to appropriate ring closure, leading to the desired intermediate,

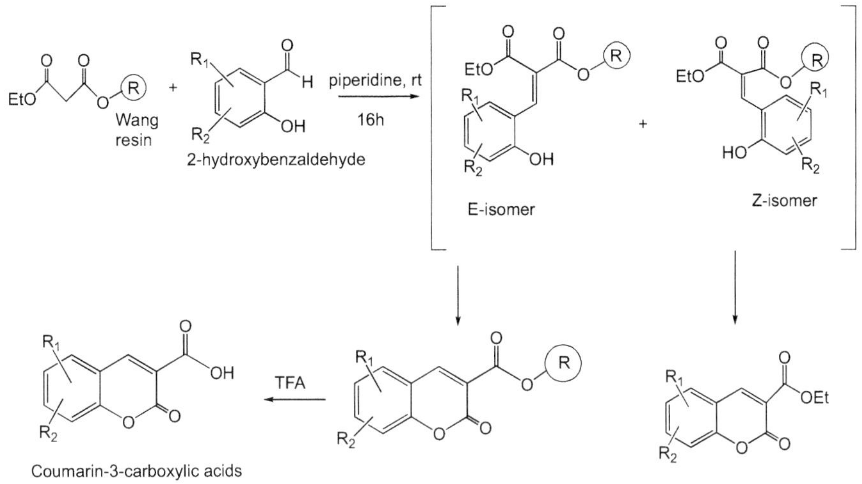

Fig. 5.23 Isomerization during Knoevenagel condensation in solid-phase chemistry to create substituted coumarin dyes. Data obtained from Watson and Christiansen (1998) with permission from Elsevier

which leads to the formation of the desired coumarin-3-carboxylate dye after resin cleavage in TFA (Watson and Christiansen 1998).

Fused coumarin analogs are also known as psoralens, used as photosensitizing agents for the treatment of psoriasis (Parrish et al. 1974), vitiligo (Perone 1972), and other skin disorders using photochemotherapy (Cimino et al. 1985). A number of recognized psoralens have been used for the clinical treatment of psoriatic patients (Langner et al. 1976) (Fig. 5.24). Particularly, 2-alkylthioimidazocoumarin family of compounds has been shown benefits for a wide range of applications, including antiulcer, antitumor, and antiviral capabilities (Cereda et al. 1987; Klimesova et al. 2002; Kugishima et al. 1994; Zarrinmayeh et al. 1999). Given the interest of these dyes, combinatorial chemistry via a solid-phase approach has been developed to diversify the chemical backbone of 2-alkylthioimidazocoumarins to improve the chemical genetics of psoralens as these molecules serve as excellent substrates for large-scale screening. The synthesis employed coumarin carboxylate as a scaffold to immobilize on Rink amide resin (Fig. 5.25). The substituted coumarin was attached to Rink amide resin using coupling reagents of DIC (1,3-diisopropylcarbodiimide) and HOBt (Song et al. 2004). In the next step, aromatic nucleophilic substitution, usually preferred a large excess of amine in the presence of an organic base, however, it is reported that the coumarins are unstable in basic conditions. Thus, optimal aromatic nucleophilic substitution on resin would use 2 equivalents of amine in 5% diisopropylethylamine (DIEA). After aromatic nitro reduction, a major transformation was achieved in solid-phase condition, whereby treating the diamino chromene on resin with 1,1′-thiocarbonyldiimidazole, followed by cyclization to form the fused coumarin product, with almost quantitative yield. The next step to diversify the chemical structures of 2-alkylthioimidazocoumarins was performed using a wide range of

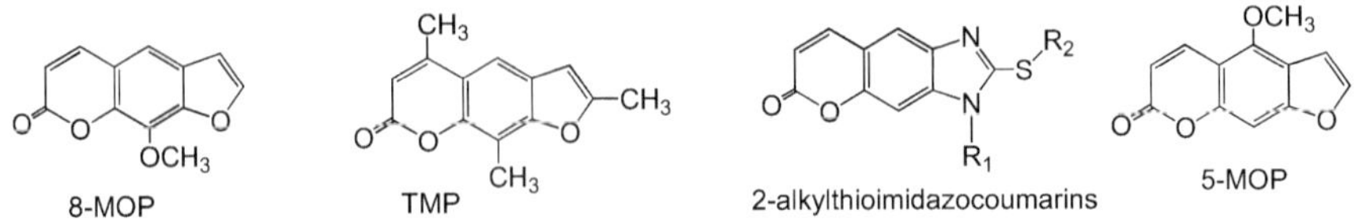

Fig. 5.24 Fused coumarins used as psoralens for photochemotherapeutic applications. Data obtained from Song et al. (2004) with permission from the American Chemical Society

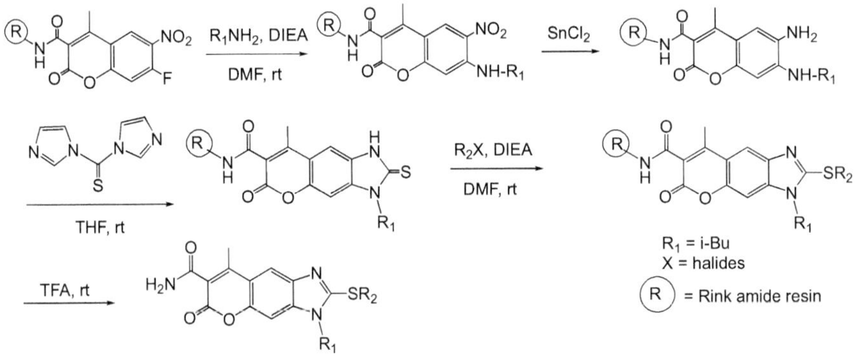

Fig. 5.25 Solid-phase synthesis of fused coumarin dyes. Data obtained from Song et al. (2004) with permission from the American Chemical Society

alkyl halides for the alkylation of benzimidazole-2-thiones. Given the strong nucleophilicity of the thiol group, this reaction can be accomplished with the nitrogen N-1 on the imidazole ring, unprotected. Finally, the desired, high-purity, and good isolated products were obtained after the resin cleavage. Since mild reaction conditions were employed during this 5-step reaction, the prospect of high-throughput synthesis via automation is within reach. Since the design enables 2 chemical mutations (n^2) per compound, it is anticipated a vast library of novel psoralens can be derived from this approach.

5.4.7 BODIPY Dyes

BODIPY dyes are another class of fluorophores that have attracted attention in imaging technology due to their high photostability. Other features that make BODIPY well recognized in the field are related to its excellent optical characteristic. The dyes usually exert a strong extinction coefficient and quantum yield. BODIPY dyes are also suitable for multichannel imaging, given their narrow excitation and emission bandwidths in the vicinity of those of quantum dots. Like every other dye, solution-phase synthesis of BODIPY faces tremendous effort during purification operation to recover the desired product; in most cases, low yield is the main discouraging issue in dye development. Specifically, this problem happened with most of the highly polar dyes. Thus, incorporation of BODIPY synthesis into SPS would overcome the purification issues, aside from the ability to diversify the chemical genetics surrounding the BODIPY scaffold. In this example, one of the first steps to anchor BODIPY dye onto the resin involved the synthesis of aminoethyl-laden BODIPY (Fig. 5.26). In this asymmetric dye design, the main intermediate 2-ketopyrrole was developed using a mild method starting from Fmoc-N-protected aminopropanoic acid, which was converted into 2-pyridylthioester using 2,2′-dipyridyldisulfide and triphenylphosphine in situ. Treating this reaction mixture with pyrrylmagnesium chloride in a one-pot sequential reaction to afford 2-ketopyrrole (Nicolaou et al. 1981). Next, the condensation between Fmoc-protected 2-ketopyrrole with 2,4-dimethylpyrrole in a typical condition using $POCl_3$, followed by in situ addition of $BF_3.OEt_2$ afforded BODIPY dye (Vendrell et al. 2011). After deprotection of the Fmoc group with a strong non-nucleophilic base, such as DBU (1,8-diazabicycloundec-7-ene), the linker with a free amino group serves as a handle for immobilizing the dye onto the CTC-PS resin that enabling modification of the dye at the C3-methyl position using Knoevenagel reaction. Finally, the modified dye can be released from the resin with brief exposure to 0.5% TFA in methylene chloride.

The compact size, stability, and high-quantum efficiency are the trademark of BODIPY dyes. There is no doubt that BODIPY dyes are excellent candidates for developing peptide-based optical probes. In one of the remarkable approaches, the dye was incorporated into an amino acid as a building block for SPS. This strategy enables the versatile synthesis of fluorescently labeled peptides without any additional effort aside from SPPS. As shown in Fig. 5.27, the retrosynthetic analysis for the synthesis of the tryptophan-BODIPY complex would start with the arylation of

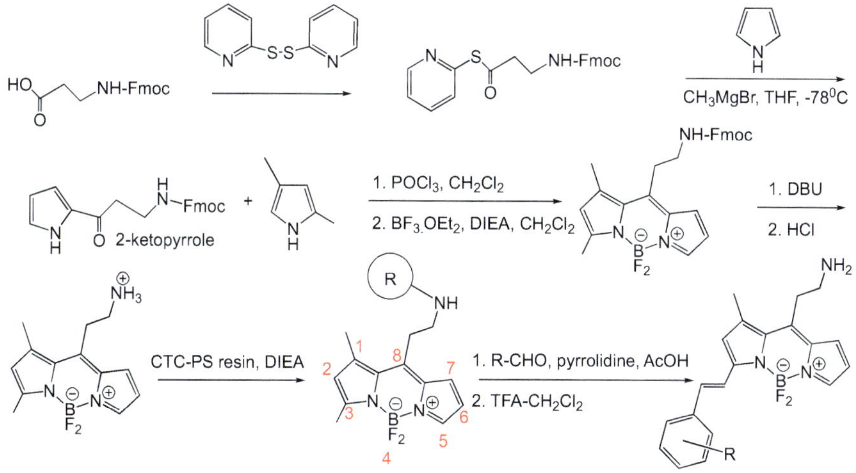

Fig. 5.26 Synthesis of BODIPY analogs via SPS. Data derived from Vendrell et al. (2011)

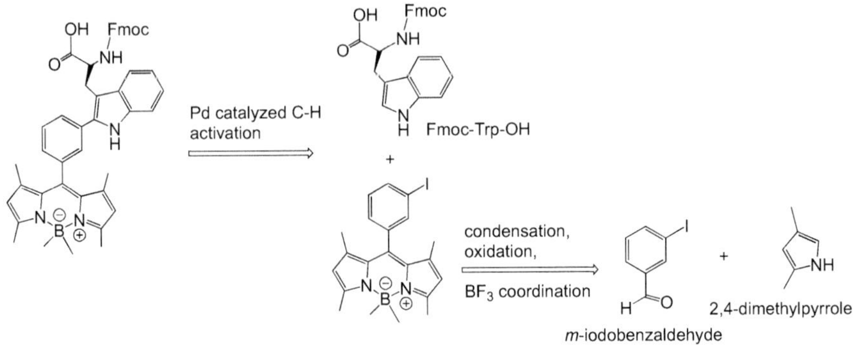

Fig. 5.27 Retrosynthetic design of a BODIPY-Fmoc-Trp-OH for SPS. Data obtained from Mendive-Tapia et al. (2017) with permission from Springer Nature

m-iodophenyl-BODIPY on the unsubstituted C_2 position of Fmoc-Trp-OH via the Pd(II)-catalyzed C–H activation/C–C cross-coupling reaction. While the BODIPY dye can be achieved through the condensation of *m*-iodobenzaldehyde with 2,4-dimethylpyrrole, followed by 2,3-dichloro-5,6-dicyano-*p*-benzoquinone oxidation and BF$_3$ coordination (Mendive-Tapia et al. 2017). The developed BODIPY-Fmoc-Trp-OH is stable as storage in solid form at room temperature over the course of 4 months. The same holds true, even when the compound is dissolved in methanol of dichloromethane. However, the Fmoc group was released from the complex if the compound is stored in DMF over the course of one month at room temperature. This issue can be overcome if storage is at 4 °C or −20 °C. Most importantly, the optical property of BODIPY dye remains stable in all conditions just mentioned. The BODIPY-Fmoc-Trp-OH was suitable for SPPS using 2-chlorotrityl polystyrene resin with Fmoc chemistry using orthogonal building blocks. The mild acidic condition (1:99 TFA:dichloromethane) did not affect the dye stability when it was used to develop a cyclic peptide (BODIPY-cPAF26) with >99% purity (Fig. 5.28).

5.4.8 Solvatochromic Dyes

So far, we have discussed the use of resin for the synthesis of molecules that required repeated procedures. The technique can be employed to overcome the purification issues. Still, derivatization of ligands or dyes onto resin may also have other applications, particularly in the area of immunosensor and surface sensing for biological and environmental applications, respectively (Liu et al. 2017, 2013). This technology relies on a unique family of dyes, usually termed fluorescence chemosensors, meaning that the emission of these fluorophores changes in response to a binding event (Mello and Finney 2005). One example of this class of compounds called solvatochromic fluorophores, whose fluorescence color would change depending on

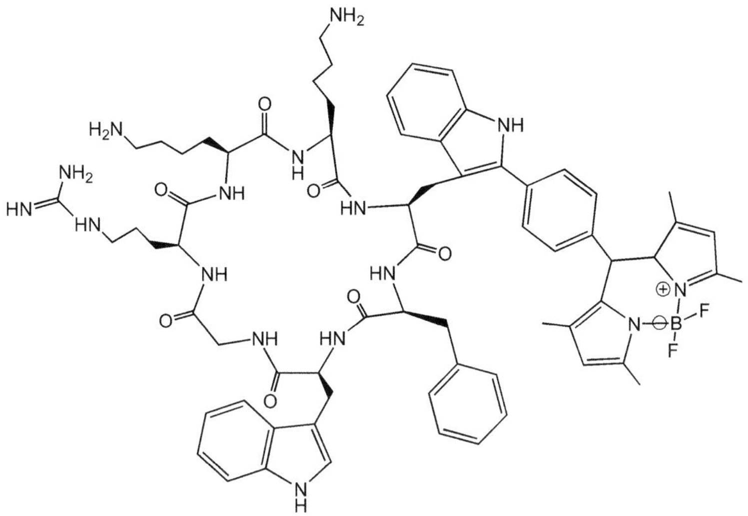

Fig. 5.28 Endogenous labeling of BODIPY dye via SPPS to make BODIPY-cPAF26 probe (Mendive-Tapia et al. 2017)

the solvent polarity (Loving et al. 2010). Solvatochromism relates to the shift of an electronic absorption band, consequently, the emission signal of a solute when there is a change in the polarity of the solvent. It occurs when there is differential solvation between the ground and the first excited state of the chromophore. Negative solvatochromism happens when the molecule undergoes a hypsochromic shift, and positive solvatochromism happens when the molecule undergoes a bathochromic shift. To study the photophysical properties of fluorescent dyes in solid-phase, a special solvatochromic heterocyclic dye was immobilized onto the modified Wang resin via the 4-iodobenzoic acid (Fig. 5.29). The key intermediate of N,N-dihexylaminophenyl thienyl boronic acid was developed based on the Suzuki cross-coupling between the thienyl boronic acid and the N,N-dihexyltetramethylboro aniline starting materials. Next, using the same Suzuki reaction between N,N-dihexylaminophenyl thienyl boronic acid with 4-iodobenzoic acid-Wang resin to create the fluorescent solvatochromic resin. The emission lambda max of the probe exerted a bathochromic shift as the concentration of the dye increased. Further, the fluorescent resin demonstrated ratiometric fluorescence response in different polarity solvent mixture (Fig. 5.30). Overall, the technology developed herein has the potential to sense bioaffinity during biomolecular interaction and conformation change, such as in the case of antigen–antibody binding. Further, this kind of solvatochromic resin can potentially be used as a versatile probe for microscale thermophoresis to characterize interactions with many biomolecules.

In a more elaborated approach, fluorescent chemosensors were developed using the combinatorial solid-phase rationale to explore new binding motifs and their subsequently new signaling mechanism (Mello and Finney 2005). In a proof-of-principle

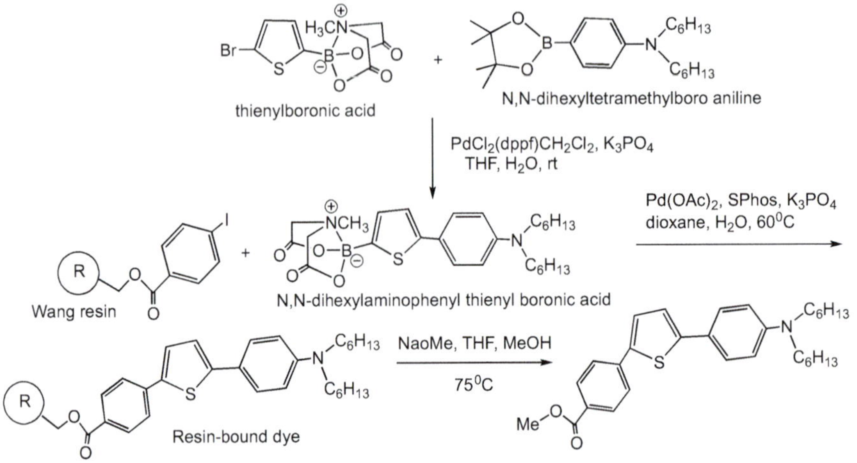

Fig. 5.29 Synthesis of fluorescent solvatochromic resin for measuring the solvent polarity on the surface. Data obtained from Otsuka et al. (2020)

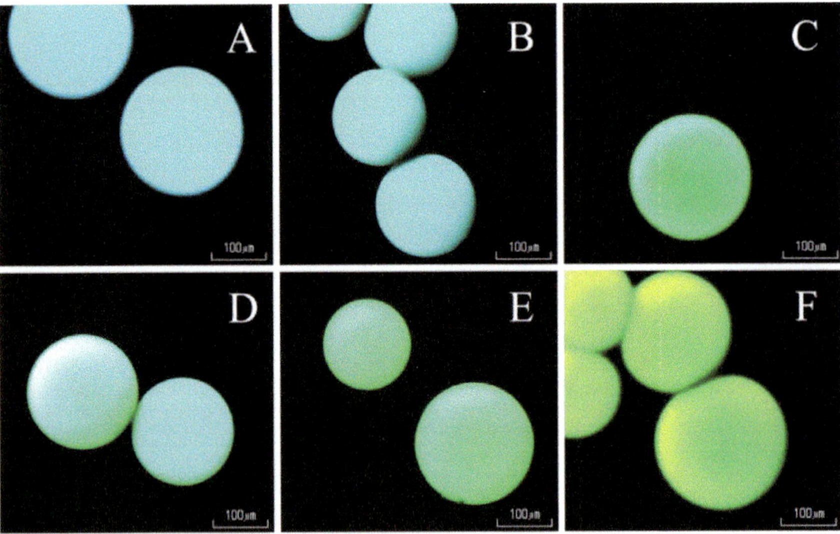

Fig. 5.30 Sensing solvent polarity using a fluorescent solvatochromic resin. **a** Toluene; **b** 1,4-dioxane; **c** ethyl acetate; **d** THF; **e** dichloromethane; **f** DMF. Data obtained from Otsuka et al. (2020)

study, a library of biarylpyridine fluorophore was developed to explore new binding characters of the probe on Hg^{2+}. A family of 198 fluorescent chemosensors was achieved using SPS based on the 2,6-biarylpyridine as the scaffold for structural modification. As shown in Fig. 5.31, the 2,6-biarylpyridine-4-carboxaldehyde, as a key intermediate, was first synthesized through multiple steps. Suzuki coupling reaction between 2,6-dichloropyrine and the boronic acid followed by a practical procedure for the catalytic reduction of nitriles to Boc-amines using catalytic quantities of nickel(II) chloride with excess sodium borohydride to facilitate the formation of the di-Boc-protected amines (Caddick et al. 2003). This step of conversion of the cyano group is reliable and tolerates some functional groups, and the amount of metal boride needs to be carefully calibrated to ensure reproducibility and selectivity. It is worth mentioning that one of the requirements for this transformation is that the starting nitrile must be present in the reaction solution prior to the formation of metal boride. Next, the tert-butyldimethylsilyl (TBS) ether was deprotected using tetra-n-butylammonium fluoride (TBAF). The alcohol product was oxidized into aldehyde using the Swern oxidation reaction to set the stage for the incorporation of the molecule onto the resin via the Horner–Wadsworth–Emmons (HWE) olefination reaction. The reason to choose HWE reaction over the Wittig reaction for the task because of the mild conditions it can offer. Specifically, the use of organic amine as a base, along with lithium salts, is compatible not only with SPS but also suitable in the presence of other functional groups. The coupling to resin was found to be very high yield and high purity (Johnson and Zhang 1995). At this stage, different amino acid sequences can be linked to the chemosensor utilizing convention Fmoc chemistry and diethoxyphosphoryloxy benzotriazine (DEPBT) as a coupling reagent. The N terminal of the compound library was acylated, capping with different acylating groups. Among the compounds within the combinatorial library, the binding affinity to Hg^{2+} with encouraging $K_a = 1.8 \times 10^{-6}$ M^{-1}. This chemistry will pave the way for further discovery of novel chemosensors, useful in environmental applications and biomedical research.

5.4.9 Styryl Dyes

Another family of dyes that has gathered significant attention in biomedical imaging and high-tech application is styryl dyes. Styryl backbone structures can be synthesized by the condensation of aldehydes with pyridinium salts under the catalytic action of a secondary amine. Traditionally, styryl dyes had a wide application in optical recording media in laserdiscs or used as photosensitizers used in photodynamic therapy, as well as in the solar cell industry. However, styryl became more recognized when it is found that this family of molecules can bind amyloid plaques, a potential biomarker in Alzheimer's disease. In fact, up to date, the styryl dye family is a subject for intensive research to develop effective probes for the early detection and assessment response to Alzheimer therapy. Amyloid occurs when protein misfolding leads to the formation of ordered secondary structures, expressed as cross-β sheets, which are present as fibrillar deposits in tissues (Sipe and Cohen 2000). The (trans,trans)-1-bromo-2,5-bis-(3-hydroxycarbonyl-4-hydroxy)styrylbenzene (BSB) was one of the first versions of styryls shown to bind to wide range of amyloid aggregations, including Abeta-rich plaques and tau-rich NFTs in situ (Schmidt et al. 2001).

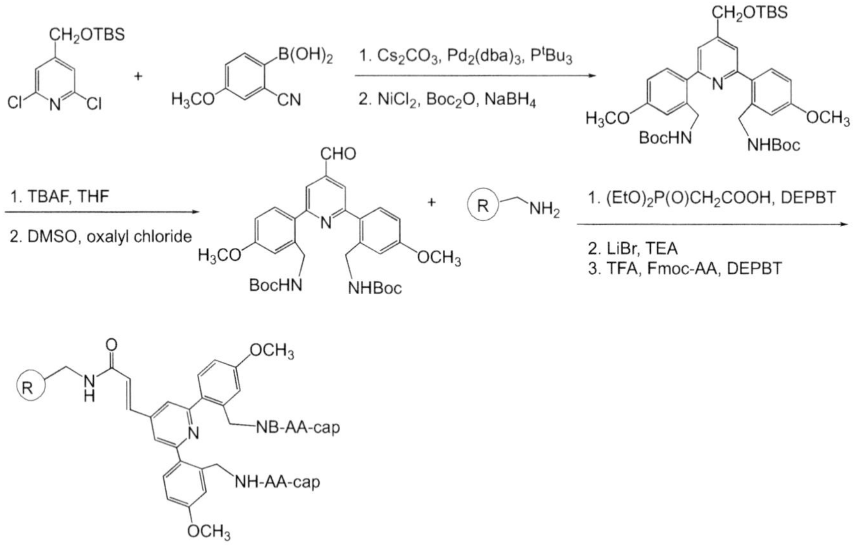

Fig. 5.31 Synthesis of a chemosensor library. Data obtained from Mello and Finney (2005) with permission from the American Chemical Society

Although no mechanisms have been determined so far to explain how styryl dyes bind to Abeta, a few notes worth mentioning can help improve the chemical design or set targets for analysis. First, there is no evidence that charged (ionic) styryl is crucial for binding (Crystal et al. 2003). Second, styrylbenzene molecules can cross the BBB, thus suitable for in vivo imaging applications. Third, styrylbenzene-based compounds bind to similar sites of the amyloid fibrils as Chrysamine G, but not with Thioflavin T (Zhuang et al. 2001). Fourth, it seems the orientation of styrylbenzene dyes has no role in dictating Abeta binding; as evidence from BSB, both cis and trans isomers showed similar binding properties (Lee et al. 2001).

Figure 5.32 shows the design of a library of styrylbenzene-based amyloid sensors (Li et al. 2004). Using 2-chlorotrityl resin, two linkers, including aminoalcohols with 2-or-6-carbon chain were loaded after the resin was activated using thionyl chloride. The alcohol end group serves as the starting point that necessitates the transformation into a mesylate group before the pyridinium salt intermediates can be achieved. The condensation between the resin-bound pyridinium salts with an aldehyde analog was assisted by microwave irradiation. Then, the dye was released from the resin under mildly acidic conditions. What is notable about this design is that two key components of the dye can be flexibly added in the construct to generate a large repertoire of novel chemical functional groups, potentially can improve Abeta-binding specificity, pharmacokinetics, and safety profiles. Most styryl dyes normally have a pyridinium or quinolinium cationic moiety conjugated with an aromatic ring at the other end in either cis or trans position separated by a methine bridge (Lee et al. 2009); with this combinatorial solid-phase chemistry, it is anticipated more creative

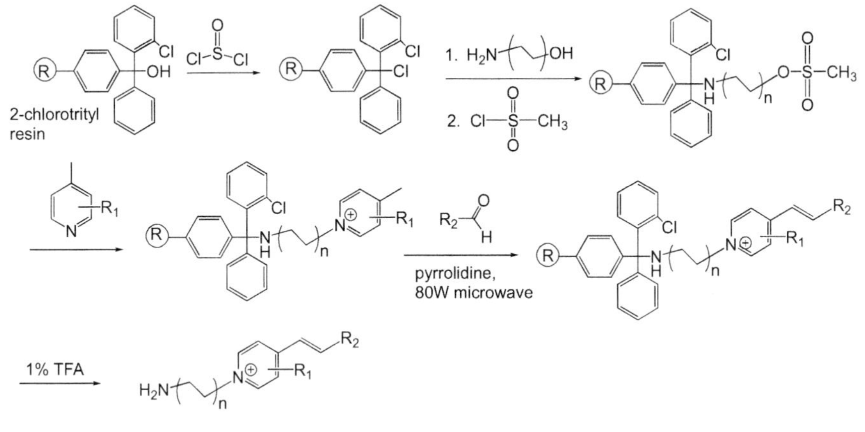

Fig. 5.32 Combinatorial solid-phase synthesis of a library of styrylbenzene as amyloid sensors. Data adapted from Li et al. (2004)

design of styryl derivatives will be developed beyond the conventional approach. In this proof-of-principle study, 320 compounds have been reported and characterized with remarkable purity (80%).

5.4.10   *Dapoxyl Dyes*

Providing activated signal by sensing the intrinsically local environment is another hallmark of fluorescence dyes. These fluorophores have been employed for the detection of intracellular protein activities and protein conformations under the platform called "smart" activatable molecular sensor (Pham et al. 2006). The dapoxyl family of dyes probably deserves recognition in this aspect. Dapoxyl's fluorescence is very low as a free dye but exhibits high-quantum yield and molar extinction coefficient depending on pH and solvent polarity. The relative quantum yield of dapoxyl dyes is influenced by solvent polarity and viscosity, with viscous solvent spheres (such as hydrophobic binding pockets) reducing molecular twists, which is responsible for non-radiative decay characters (Yates et al. 2016). This is attributed to the greater dipole moment of the molecules in the S_1 state compared to those in the S_0. On average, a 30-fold increase in relative quantum yield has been reported upon binding to the β-sheet conformer of amyloid (Yates et al. 2016). While other work reported 30-fold increase when the peptide-bound dapoxyl is in the vicinity of the hydrophobic group, transferred by farnesyl protein transferase (Pham et al. 2006).

The SPS of dapoxyl dyes using 2-chlorotrityl chloride resin, which is conjugated to the Fmoc version of the first building block, can be found in Fig. 5.33 (Min et al. 2007). The coupling reaction of the next building block was carried out under HATU/DIEA or pyridine conditions. The cyclodehydration ring closure can be achieved by treating

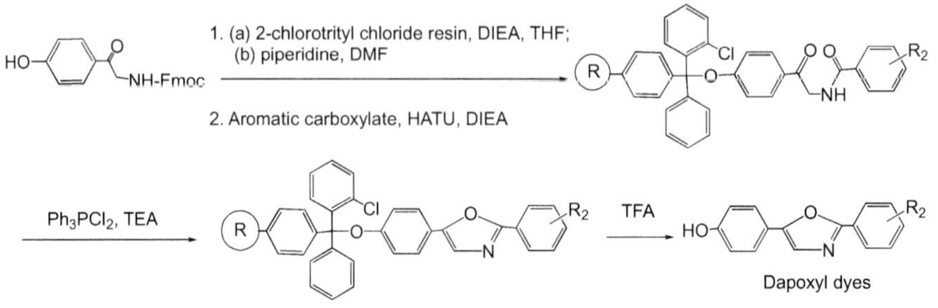

Fig. 5.33 Solid-phase synthesis of dapoxyl dye library. Data obtained from Min et al. (2007) with permission from the American Chemical Society

Ph_3PCl_2 in the presence of triethylamine. While Burgess reagent can also be used for the same task, albeit there is a high chance of cleavage from the resin. The dapoxyl dye products can be retrieved and characterized after treating the resin-bound dyes with mild TFA solution.

5.5   Conclusion

Solid-phase peptide chemistry was originally designed for peptide synthesis. Over time, this revolutionary technology was translated to synthesize dyes since it resonates very well with the needs in dye chemistry. The most challenging issue in dye chemistry, particularly the soluble dyes, is how to purify them in the most time- and cost-effective way. This chapter just describes a few exemplary cases to demonstrate the potential of SPS in answering these problems. And that it is likely the future of dye chemistry will entangle more with SPS; what we see so far is just the beginning of what could be an exciting and long implication. We are now moving into the last chapter to discuss the applications of the probes we discussed so far and imaging methods.

References

O. Al Musaimi, A. El-Faham, Z. Almarhoon, A. Basso, B.G. de la Torre, F. Albericio, Bypassing osmotic shock dilemma in a polystyrene resin using the green solvent cyclopentyl methyl ether (CPME): A morphological perspective. Polymers (Basel). **11** (2019)

M.F. Aly, M.I. Younes, S.A.M. Metwally, Non-decarboxylative 1,3-dipolar cycloadditions of imines of α-amino acids as a route to proline derivatives. Tetrahedron **50**, 3159–3168 (1994)

C. Amadi-Kamalu, H. Clarke, M. McRobie, J. Mortimer, M. North, Y. Ran, A. Routledge, D. Sibbald, M. Tickias, K. Tse, H. Willway, Investigation of parameters that affect resin swelling in green solvents. ChemistryOpen **9**, 431–441 (2020)

K. Barlos, O. Chatzi, D. Gatos, G. Stavropoulos, 2-Chlorotrityl chloride resin Studies on Anchoring of Fmoc-Amino Acids and Peptide Cleavage. Int. J. Pept. Protein Res. **37**, 513–520 (1991a)

K. Barlos, D. Gatos, S. Kapolos, C. Poulos, W. Schafer, W.Q. Yao, Application of 2-chlorotrityl resin in solid phase synthesis of (Leu15)-gastrin I and unsulfated cholecystokinin octapeptide. Selective O-deprotection of tyrosine. Int. J. Pept. Protein Res. **38**, 555–561 (1991b)

A. Basha, M. Lipton, S.M. Weinred, A mild, general method for conversion of esters to amides. Tetrahedron Lett. **48**, 4171 (1977)

R. Bollhagen, M. Schmiedberger, K. Barlos, E. Grell, J. Chem. Soc. Chem. Comm. 2559 (1994)

M. Bossi, V. Belov, S. Polyakova, S.W. Hell, Reversible red fluorescent molecular switches. Angew. Chem. Int. Ed. Engl. **45**, 7462–7465 (2006)

L.G.S. Brooker, F.L. White, G.H. Keyes, E.P. Smyth, P.F. Oesper, JaCS **63**, 3192–3203 (1941)

S. Caddick, D.B. Judd, A.K.K. Lewis, M.T. Reich, M.R.V. Williams, A generic approach for the catalytic reduction of nitriles. Tetrahedron **59**, 5417–5423 (2003)

E. Cereda, M. Turconi, A. Ezhaya, E. Bellora, A. Brambilla, F. Pagani, A. Donetti, Anti-secretory and anti-ulcer activities of some new 2-(2-pyridylmethyl-sulfinyl)-benzimidazoles. Eur. J. Med. Chem. **22**, 527–537 (1987)

J.W. Christensen, in *Resins for Solid Phase Chemistry*. Advanced Chemtech Handbook (1998), 101

G.D. Cimino, H.B. Gamper, S.T. Isaacs, J.E. Hearst, Psoralens as photoactive probes of nucleic acid structure and function: organic chemistry, photochemistry, and biochemistry. Annu. Rev. Biochem. **54**, 1151–1193 (1985)

A.S. Crystal, B.I. Giasson, A. Crowe, M.P. Kung, Z.P. Zhuang, J.Q. Trojanowski, V.M. Lee, A comparison of amyloid fibrillogenesis using the novel fluorescent compound K114. J. Neurochem. **86**, 1359–1368 (2003)

S. Eissler, M. Kley, D. Bachle, G. Loidl, T. Meier, D. Samson, Substitution determination of Fmoc-substituted resins at different wavelengths. J. Pept. Sci. **23**, 757–762 (2017)

B.F. Gisin, Helv. Chim. Acta. **56**, 1476 (1973)

B.C. Hamper, D.R. Dukesherer, M.S. South, Solid-phase synthesis of proline analogs via a three component 1,3-dipolar cycloaddition. Tetrahedron Lett. **37**, 3671–3674 (1996)

B.C. Hamper, S.A. Kolodziej, A.M. Scates, R.G. Smith, E. Cortez, Solid phase synthesis of beta-peptoids: N-substituted beta-aminopropionic acid oligomers. J. Org. Chem. **63**, 708–718 (1998)

I.A. Hemmila, Appl. Fluoresc. Technol. **1**, 1–8 (1989)

C.R. Johnson, B. Zhang, Solid phase synthesis of alkenes using the Horner-Wadsworth-Emmons reaction and monitoring by gel phase 31P NMR. Tetrahedron Lett. **36**, 9253–9256 (1995)

V. Klimesova, J. Koci, M. Pour, J. Stachel, K. Waisser, J. Kaustova, Synthesis and preliminary evaluation of benzimidazole derivatives as antimicrobial agents. Eur. J. Med. Chem. **37**, 409–418 (2002)

J. Kowalski, M.A. Lipton, Solid phase synthesis of a diketopiperazine catalyst containing the unnatural amino acid (S)-norarginine. Tetrahedron Lett. **37**, 5839–5840 (1996)

H. Kugishima, T. Horie, K. Imafuku, Synthesis of 3-[[(2-benzimidazolyl)thio]methyl]-1-methyl-1,8-dihydrocycloheptapyrazol-8-ones. J. Heterocycl. Chem. **31**, 1557–1559 (1994)

H. Kumagai, W. Pham, M. Kataoka, K. Hiwatari, J. McBride, K.J. Wilson, H. Tachikawa, R. Kimura, K. Nakamura, E.H. Liu, J.C. Gore, S. Sakuma, Multifunctional nanobeacon for imaging Thomsen-Friedenreich antigen-associated colorectal cancer. Int. J. Cancer **132**, 2107–2117 (2013)

A. Langner, H. Wolska, J. Kowalski, H. Duralska, E. Murawska, Photochemotherapy (PUVA) and psoriasis: Comparison of 8-MOP and 8-MOP/5-MOP. Int. J. Dermatol. **15**, 688–689 (1976)

C.W. Lee, Z.P. Zhuang, M.P. Kung, K. Plossl, D. Skovronsky, T. Gur, C. Hou, J.Q. Trojanowski, V.M. Lee, H.F. Kung, Isomerization of (Z, Z) to (E, E)1-bromo-2,5-bis-(3-hydroxycarbonyl-4-hydroxy)styrylbenzene in strong base: Probes for amyloid plaques in the brain. J. Med. Chem. **44**, 2270–2275 (2001)

J.S. Lee, Y.K. Kim, M. Vendrell, Y.T. Chang, Diversity-oriented fluorescence library approach for the discovery of sensors and probes. Mol. Biosyst. **5**, 411–421 (2009)

Q. Li, J.S. Lee, C. Ha, C.B. Park, G. Yang, W.B. Gan, Y.T. Chang, Solid-phase synthesis of styryl dyes and their application as amyloid sensors. Angew. Chem. Int. Ed. Engl. **43**, 6331–6335 (2004)

Z. Liu, W. He, Z. Guo, Metal coordination in photoluminescent sensing. Chem. Soc. Rev. **42**, 1568–1600 (2013)

L. Liu, X. Zhou, J.S. Wilkinson, P. Hua, B. Song, H. Shi, Integrated optical waveguide-based fluorescent immunosensor for fast and sensitive detection of microcystin-LR in lakes: Optimization and analysis. Sci. Rep. **7**, 3655 (2017)

A. Loffet, Improvement to the esterification procedure used in solid phase peptide synthesis. Int. J. Protein Res. **3**, 297–299 (1971)

M. Loncaric, M. Susjenka, M. Molnar, An extensive study of coumarin synthesis via Knoevenagel condensation in choline chloride based deep eutectic solvents. Curr. Org. Synth. **17**, 98–108 (2020)

G.S. Loving, M. Sainlos, B. Imperiali, Monitoring protein interactions and dynamics with solvatochromic fluorophores. Trends Biotechnol. **28**, 73–83 (2010)

S.J. Mason, S. Balasubramanian, Solid-phase catch, activate, and release synthesis of cyanine dyes. Org. Lett. **4**, 4261–4264 (2002)

S.J. Mason, J.L. Hake, J. Nairne, W.J. Cummins, S. Balasubramanian, Solid-phase methods for the synthesis of cyanine dyes. J. Org. Chem. **70**, 2939–2949 (2005)

G.R. Matsueda, J.M. Stewart, A p-methylbenzhydrylamine resin for improved solid-phase synthesis of peptide amides. Peptides **2**, 45–50 (1981)

J. Meienhofer, M. Waki, E.P. Heimer, T.J. Lambros, R.C. Makofske, C.D. Chang, Solid phase synthesis without repetitive acidolysis. Preparation of leucyl-alanyl-glycyl-valine using 9-fluorenylmethyloxycarbonylamino acids. Int. J. Pept. Protein Res. **13**, 35–42 (1979)

J.V. Mello, N.S. Finney, Reversing the discovery paradigm: A new approach to the combinatorial discovery of fluorescent chemosensors. J. Am. Chem. Soc. **127**, 10124–10125 (2005)

L. Mendive-Tapia, R. Subiros-Funosas, C. Zhao, F. Albericio, N.D. Read, R. Lavilla, M. Vendrell, Preparation of a Trp-BODIPY fluorogenic amino acid to label peptides for enhanced live-cell fluorescence imaging. Nat. Protoc. **12**, 1588–1619 (2017)

J. Min, J.W. Lee, Y.H. Ahn, Y.T. Chang, Combinatorial dapoxyl dye library and its application to site selective probe for human serum albumin. J. Comb. Chem. **9**, 1079–1083 (2007)

A.R. Mitchell, B.W. Erickson, M.N. Ryabtsev, R.S. Hodges, R.B. Merrifield, Tert-butoxycarbonylaminoacyl-4-(oxymethyl)-phenylacetamidomethyl-resin, a more acid-resistant support for solid-phase peptide synthesis. J. Am. Chem. Soc. **98**, 7357–7362 (1976)

A. Nefzi, J.M. Ostresh, R.A. Houghten, Solid phase synthesis of 1,3,4,7-tetrasubstituted perhydro-1,4-diazepine-2,5-diones. Tetrahedron Lett. **38**, 4943–4946 (1997)

T. Nguyen, M.B. Francis, Practical synthetic route to functionalized rhodamine dyes. Org. Lett. **5**, 3245–3248 (2003)

K.C. Nicolaou, D.A. Claremon, D.P. Papahatjis, A mild method for the synthesis of 2-ketopyrroles from carboxylic acids. Tetrahedron Lett. **22**, 4647 (1981)

R. O'Kennedy, R.D. Thornes, *Coumarins: Biology, Applications and Mode of Action* (Wiley, Chichester, U.K., 1997)

G. Orosz, L.P. Kiss, Simple and efficient synthesis of 2-chlorotritylchloride resin. Tetrahedron Lett. **39**, 3241–3242 (1998)

Y. Otsuka, G. Li, H. Takahashi, H. Satoh, K. Yamada, Synthesis of a fluorescent solvatochromic resin using Suzuki-Miyaura cross-coupling and its optical waveguide spectra to measure the solvent polarity on the surface. Materials (Basel) **13** (2020)

J.A. Parrish, T.B. Fitzpatrick, L. Tanenbaum, M.A. Pathak, Photochemotherapy of psoriasis with oral methoxsalen and longwave ultraviolet light. N Engl. J. Med. **291**, 1207–1211 (1974)

V.B. Perone, Microb. Toxins **8**, 71–81 (1972)

W. Pham, P. Pantazopoulos, A. Moore, Imaging farnesyl protein transferase using a topologically activated probe. J. Am. Chem. Soc. **128**, 11736–11737 (2006)

K.C. Pugh, E.J. York, J.M. Stewart, Effects of resin swelling and substitution on solid phase synthesis. Int. J. Pept. Protein Res **40**, 208–213 (1992)

S. Sakuma, H. Kumagai, M. Shimosato, T. Kitamura, K. Mohri, T. Ikejima, K.I. Hiwatari, S. Koike, E. Tobita, R. McClure, J.C. Gore, W. Pham, Toxicity studies of coumarin 6-encapsulated polystyrene nanospheres conjugated with peanut agglutinin and poly(N-vinylacetamide) as a colonoscopic imaging agent in rats. Nanomedicine (2015)

M.L. Schmidt, T. Schuck, S. Sheridan, M.P. Kung, H. Kung, Z.P. Zhuang, C. Bergeron, J.S. Lamarche, D. Skovronsky, B.I. Giasson, V.M. Lee, J.Q. Trojanowski, The fluorescent Congo red derivative, (trans, trans)-1-bromo-2,5-bis-(3-hydroxycarbonyl-4-hydroxy)styrylbenzene (BSB), labels diverse beta-pleated sheet structures in postmortem human neurodegenerative disease brains. Am. J. Pathol. **159**, 937–943 (2001)

J.D. Sipe, A.S. Cohen, Review: History of the amyloid fibril. J. Struct. Biol. **130**, 88–98 (2000)

A. Song, J. Zhang, C.B. Lebrilla, K.S. Lam, Solid-phase synthesis and spectral properties of 2-alkylthio-6H-pyrano[2,3-f]benzimidazole-6-ones: A combinatorial approach for 2-alkylthioimidazocoumarins. J. Comb. Chem. **6**, 604–610 (2004)

X.L. Tang, Y. Wang, D.L. Li, J. Luo, M.Y. Liu, Orphan G protein-coupled receptors (GPCRs): Biological functions and potential drug targets. Acta Pharmacol. Sin. **33**, 363–371 (2012)

V. Uryga-Polowy, D. Kosslick, C. Freund, J. Rademann, Resin-bound aminofluorescein for C-terminal labeling of peptides: High-affinity polarization probes binding to polyproline-specific GYF domains. ChemBioChem **9**, 2452–2462 (2008)

M. Vendrell, G.G. Krishna, K.K. Ghosh, D. Zhai, J.S. Lee, Q. Zhu, Y.H. Yau, S.G. Shochat, H. Kim, J. Chung, Y.T. Chang, Solid-phase synthesis of BODIPY dyes and development of an immunoglobulin fluorescent sensor. Chem. Commun. (Camb) **47**, 8424–8426 (2011)

S.S. Wang, p-alkoxybenzyl alcohol resin and p-alkoxybenzyloxycarbonylhydrazide resin for solid phase synthesis of protected peptide fragments. J. Am. Chem. Soc. **95**, 1328–1333 (1973)

S. Wang, Y.T. Chang, Combinatorial synthesis of benzimidazolium dyes and its diversity directed application toward GTP-selective fluorescent chemosensors. J. Am. Chem. Soc. **128**, 10380–10381 (2006)

S.S. Wang, B.S. Wang, J.L. Hughes, E.J. Leopold, C.R. Wu, J.P. Tam, Cleavage and deprotection of peptides on MBHA-resin with hydrogen bromide. Int. J. Pept. Protein Res. **40**, 344–349 (1992)

B.T. Watson, G.E. Christiansen, Solid phase synthesis of substituted coumarin-3-carboxylic acids via the Knoevenagel condensation. Tetrahedron Lett. **39**, 6087–6090 (1998)

D. Wootten, A. Christopoulos, M. Marti-Solano, M.M. Babu, P.M. Sexton, Mechanisms of signalling and biased agonism in G protein-coupled receptors. Nat. Rev. Mol. Cell Biol. **19**, 638–653 (2018)

T.Y. Wu, S. Ding, N.S. Gray, P.G. Schultz, Solid-phase synthesis of 2,3,5-trisubstituted indoles. Org. Lett. **3**, 3827–3830 (2001)

E.V. Yates, G. Meisl, T.P. Knowles, C.M. Dobson, An environmentally sensitive fluorescent dye as a multidimensional probe of amyloid formation. J. Phys. Chem. B **120**, 2087–2094 (2016)

H. Zarrinmayeh, D.M. Zimmerman, B.E. Cantrell, D.A. Schober, R.F. Bruns, S.L. Gackenheimer, P.L. Ornstein, P.A. Hipskind, T.C. Britton, D.R. Gehlert, Structure-activity relationship of a series of diaminoalkyl substituted benzimidazole as neuropeptide Y Y1 receptor antagonists. Bioorg. Med. Chem. Lett. **9**, 647–652 (1999)

H.C. Zhang, K.K. Brumfield, L. Jaroskova, B.E. Maryanoff, Facile substitution of resin-bound indoles via the Mannich reaction. Tetrahedron Lett. **39**, 4449–4452 (1998)

H.C. Zhang, H. Ye, A.F. Moretto, K.K. Brumfield, B.E. Maryanoff, Facile solid-phase construction of indole derivatives based on a traceless, activating sulfonyl linker. Org. Lett. **2**, 89–92 (2000)

X.F. Zhang, J. Zhang, L. Liu, Fluorescence properties of twenty fluorescein derivatives: Lifetime, quantum yield, absorption and emission spectra. J. Fluoresc. **24**, 819–826 (2014)

Z.P. Zhuang, M.P. Kung, C. Hou, D.M. Skovronsky, T.L. Gur, K. Plossl, J.Q. Trojanowski, V.M. Lee, H.F. Kung, Radioiodinated styrylbenzenes and thioflavins as probes for amyloid aggregates. J. Med. Chem. **44**, 1905–1914 (2001)

Chapter 6
Construction of Molecular Probes and Imaging Applications

6.1 Introduction

After discussing methods to design contrast agents, we will examine how these agents can be assembled in a meaningful way for imaging applications. Thus, this last chapter will focus on discussing different approaches for the design and application of molecular probes. The past chapters described much of the probe development for MRI and PET imaging. However, we covered the design of fluorescence dyes, but we have not delved into more detail on the probe synthesis. Therefore, this chapter will see a complete picture of how optical probes are designed, mostly for bioassay and preclinical applications. Increasing breakthrough in optical probe design in the past decade is evident from recent reports on clinical translation. Optical imaging is an emerging clinical favorite given the simplicity, safety, and cost of operation. Further, the switching of signal on/off afforded by the dynamic interaction between excited photons and the surrounding environment makes this imaging modality the most abundant source of creative probes.

Several approaches have been undertaken for the creation of molecular imaging probes. In most cases, probes are derived from known drug molecules; thus, the process is an extension of drug discovery, which can be achieved via (i) rational design; (ii) high-throughput screening, and (iii) natural products pathways. Sometimes, the combination of more than one approach is necessary; it is no longer an option; rather, it is imperative to guarantee successful probe development. This lecture will walk through each of these methods during the following discussions.

The rational design has the advantage of tailoring the final product with desired attributes through a number of precision manipulations. In medicinal chemistry, the rational design of drugs relies on the structure–activity relationship (SAR); similarly, in probe chemistry, the notion of SAR is taken to another level. It is not only the sensitivity, specificity, safety, and pharmacokinetics of the probe that will be after but other parameters such as stability and efficiency (e.g., quantum efficiency for fluorescent probes) are also considered. When the probes are linked to a linker, and

© The Author(s), under exclusive license to Springer Nature Singapore Pte Ltd. 2023   239
W. Pham, *Principles of Molecular Probe Design and Applications*,
https://doi.org/10.1007/978-981-19-5739-0_6

thus its physical property, length, and composition will also be assessed as well. Overall, these are essential elements in the overall design equation.

High-throughput screening can potentially overcome the limitation of the conventional approach, given the nature of this work is to identify a large number of lead compounds with desired attributes in a speedy and economical manner. Aside from traditional targets, such as kinases, nuclear receptors, G-protein coupled receptors, ion channels, and proteases (Attene-Ramos et al. 2014), other protein–protein, protein-DNA, and protein-RNA interaction targets are also good candidates for high-throughput screening (Attene-Ramos et al. 2014; McClure et al. 2019). Before the screening, an assay targeting a particular mechanism of interest must be vigorously developed and and confirmed with appropriate controls and sample size for quality assessment through the calculation of Z' values. In principle, Z' prime helps to justify whether the assay is robust and yields a large response of the tested over the control groups to warrant further analysis. Plates that do not have a Z' > 0.5 are considered suboptimal and need further optimization or rejected. Compounds identified as hits in the initial screen should be confirmed in duplicate in a confirmation screen. Further, counterscreen should be performed in parallel. The confirmed hits can also be run in the assay where concentration–response mode using 10-point concentration curves can be performed. Another advantage of HTS is that a selective library of compounds can be selected for the screening. For instance, if the targeted imaging probe is designed for neurological disorders, the screening compounds can be cherry-picked with appropriate logP values to ensure hit compounds have the capability for BBB penetration.

Natural product research has become an indispensable tool in drug discovery, and it has been implicated in pharmaceutical research and, later, molecular probe development for many years. Natural molecules exhibit the most sophisticated and high degree of structural orientation through natural selection; thus it is an invaluable target for probe development. Natural products are recognized by enzymes or other biological systems as specific as a "lock-and-key" mechanism. Specifically, natural product-based fluorescent dyes have been used in many aspects of life for hundreds of years and recently for imaging technology. These dyes cover the whole spectrum ranging from visible to near-infrared windows. For example, the anthranilates structures, coumarins, and quinolines emit fluorescence from the violet-blue range. Many compounds belong to the family of flavonoids, curcumin, and alkaloids from the green-yellow fluorescent range. In contrast, polycyclic aromatic quinones, porphyrins, and chlorophylls can extend the emission wavelength in the red and near-infrared window. Modification of these dyes have great implications in molecular imaging. For example, recently, synthetic curcumin analogs have been demonstrated excellent contrast agent to report amyloid plaques in Alzheimer's disease (McClure et al. 2015; Yanagisawa et al. 2011, 2015, 2021, 2010).

When it comes to the design of molecular probes, it is advised to adopt the strategy that has proved to be very successful in natural product synthesis, meaning that the design should first be started with the final products. One would need to define what kind of molecules we need and what then the process can be traced back to the precursors and eventually the starting materials; to see whether there is chemical feasibility to obtain them. Multiple criteria would be factored in the decision in terms of how many reaction steps need to be performed, the chemical yield, purification issues, whether the intermediates are stable, cost, and so on. By the same token, before designing the imaging probes, one would know whether the probe will be used for diagnosis, detection, staging, grading, or assessment of response to treatment. And what are the targets? Whether they are located in deep tissue and what are the anticipated kinetics and washout; whether they are located in the extracellular matrix, or they belong to the intracellular metabolic pathway? or whether the goals are purely for detection of isolated targets through biopsied or blood samples without the need to concern about the distribution, pharmacokinetics or any toxicity as the result of side effects to other peripheral organs. Or the target aims to image anatomy. Once the answers to these questions are clearly defined, an appropriate imaging modality will emerge with subsequent suggestions for synthesizing the corresponding probes.

As mentioned in Chap. 3, small organic-based PET probes are the only entities that can be developed without using a linker; otherwise, the majority of the probes need a prosthetic linker to assemble the ligand and the emitter together. Overall, this suggests that aside from consideration of the targeted molecule/ligand and a contrasting source during the design, there will be a selection number of available linkers for assembling these agents together. As a matter of fact, wet-lab chemistry employing bioconjugation synthesis is another essential operation in probe development. These state-of-the-art linkers are designed to conjugate an array of different bioactive functional groups with reliable chemical transformations. While no toxicity or interference with the function of the probes occurs. They can conjugate ligands with a low energy barrier; a reaction can occur at low temperature in a neutral buffer and fast reaction kinetics even without a catalyst. Some design enables conjugation to happen in vivo upon injection of two separate components. Notably, some linkers are equipped with a chromophore that can be activated post-conjugation, providing quantifiable parameters to account for conjugation efficiency via spectrophotometry.

Although there are many different types of imaging works, they can be grouped into 3 strategies: (i) targeted imaging; (ii) protease-activated; and (iii) environment-activated imaging. Targeted imaging comprises majority of the design. The probes derived from this approach have been used in preclinical and clinical work for a variety of disease detection. While the development of protease-activated probes is on the rise. This thanks to the presence of an extensive repertoire of proteases, which serves as a great asset for designing more advanced probes, sometimes called "smart-activated" probes. The different expression levels of the proteases in healthy versus pathological conditions can be detected and quantified by these probes. The third approach involves the use of environmentally sensitive probes to detect the morphology associated with pathological progress of diseases. All of these approaches will be the subject of discussion in this chapter.

6.2 Physical Information of Different Imaging Modalities

Several imaging modalities are available for the molecular detection of biomarkers related to pathological diseases or metabolic pathways. This lecture discussed only the imaging technologies that use contrast agents, including PET, MRI, and optical modalities. Each has pros and cons regarding the detection limit, resolution, sensitivity, and application. Understanding these features enables the appropriate design of the probes. For example, at least in preclinical animal models, when MRI is incorporated with exogenous contrasting agents, prescanning of untreated animals is necessary to generate knowledge about the background to the tissue of interest. After injecting the contrast agent, one of the most important and challenging tasks involves finding and matching that tissue to ensure appropriate comparison between pre- and post-contrast tissues.

In the PET imaging case, although the method provides excellent sensitivity, it lacks anatomical information. Therefore, after PET imaging, the subject must be scanned with computed tomography (CT). The availability of PET-CT scanners provide the ability to accurately register molecular and metabolic aspects of the target of interest with anatomical and morphological information (Seemann 2004). PET-MRI systems are also available, providing state-of-the-art imaging capability. Especially, MRI offers high resolution, so the hybrid system is designed for quantitative imaging of small and large subjects.

As this lecture focuses on preclinical validation of the imaging biomarker probes, aside from in vivo imaging data, it is crucial to perform additional ex vivo work to corroborate the information. And thus, ex vivo autoradiography and immunohisto-chemistry are inseparable operations during this validation process. The list below shows some of the major features of popular imaging technology.

PET imaging
Tissue penetration: unlimited, suitable for any target.
Spatial Resolution: Many factors affect the resolution, including the detectors and the radionuclides. In general, the resolution of preclinical and clinical PET is 0.75–1.0 mm and 2–3 mm, respectively (Moses 2011).
Temporal resolution: several minutes.
Sensitivity: 10^{-12} M.
Imaging platform: cell, preclinical animals, human.
Source of production: cyclotron.

Magnetic resonance imaging
Tissue penetration: Excellent for soft tissue, limitless.
Spatial Resolution: 25–100 μm.
Temporal resolution: minutes-hours.
Sensitivity: $10^{-3}-10^{-5}$ M.
Imaging platform: cell, preclinical animals, human.
Source of production: magnet.

Optical Fluorescence imaging

Tissue penetration: ~1 cm (reflectance imaging), 500 μm (OCT), 100 μm (confocal microscopy).

Spatial resolution: 2–3 mm (reflectance imaging), and 3–10 μm (OCT), 0.8–1 μm (confocal microscopy).

Temporal resolution: seconds (reflectance imaging), millisecond (OCT), millisecond (confocal microscopy).

Sensitivity: $10^{-9} - 10^{-12}$

Imaging platform: cell, preclinical animals, some limited use in human.

Source of production: fluorescent dyes can be developed in the laboratory. Several activated dyes cover a whole spectrum of different wavelengths are commercially available. These dyes are designed to label any functional groups available in biological ligands.

Optical bioluminescence imaging

Tissue penetration: 1–2 cm.

Spatial Resolution: 3–5 mm.

Temporal resolution: minutes.

Sensitivity: $10^{-15} - 10^{-20}$ M.

Imaging platform: preclinical animals.

Source of production: luciferin.

6.3 Targeted Imaging Approach

The majority of the molecular imaging probes belong to this category. In this approach, specific ligands can be synthesized and labeled with an emitter, depending on what imaging modality will be used, so the emitter can be incorporated directly onto the structure or tethered through a linker. Practically, a drug molecule is an ideal ligand for a targeted imaging approach, given the known specificity, pharmacokinetics, and toxicity profile of the compound; labeling known drugs and converting them into imaging probes can speed up the process. Some efforts, though, are required to ensure the physical property of the drug remains unchanged, or at least not much altered after labeling with an emitter. As mentioned before, most [^{11}C] labeling would retain an identical chemical structure to the ligand. In contrast, MRI and optical probes need some modifications, such as incorporating a bifunctional linker to attach the emitter to the drug molecule. In many cases, careful designing of a spacer linker helps to prevent interference with the drug binding. Overall, the goal of targeted imaging is that the probe can directly detect the designated targets after exposure to cells or injected in vivo. In either case, the emitted signal always presents in the peripheral, it needs some time to resolute the signal before imaging commences. To remove the unspecific signal from the background, in a cell-based study, this can be achieved by repeated washing several times to remove the unspecific probe from the media. However, for in vivo cases, this clearance process relies on the physiological actions in the body. After this clearance from the blood pool and the peripheral, the signal will be resolute for better imaging. Usually, for a specific probe, the ideal imaging window would be somewhere $1-2$ h post intravenous injection. As depicted in Fig. 6.1, the probe will quickly circulate in the blood pool upon intravenous injection. Several minutes post-injection, the probe would extravasate from the vascular system and bind to the specific target and help provide information on the mechanism of investigation. For each newly developed probe, the pharmacokinetics profile needs to be assessed, and the optimal data can be obtained and defined where the prolonged binding of the probe to the intended target provides good target-to-background ratios. The imaging process best starts when the background is totally subdued. This condition provides a possibility of imaging subtle differences between normal and pathological conditions. The ideal imaging probes would be cleared from the target site and washed out via the kidneys and bladder after doing what they are designed to do. Targeted imaging was also used to validate potential biomarkers to inform disease detection and characterization of treatment response (deSouza et al. 2019). Another potential forte of targeted imaging is that it provides objective decision-making for the management of patient care.

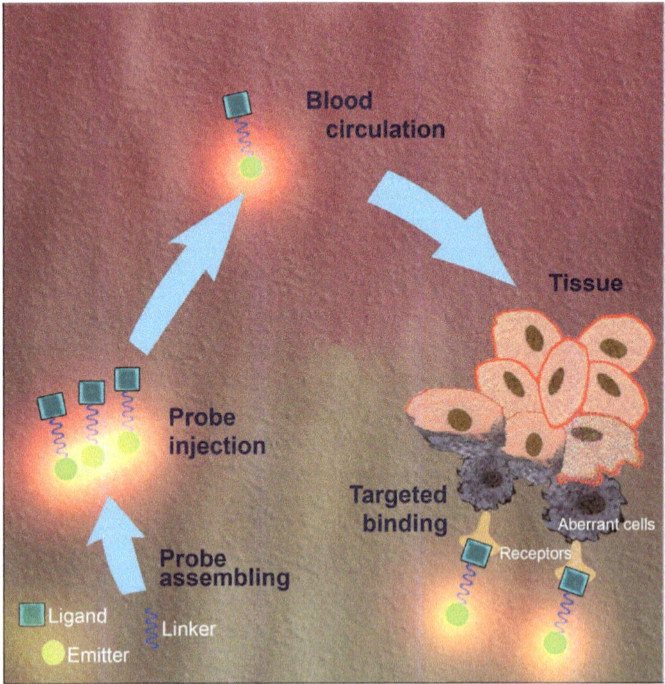

Fig. 6.1 In vivo targeted imaging strategy. Distribution kinetic designed to optimize the signal detection when non-specific blood-pool signal fades away

The ^{18}F-labeled pyridine-based benzamide derivative [^{18}F]DMPY2 designed for the detection of malignant melanoma, exemplifies the rationale of optimizing pharmacokinetics to enhance signal-to-noise ratios for targeted imaging (Pyo et al. 2020). With the emergence of new therapeutic drugs, the diagnosis of melanoma at the earliest possible stage has become more important for improving the survival of patients (Liu et al. 2013). Early detection enables successful treatment via surgical removal of cancer. Statistically, up to 85% of patients are cured with this treatment following diagnosis (Zhu et al. 2011). Although a number of PET probes have been tested for clinical detection of melanoma, only [^{18}F]FDG is currently used for clinical detection of melanoma (Krug et al. 2008). FDG imaging focuses on metabolic assessment to compare normal tissue from disease conditions. For decades [^{18}F]FDG has been used clinically for tumor imaging due to increased glucose metabolism in most types of tumors, and has been shown to improve the diagnosis and subsequent treatment of cancers (Zhu et al. 2011). However, this metabolic-based imaging is incapable of staging and restaging cutaneous melanoma since it does not detect microscopic disease with acceptable sensitivity. Further, the low signal-to-noise ratio as the result of using this metabolic probe resulted in high background images (Fig. 6.2). Further, FDG-PET sensitivity for melanoma is dependent on tumor volume (Wagner et al. 2001). There is an emerging clinical need to develop more specific probes for melanoma. [^{18}F]DMPY2 was developed based on the backbone of the benzamide family of molecules that have a high affinity for melanin (Rouanet et al. 2021). Melanin is present in most primary melanoma tumors (Koch and Lange 2000), working as a protective element. For example, amelanotic melanomas (skin cancers in which the cells do not have melanin) are usually associated with a lower survival rate than pigmented melanomas (Wee et al. 2018). Therefore, melanin is an obstacle when it comes to treatment using photodynamic therapy, radiotherapy, and chemotherapy because of its photo-protection and scavenging capabilities. However, the presence of melanin in melanoma serves as a perfect biomarker from which targeted molecular probes can be developed for specific tumor targeting (Rouanet et al. 2021). The [^{18}F]DMPY2 was designed with 3 major components, which were modified, including the pyridine-based benzamide, an aliphatic linker, and an aliphatic amine moiety, to enhance binding specificity and with enhanced bioavailability.

The labeling for making [^{18}F]DMPY2 is about 50 min, with the overall decay-corrected RCY approximately 15–20% (Fig. 6.3). The product was confirmed by HPLC, and the molar activity was greater than 7.6 GBq/μmol. No metabolites were detected when incubating the probe in human serum for 2 h at 37 °C. The same observation was reported when the probe was assessed via mouse serum after by intravenous injection for 60 and 120 min. The PET data demonstrated that if imaging occurred 60 min post intravenous injection, high tumor uptake was observed with very minimal signal interference from the background noise compared to imaging the same animal 30 min earlier.

Aside from using a specifically targeted probe with optimized kinetics to reduce the background signal and improve image quality, a few other parameters should also be examined and scrutinized in a typical imaging workflow to improve tracer biodistribution and image resolution. For example, for metabolic imaging using [^{18}F]FDG,

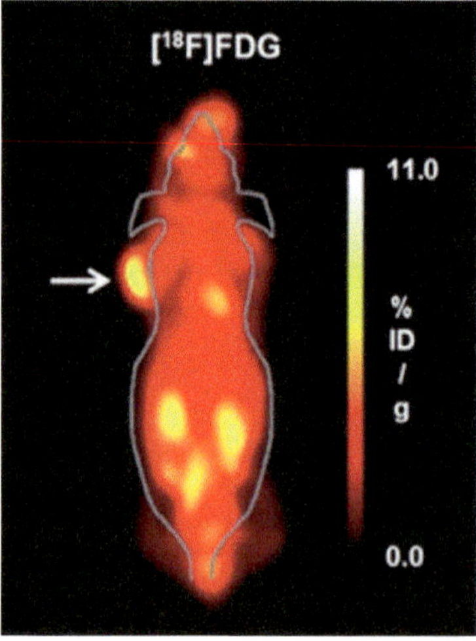

Fig. 6.2 High background signal using a metabolic probe, such as [^{18}F]FDG. Coronal images of B16F10-bearing mice at 60 min post intravenous injection of the radiotracer. The tumor is indicated by a white arrow. Data obtained from Pyo et al. (2020)

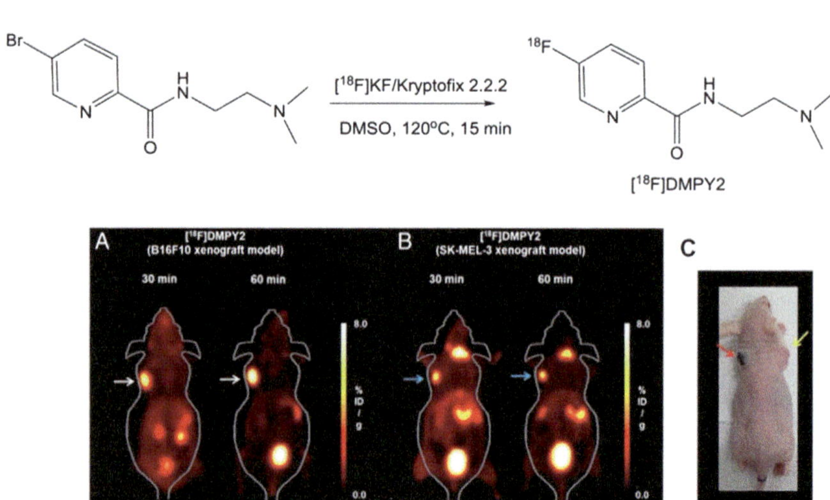

Fig. 6.3 Time-course imaging of tumor-bearing nude mice using the [^{18}F]DMPY2. Data obtained from Pyo et al. (2020)

animal handling directly dictates [^{18}F]FDG biodistribution and influences the results of PET imaging data in tumor-bearing mice (Fueger et al. 2006). It is found that warming and fasting animals prior to tracer injection reduced brown adipose tissue uptake and improved tumor visualization. Further, anesthesia used in [^{18}F]FDG imaging might contribute to artifacts since ketamine/xylazine induce hyperglycemia (Fueger et al. 2006). In contrast, isoflurane is compatible with [^{18}F]FDG imaging on tumor-bearing mice. It is also worth mentioning that intravenous injection (IV) (the process of IV injection was described in the bonus section at the end of this book) is the best route for PET imaging. Other routes of distribution, such as intraperitoneal, retroorbital or subcutaneous injection, resulted in slow tracer absorption into the circulation and signification tracer retention at the injection site, which may interfere target visualization (Kelada et al. 2018).

Diagnostic accuracy using metabolic [^{18}F]FDG probe can improve malignant foci and lesion detection in approximately 50% of patients (Delbeke and Martin 2004). Nevertheless, there is an unmet clinical need to develop more specific probes to improve cancer detection capability. In one particular case, to exemplify this notion that the current clinical standard using metabolic [^{18}F]FDG probe often cannot distinguish inflamed from malignant lymph nodes. Thus, it is challenging for detecting diffuse large B-cell lymphoma (DLBCL). Due to this issue, there is a penchant for replacing [^{18}F]FDG. One of the biomarkers emerged as a good candidate for imaging DLBCL; Poly(ADP-ribose) polymerase (PARP) enzymes belong to a family that has been identified with 17 members. PARPs play an essential role in a number of cellular processes, including modulation of chromatin structure, transcription, replication, recombination, and DNA repair (Morales et al. 2014). Due to its primary function as a DNA damage sensor, PARP1 is a target for therapeutic intervention in cancer (Ali et al. 2012). Particularly, a high expression of PARP1 mRNA, not others, was high in primary tumors of the breast, endometrium, lung, ovary, skin, and in DLBCL (Ossovskaya et al. 2010; Tang et al. 2017). The overexpression of PARP1 is found to associate with deterioration, metastasis, and angiogenesis in tumors (Wielgos et al. 2017). Taken altogether, PARP1 can serve as a good therapeutic target for human malignancies and a biomarker for imaging cancer. The imaging information obtained from this probe can also serve as a biomarker to detect response to therapy, another forte of the targeted probe. In response to that need, a PARP1-targeted imaging probe was developed by labeling the PARP inhibitor, Olaparib, which is an FDA-approved drug for the treatment of BRCA-mutated advanced ovarian cancer in adults, with a BODIPY fluorescence dye (PARPi-FL) (Reiner et al. 2012). The binding study showed that the EC$_{50}$ of the PARPi-FL probe (12.2 nM) fluctuates just slightly compared to that of Olaparib (5 nM) (Fig. 6.4)(Irwin et al. 2014). Further, the probe exhibited high stability in vivo and targeted glioblastomas with great selective and low toxicity.

In an extension to appreciate the potential of PARP-based imaging probe for non-invasive detection of tumor in vivo and future clinical work, a [^{18}F]PARPi version was developed, in which the precursor was labeled with [^{18}F] radioisotope using N-succinimidyl-4-[^{18}F]fluorobenzoic acid as a prosthetic linker.

To test the probe, first DLBCL mouse model was developed by subcutaneous injection of hematopoietic precursor cells (HPCs) that have the overexpression of Myc and Bcl2 into C57BL/6 (B6) mice. It has been demonstrated in the past that MYC and BCL2 are two main oncogenes that drive the pathogenesis of DLBCL (Aukema et al. 2011; Hu et al. 2013; Savage et al. 2009). Then, the specificity of the [^{18}F]PARPi was assessed by intravenous injection of the probe (300 µCi/mouse) in either DLBCL mice, B6 mice, or DLBCL mice pretreated with excess Olaparib (500 µg/mouse). The PET signal depicted from the lymph nodes of DLBCL mice was 5.6-fold more

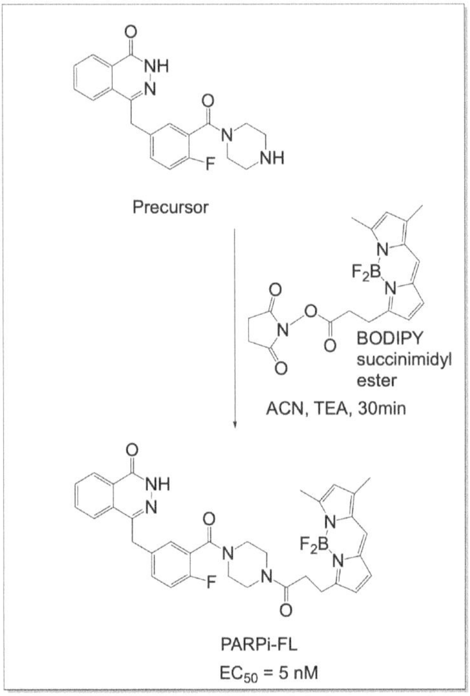

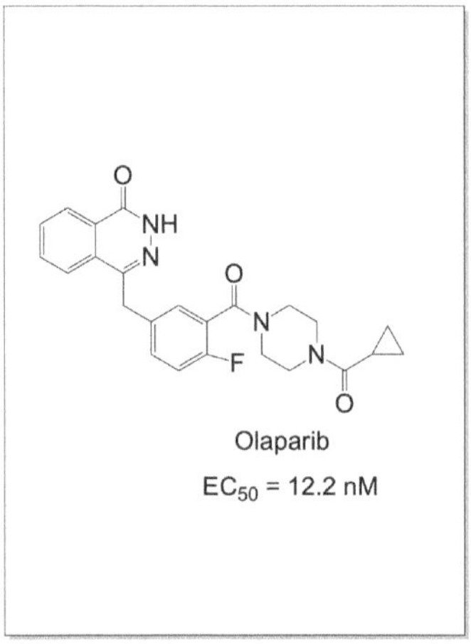

Fig. 6.4 Synthesis of PARPi-FL probe and comparing the EC_{50} of the new probe with the parent drug, Olaparib. Data obtained from Irwin et al. (2014), (Reiner et al. 2012) with permission from Elsevier

than in those of control B6 mice. While the signal decreased by approximately 87% if pretreated with Olaparib, suggesting the distribution of [^{18}F]PARPi was specific to PARP1. This in vivo data was consistent with the biodistribution study.

To demonstrate the probe could be used to differentiate inflamed from malignant lymph nodes, a mouse model of inflamed lymph nodes was generated by the injection of Flt3L. In animal models, administration of Flt3L to mice induces a significant increase in the number of functionally active DCs in the circulation and in all organs (Pham et al. 2009). Further, locally injection of Poly-IC in the lymph nodes of B6 mice to induce inflammation. As shown in Fig. 6.5, the PET/CT imaging using [^{18}F]PARPi in DLBCL mice, B6 with inflamed lymph nodes, and normal B6 mice as controls demonstrated that the probes could report the difference between inflamed versus malignant lymph nodes in these mouse models. Quantitatively, the PET signal registered at 76% higher intensity in the malignant lymph nodes than in the inflamed counterparts and 152% higher signal than in normal lymph nodes. The data also shows a mild increase in [^{18}F]PARPi PET signal in inflamed versus normal lymph nodes, albeit not significant. Further, immunohistochemistry of the lymph nodes in these animals also revealed consistently high expression of PARP1 across the tissue, while much lower and only in the germinal center expressing in inflamed and normal lymph nodes.

As mentioned earlier, the tissue penetration of optical imaging is limited to 1 cm; it is suitable for only surface imaging, even in preclinical animal models. Nevertheless, the proclivity of optical imaging focuses on its strength by integrating it with optimal surface detection modules, such as laparoscopy and endoscopy, along with a robust engineering approach and image reconstruction software in a new technology called intraoperative fluorescence imaging. Recently, the hybrid optical-confocal system enables high-resolution imaging using the endomicroscopy technique (Bajbouj et al.

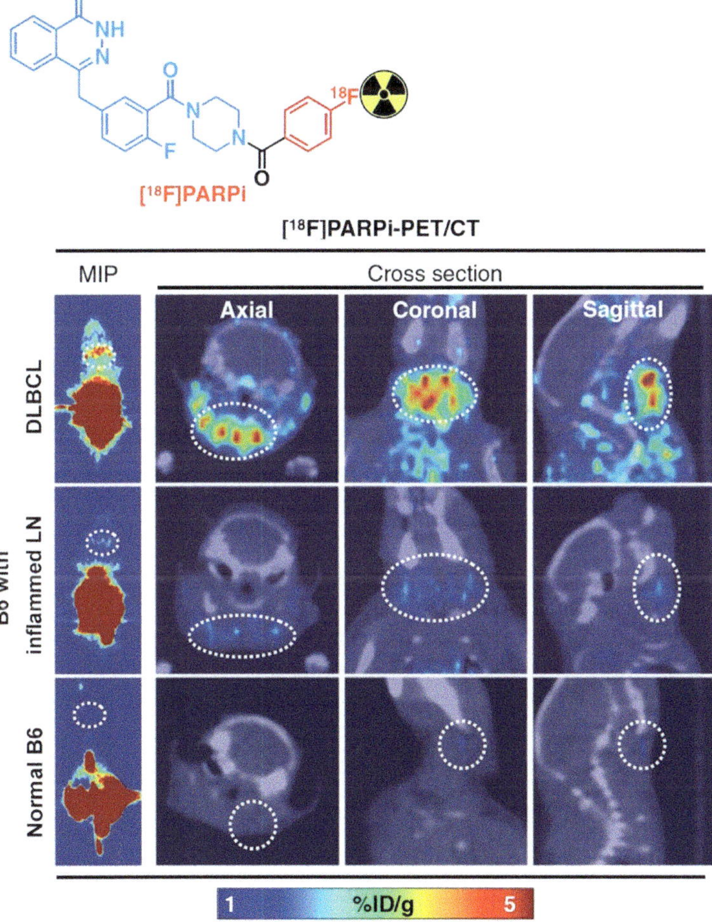

Fig. 6.5 Targeted PET imaging to discern inflamed lymph nodes from malignant counterparts in diffuse large B-cell lymphoma (DLBCL). Data obtained from Tang et al. (2017) with permission from the PNAS

2010). Other applications for optical imaging exploit the availability of the surface-located targets on the surface, found in melanoma, breast cancer, throat and mouth disease.

The use of targeted fluorescent probes is essential in this paradigm-shift approach for the diagnosis, detection, staging, and chasing of metastatic tumors, all with the direct intervention of a surgeon. One of the methods to enhance the specific detection of probes relies on the receptor recognition of the ligand. For example, folic acid, a small molecule, has been used for such an approach; given cancer cells are overexpressed with folic receptors significantly higher than normal cells, there is a potential to create high contrast image. To label folate with a fluorescent dye is not trivial. By nature, it is conceivable that the carboxylic acid group present in folate can be activated and coupled with the aminated ligand via the amide bond. Although these carboxylic groups can react with coupling reagents at a different rate, they are close enough to generate undesired products, including inactive α-conjugate and the active γ-conjugate, accompanied by the bis-functionalized product (Luo et al. 1997). Separation of these mixtures is often challenging. To overcome this problem, the rationale approach for labeling folate would go to an opposite but meaningful approach. It would focus on the regiospecific functionalization of the γ-carboxylate of glutamic acid with a primary amine. The first chemical report toward the development of folate (γ)-deferoxamine by direct DCC coupling of deferoxamine with folate followed by exhausted characterization and purification using ion-exchange chromatography and HPLC to afford the desired products with 10–15% yield (Wang et al. 1996).

An improved chemistry had shown a more robust modification of folate for bioconjugation, starting with γ-methyl folate. Treating this starting material with excess and neat ethylenediamine to furnish the folate ethylenediamine with up to 87% yield. Most impressively, without the need for purification via chromatography (Luo et al. 1997). With the availability of a primary amine, now the folate can conjugate with any activated carboxylate. Figure 6.6 shows how FITC-labeled folate was generated. The art of chemical optimization is demonstrated in this chemistry. There is no evidence of chemical inference during these reaction steps given by guanidine even it is considered as one of the most powerful organo-bases. If the reaction or purification conditions happen to have other reagents in the system, for example, an aldehyde, the outcome will be different. There are reports showing that if guanidine occurs in the presence of aldehyde, aziridine will be formed consequently (Ishikawa 2010). A near-infrared version of folate, OTL38, has been reported, in which folate was labeled with indocyanine green dye for clinical study (Randall et al. 2019). Aside from these optical probes, a chelator, like DTPA for indium-111, can also be conjugated to convert the probe for PET imaging.

For the preclinical study, FITC-folate can be injected intravenously into tumor-bearing mice (10 nmol, 100 μL), and the fluorescent intensity against the background signal was best at 2 h post-injection (Fig. 6.7). The specificity of the probe was demonstrated in the L1210 tumor model in which one of the mice was pre-injected with over 100-fold molar excess of folate before treating with the probe, resulting in quenching of the fluorescent signal compared to other mice. Further, the illumination

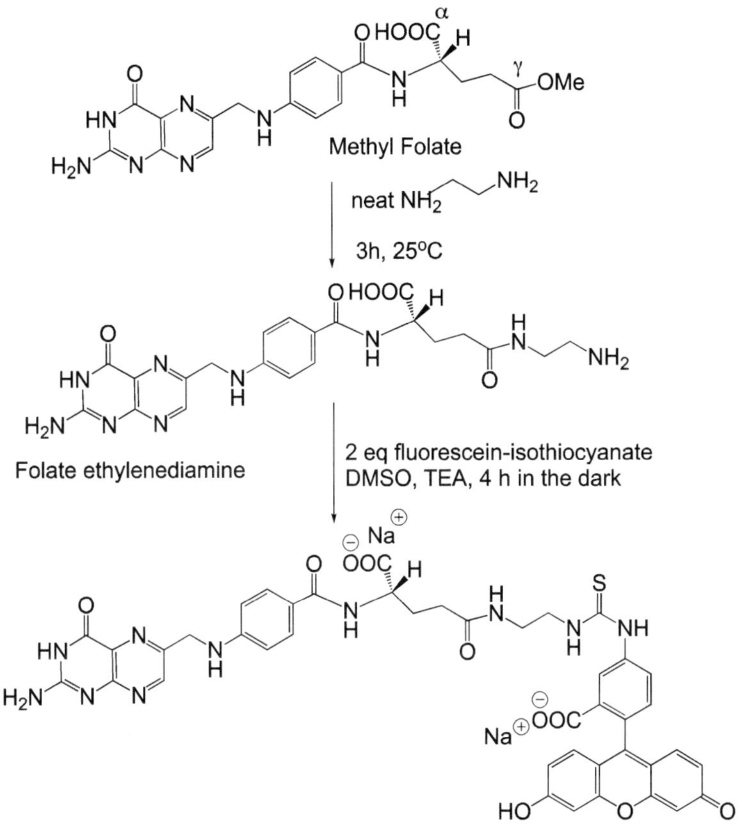

Fig. 6.6 Synthesis FITC-labeled folate as a targeted probe for optical imaging. Data obtained from Luo et al. (1997) with permission from the American Chemical Society

provides a clear tumor margin than the adjacent normal tissues. In the metastatic M109 tumor model, the probe could report tumor nodules in the lungs as small as 0.5 mm with high intensity compared with adjacent non-malignant tissue. Based on these observations, this folate-based optical imaging has two important clinical applications. First, the high specificity afforded by folate enables the probe to discern tumor margin from healthy tissues in the vicinity, which will assist physicians while performing tumor resection. Second, the detection of small and metastatic tumors of any size smaller than 1 cm for any kind of tumor, either during laparoscopy or colonoscopy, is a significant challenge; this is where this probe will find its niche.

Another excellent targeted imaging probe worth mentioning is the heat-shock protein 90 (Hsp90). Hsp90works as a molecular chaperone that aids the folding, maturating, transport, and maintenance of conformational stability of many proteins, including Her2, c-Met, and Cdk-4 (Pratt and Toft 2003; Richter and Buchner 2001; Sato et al. 2000; Schulte et al. 1996). Regarding the targeted probe's design, the target molecule should be differentially expressed in pathological cells versus control

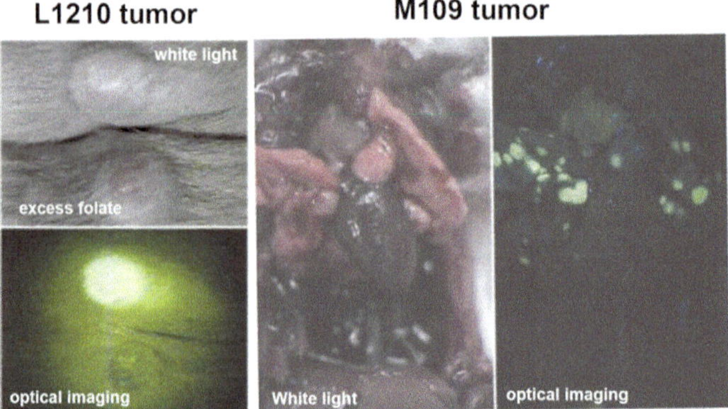

Fig. 6.7 Application of FITC-folate for imaging using lymphocyte-derived L1210 on DBA mice and metastatic lung M109 tumor on BALB/C mice. The probe was injected via intravenous route, and imaging started 2 h post-injection. Data obtained from Kennedy et al. (2003) with permission from the author

counterparts. And they must have much higher expression, and Hsp90 fits well in this respect. It is overexpressed in malignant cells since these cells rely on Hsp90 for chaperoning function. Hsp90 comprises three associated domains: (i) the N terminal domain, which is responsible for the protein's ATPase activity; (ii) the C terminal domain, which contains an ATP site and (iii) a spacer connecting the N and C terminal. Understanding the architecture of Hsp90 enables the rationale design of the inhibitors. As a matter of fact, evidence showed the direct involvement of Hsp90 in oncogenic protein folding/stability by inhibition of the ATPase function would lead to the disruption of the folding cycle, subsequently leading to destabilization, ubiquitination, and ultimately proteasomal degradation of the client proteins (Workman 2004; Zhang and Burrows 2004). Through molecular modeling and medicinal chemistry methods, several compound libraries focused on ATP-binding proteins, and the lead compound was identified as 4-(2,6,6-trimethyl-4-oxo-4,5,6,7-tetrahydro-1H-indole-1-yl)benzamide, a moderate Hsp90 inhibitor with K_d of 3.7 μM (Fig. 6.8) (Huang et al. 2009). Further SAR studies delivered a novel class of Hsp90 inhibitor (SNX-2112) with an IC_{50} for Her-2 degradation of 10 nM, and the drug has remarkable Hsp90 affinity with a K_d of 16 nM.

The SNX-2112 scaffold was further modified to carry an optical and radioiodinated emitter via a polyethylene glycol spacer (Barrott et al. 2013). There are no golden rules of what type of fluorescent dyes or linkers should be used for developing targeted probes. The general approach is to create a number of probes where the targeted ligand is labeled with different dyes and linked with spacers of different lengths. Probe optimization will then be screened via the in vitro binding assays, in this case, against the immobilized ATP. Like every other probe, tethering an inhibitor with a spacer, as shown in probe HS-27 (Fig. 6.9), would experience a reduced binding affinity for the native Hsp90. Nevertheless, not every modification leads to bad trends, the good sign is that the presence of a tether enhances specificity

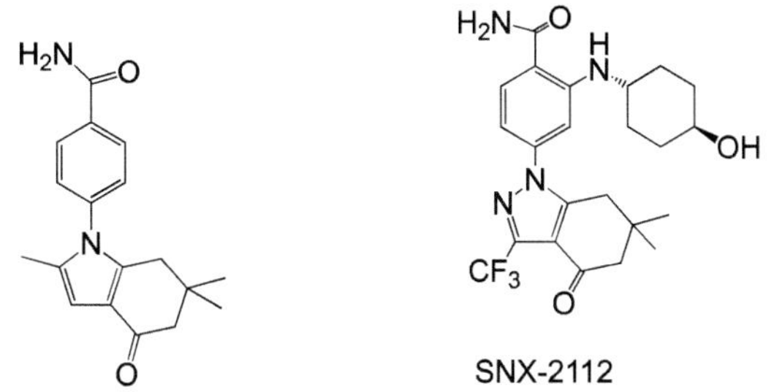

Fig. 6.8 Rational design of Hsp90 inhibitors. Data obtained from Huang et al. (2009) with permission from the American Chemical Society

by eliminating binding to Grp94, probably caused by the bulkiness of the tether, as well as the dye, blocking the access to the ATP-binding site of Grp94.

After achieving acceptable specificity, the next step in the probe validation assay would focus on cell study. Probably, cell work is one of the most reliable and robust works before in vivo translation. Asides from a semi-high-throughput manner, suitable for screening a new probe library across different cell types, numerous information about the probe can be obtained from cell works, such as toxicity, uptake kinetic, stability of the probe, and more. This work is sometimes considered a must since it can guide the next in vivo study. For instance, it is found that HS-27 is internalized readily by most of breast cancer cell lines, albeit with different degree of uptake: MCF10 <<MCF7<MDA-MB-468<BT474. But surprisingly, the probe

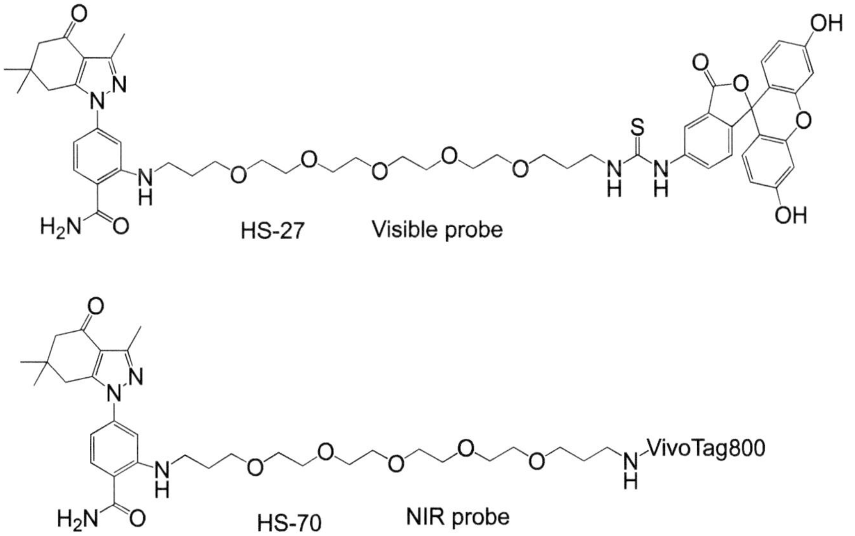

Fig. 6.9 Hsp90 optical probe. Data obtained from Barrott et al. (2013) with permission from Elsevier

uptake is selective for some cell types but not others. Even though identical conditions were used in breast cancer cells, the fluorescent signal was nearly negligible for Huh7 cells, a hepatocarcinoma cell line, which also has a high expression of Hsp90. Competition assays with the ligand alone demonstrate that HS-27 uptake and selectivity for breast cancer are Hsp90-dependent.

To understand the mechanism why HS-27 probe reports Hsp90 in breast cancer but not in hepatocarcinoma cells, MDA-MB 468 and Huh7 cells were treated with HS-27 in the presence of permeabilizing agents, such as β-escin or Triton X-100. Both cell types showed remarkable fluorescent signals intracellularly. And the signal depicted inside these two cells from HS-27 is specific for Hsp90, evidenced by the quenching of the fluorescence when these cells were treated with excess unlabeled Hsp90 inhibitor. Taken together, the data support the notion that the probe displays selectivity for breast cancer cells, probably influenced by ectopically expressed Hsp90. While Huh7 cells do not express ectopic Hsp90 (Barrott et al. 2013).

Guided by the cell imaging data, a preclinical mouse model of breast cancer was generated by injection of the MDA-MB-468 subcutaneously to generate a xenograft model. Intravenous injection of HS-27 or HS-70 probe can detect the malignancies non-invasively using the IVIS optical imaging system. The tumor mass can be detected as early as 5 min post-injection, but better resolution requires a much longer time for clearing non-specific distribution of the probe in the blood pool. The in vivo imaging data shown in Fig. 6.10 was obtained 1 h post-injection, and the retention of the targeted probe in the tumor remained strong even 24 h post-injection. This work also demonstrates the idea mentioned in Chap. 2 regarding the advantage of using a near-infrared probe for deep tissue imaging. In the visible probe, HS-27, a reported eightfold enhanced signal compared to the control tumor-bearing mice without receiving the probe. However, if using a near-infrared probe, the registered enhancement of 150-fold was observed (Fig. 6.10).

It would be a mistake not to discuss MRI-based probes when covering a chapter on targeted imaging. Most MRI contrast agents are designed based on the premise that differential enhancement of pathological tissues relies on residual differences in water content. These probes could provide information on disease initiation and progress. But it is widely recognized that the signals depicted from these situations are too low and probably too late for early detection purposes, nor can this difference be detected in time for imaging response to therapy. Since the emergence of new contrast agents, which have changed this landscape. As mentioned in Chap. 4, two major magnetic resonance contrast agents, namely gadolinium and superparamagnetic iron oxide nanoparticles for T1 and T2-mediated imaging, respectively, have been proven effective for molecular imaging of several clinical-relevant targets. Given the low detection sensitivity of MRI, the design of these probes must have a different approach compared to those from other modalities. Multivalency plays an essential role in the design of the MRI contrast agents. For T1-mediated imaging, the targeting molecules should have multiple and identical active groups for bioconjugation with gadolinium-chelate complex. And the opposite way, in the case of iron oxide nanoparticles for T2-mediating MRI. With an average size from 5 to 30 nm, these particles are fabricated with a large number of functional groups to accommodate several targeted

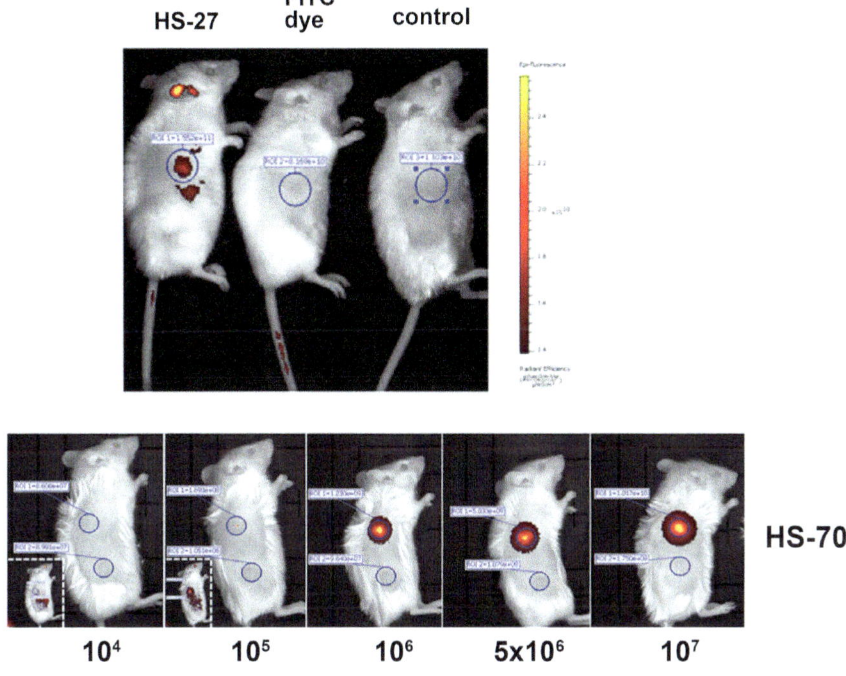

Fig. 6.10 Assessment of sensitivity and specificity of the HS-27 (top) and HS-70 (bottom) on MDA-MB-468 xenograft mouse model of breast cancer using the IVIS. Data obtained from Barrott et al. (2013) with permission from Elsevier

molecules through surface fabrication. Therefore, it is not surprising, MRI probes are usually associated with large molecules, like antibodies, proteins, graft-copolymers, and nanoparticles.

Albumin labeled with DTPA(Gd) was one of the first gadolinium-based probes for intravascular and blood-pool-enhancing MRI agents (Schmiedl et al. 1987). The design of large-sized probes aims to limit their circulation in the intravascular space for MR angiography. The conjugate was achieved via a 2-step synthesis, with the DTPA conjugated to aminated albumin, followed adding an excess of Gd^{3+} ions and purification by dialysis. Although this probe did not target any particular tissues or pathological targets, the use of macromolecules as a carrier demonstrated in this work was a motivation for many following designs where targeted probes were developed. This work also revealed the opportunity and implications of macromolecule chemistry in probe development. For instance, depending on the chemical procedure, the number of DTPA(Gd) fabricated on the albumin varies significantly; it could be 7–18. And this fluctuation resulted in the in vitro T1 relaxivities (R1) of the albumin-DTPA(Gd) in the range of 110–313 mM^{-1} s^{-1}. The concentration of this probe was relatively high for in vivo study using rats; it is reported the range from 140–340 mg/Kg based on albumin molecular weight.

The modification of macromolecular-Gd for MR imaging employing antibodies is a great move for enhanced targeted imaging. In one approach, the DTPA anhydride and 9L monoclonal antibodies were first conjugated, followed by the removal of the excess starting materials using gel filtration via a Sephadex G-50 (Matsumura et al. 1994). Then, gadolinium acetate was added to label the Ab-DTPA. No report from this work about the labeling ratio, but in a similar work, it has been demonstrated about 20 DTPA per an antibody can be achieved through this chemistry (Shahbazi-Gahrouei et al. 2001).

Another notable work aimed to reduce the concentration of the probe to benefit in vivo study while enhancing the signal intensity by loading a large amount of Gd via the polyamidoamine PAMAM™ dendrimers. Depending on the number of branching of these spherical nanosized particles, generation-3 (G3D), -4 (G4D), -5(G5D), and -6 (G6D) contain about 32, 64, 96 and 192 surface primary amino groups, respectively. With a defined structure and reproducible starting materials, PAMAM™ macromolecules are the ideal platform when it comes to multivalency. And this work converted the amino-rich PAMAM™ dendrimers into carboxylates via the 2-(p-isothiocyanatobenzyl)-6-methyl-diethylenetriaminepentaacetic acid for use as a super chelator for Gd^{3+} ions. As shown in Fig. 6.11, the homogenous coupling of PAMAM with p-SCN-Bz-DTPA through the isothiocyanate group in an optimized reaction condition using strong basic condition with slightly heated temperature. After the reaction was completed, the starting materials were removed using a filtration membrane (Centricon). For the G3D dendrimer, Centricon-10 filters are suitable for this operation, and Centricon-30 for G4D, G5D, and G6D. In this construct, the isothiocyanate group was designed as an activated group for labeling. Although not the best, it is a reliable and reproducible reaction, and often this function group is very stable, altogether contributing to a high reaction yield. Through this step, almost 98% and the amine groups on the dendrimers have been reacted and eventually converted to carboxylates. To label the PAMAM-[(DTPA)], $GdCl_3$ was added to the reaction solution in citrate buffer to afford the desired product. Overall, this dendrimer-based MRI probe is suitable for imaging tumor microvasculature during anti-angiogenesis therapy of cancer. The control synthesis for a desired branching of dendrimer enables maximizing the gadolinium payload, thus reducing the concentration of the probe and consequent averted toxicity.

Heterogenous and sequential modifications of the polyamidoamine PAMAM dendrimers took this field to another level for targeted imaging. In this work, folic acid was fabricated on the surface of the G5 PAMAM dendrimer via a conventional amide coupling reaction. Then, the rest of the terminally free amino groups were conjugated with DOTA (Swanson et al. 2008). Based on the design, a typical water-soluble coupling reagent like EDC (1-ethyl-3-(3-dimethylaminopropyl)carbodiimide) could easily work to mediate the conjugation of materials of different polarity, such as in this case, folic acid and the dendrimer. On average, 4.5 folic acid-targeted molecules were incorporated onto the G5 dendrimer (Fig. 6.12). To control the amount of DOTA-Gd attached to the dendrimer, approximately half of the amino groups were acetylated via controlled chemistry. The rest of the amines were let to react with NCS-DOTA

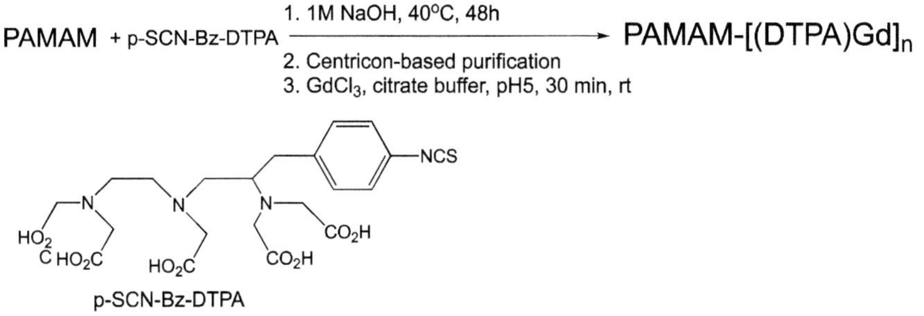

Fig. 6.11 Design of PAMAM-[(DTPA)Gd]$_n$. Data deduced from Kobayashi et al. (2001)

at high pH conditions, followed by coordination with Gd^{3+} ions to afford the folic acid-targeted dendritic chelate.

Compared to the gadolinium complex of diethylenetriamine pentaacetic acid bismethylamide (Omniscan™), the relaxivity Gd-DOTA complex associated with dendrimer is 4 times greater than that of Omniscan™, making this Gd(III)-DOTA-G5-folic acid an ideal contrast agent MRI. The specificity and signal enhancement of the targeted probe was tested by intravenous injection into the tail veins of folic acid receptor-expressing human epithelial cancer (KB) cells, implanted on immune-compromised SCID mice. On the other flank of the same mouse, xenograft tumor using MCA207 cells, which is negative for folate receptors, was used as control. With an injected dose of approximately 0.03 mmol of Gd/Kg for both the Gd(III)-DOTA-G5-folic acid and the non-targeted control probe Gd(III)-DOTA-G5, the former showed high signal intensity in the tumor, liver, kidney, and gut. The T1-mediated signal started to pick up 1 h post-injection and remained in the tumor 48 h. While

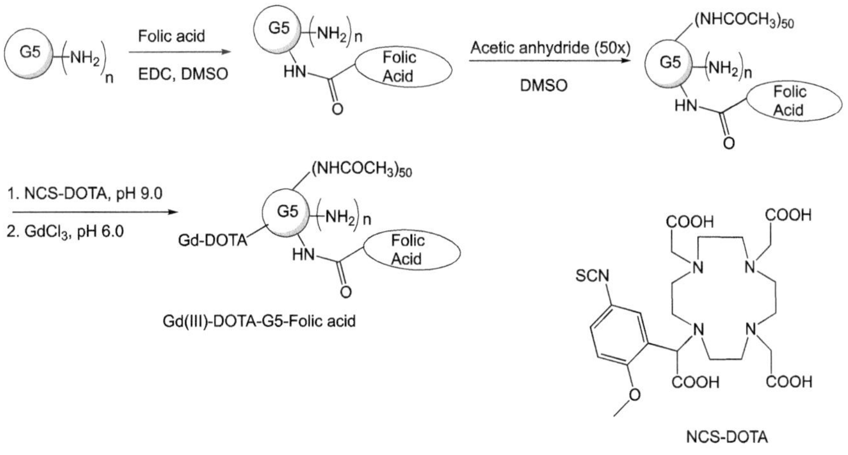

Fig. 6.12 Folic acid-targeted MRI probe employing heterogenous coupling strategy on polyamidoamine dendrimer. Data deduced from Swanson et al. (2008)

the latter showed the signal in the tumor 1 h post-injection, then washed out through the kidneys, as evidenced by the bright signal intensity in the bladder obtained from the 1 h and 4 h images.

Another way to amplify the MR signal is by exploiting the impeccably intrinsic machinery in the biological milieu to accomplish the task. One example of this category involves myeloperoxidase (MPO), a heme-containing enzyme. MPO is found mostly in the granules of polymorphonuclear leukocytes and monocytes (Schultz and Kaminker 1962), and it is released into the cell vacuoles during phagocytosis (Tobler and Koeffler 1991). This enzyme catalyzes substrates using a typical peroxidase cycle (Dunford and Hsuanyu 1999), as shown in steps 1–3 of Fig. 6.13.

Basically, hydrogen peroxide available from the oxidative burst of neutrophils oxidizes native MPO to form the intermediate compound I (MPO-I), followed by two successive reduction steps to regenerate the native enzyme; as shown in reactions (Chen et al. 2004) by interaction with the aromatic electron donor (AH_2) via intermediate compound II (MPO-II) (Fig. 6.13). Given the natural abundance of chloride, the peroxidase compound I (MPO-I) can also oxidize chloride ions to regenerate MPO (Chen et al. 2004). The product of such a reaction generates hypochlorous acid (HOCl), a potent oxidant, which acts against invading bacteria, viruses, and tumor cells. However, the protective mechanism can also cause a deleterious effect on the host tissues. And it seems this is the major reaction in the biological system compared to pathways (Chen et al. 2004) via electron donor AH_2. However, in some cases, if the donor of electron AH_2 is appropriately designed, as the way nature always does, it will outpace the MPO-H_2O_2-Cl^- system. For example, it has been reported that the rate constant for MPO-I reaction with serotonin (5-hydroxytryptamine (5-HT)) at neutral pH is approximately 10 times as fast as that for its reaction with chloride, even the concentration of chloride was over 5000 times higher (Dunford and Hsuanyu 1999). This observation leads to a hypothesis that it is possible to image MPO if the substrate 5-HT is conjugated with a paramagnetic gadolinium probe. The oxidized products of reactions could form polymeric products initiated by a spontaneous radical chain reaction (Chen et al. 2004). The rationale is that these polymers would offer higher magnetic signals than their monomeric counterparts. Three different AH_2 substrates attached to GlyMetDOTA-Gd complex were developed as shown in Fig. 6.14. First, the MPO substrates were conjugated into GlyMetDOTA tert-Butyl ester using a conventional coupling procedure, followed by deprotection of the ester using 90% TFA. The free carboxylate products were purified by repeated precipitation in ether. Next, the association of the chelator with gadolinium chloride occurred in optimized conditions using triethylammonium acetate at pH5 at room temperature. All three final products were purified by recrystallization, followed by HPLC to afford high-grade materials suitable for biological assay.

Among the three probes, 5-HT-DOTA(Gd) demonstrated ideal for MPO with higher relaxivity and fast kinetics. Size-exclusion HPLC confirmed the direct evidence of polymerization of the probe upon exposure to MPO. Two retention-time peaks were detected in the chromatogram corresponding to the starting material and the associated polymer product. The MPO-converted product, with a much larger size, resulting from polymerization, eluted out first at 20.8 min. While the

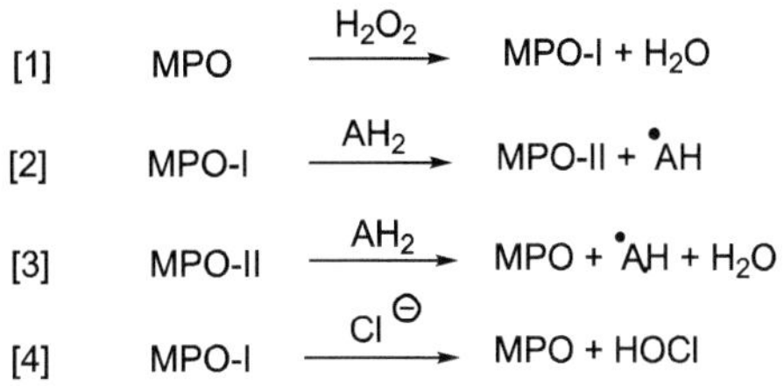

Fig. 6.13 Myeloperoxidase reaction via the peroxidase cycle. Data deduced from Dunford and Hsuanyu (1999)

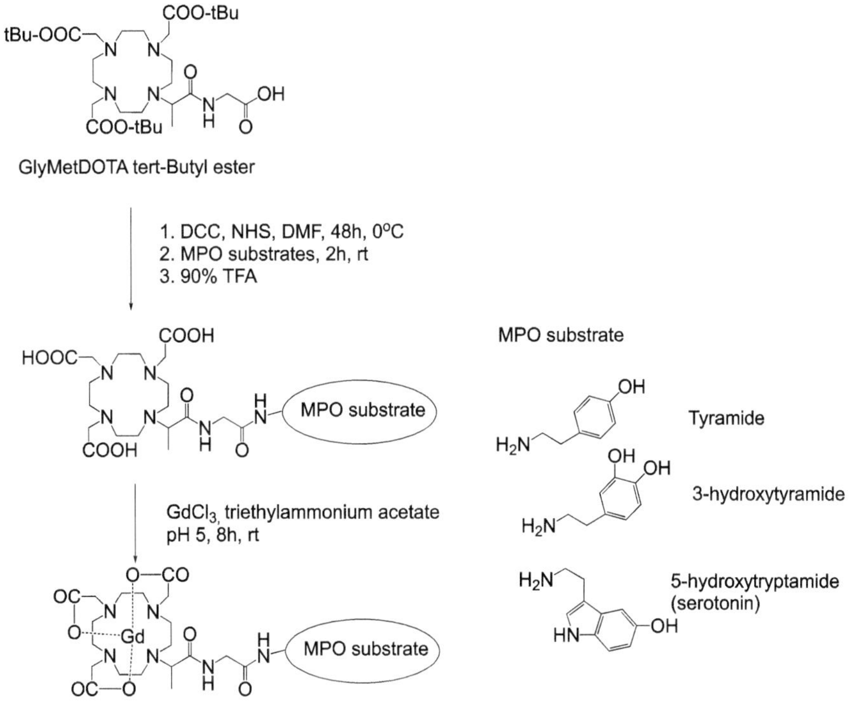

[1]   MPO $\xrightarrow{H_2O_2}$ MPO-I + H_2O

[2]   MPO-I $\xrightarrow{AH_2}$ MPO-II + $^\bullet$AH

[3]   MPO-II $\xrightarrow{AH_2}$ MPO + $^\bullet$AH + H_2O

[4]   MPO-I $\xrightarrow{Cl^\ominus}$ MPO + HOCl

GlyMetDOTA tert-Butyl ester

1. DCC, NHS, DMF, 48h, 0°C
2. MPO substrates, 2h, rt
3. 90% TFA

GdCl$_3$, triethylammonium acetate
pH 5, 8h, rt

MPO substrate

Tyramide

3-hydroxytyramide

5-hydroxytryptamide (serotonin)

Fig. 6.14 Myeloperoxidase-mediated polymerized MRI probe. Data obtained from Chen et al. (2004)

enzyme-free substrate, with a much smaller size, was eluted at 23.9 min. As the result of MPO-mediated polymerization 5-HT-DOTA(Gd), a largely enhanced relaxivity was observed compared to other substrates for an equivalent amount of MPO. Visible signal changes were detected in a low-field MRI scanner. The T_1-weighted images exhibit a linear regression between signal intensity and the MPO concentration (Fig. 6.15). The probe can report the presence of MPO as small as 250–650 U. further, it can unequivocally distinguish a minute fluctuation of MPO. For example, by increasing the MPO from 1300 to 2000 U, a recorded 40% signal difference was observed (Fig. 6.15c,d).

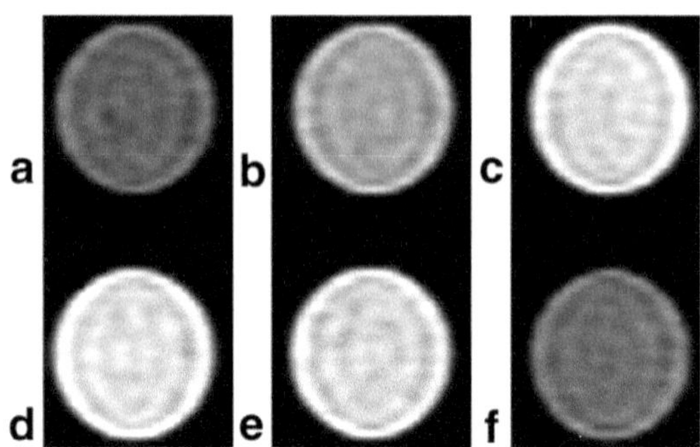

Fig. 6.15 Specificity of the 5-HT-DOTA(Gd) probe in the presence of **a** no MPO; **b** 650 U MPO; **c** 1300 U MPO; **d** 2000 U MPO; **e** 1400 U horseradish peroxidase; and **f** 650 U MPO without H_2O_2. Data obtained from Chen et al. (2004)

Targeted imaging with SPIO nanoparticles is another area that has attracted significant attention in the past two decades. SPIO nanoparticles alone are the non-targeted probe, but their large size and the chemical constituent of the coating materials are often singled out as a target for immune cells, such as dendritic cells or macrophages, and thus this approach makes SPIOs targeted probes. One of the most notable experiments that exemplifies this notion is work focusing on the use of SPIO nanoparticles for targeted imaging of dendritic cells (DCs). DCs are the primary initiators and modulators of both innate and adaptive immune responses (Pham et al. 2009). Upon uptake of endogenous and/or exogenous substances, DCs undergo a cascade steps of maturation, in a process described in Fig. 6.16. This manifests a major turning point in the phenotypical expression, particularly with the hallmarks of upregulation of MHC-I, MHC-II, and the costimulatory molecules, such as CD80 and CD86, along with the secretion of Il-12 (Cella et al. 1996; Koch et al. 1996). The loss of antigen-uptake capacity also accompanies this maturation process, a change in cell morphology, and the increased expression of the chemokine receptors, such as CCR7, which guides the migration of DCs into the secondary lymphoid tissues (Pham et al. 2009). It is the most important process in a healthy individual to guard off infection and other major illnesses, including cancer. However, existing growth factors and cytokines in the tumor microenvironment compromise the function of endogenous DCs, and thus, the tumor escapes immune surveillance. In clinical cell-mediated immunotherapy, DCs are isolated from the hosts and pulsed with exogenous and specific tumor antigens in a culture dish. The specific antigen-activated DCs are then being transferred back to the hosts to enhance the immune response against tumor targets. In this therapy, the recruitment of DCs from the injection site to the secondary lymphoid tissues, such as lymph nodes, is critical for the induction of adaptive immune response. Information regarding DC migration, fate, and how DCs interact with other cells in the lymph nodes remains a subject of intensive study.

Imaging DC migration with SPIO nanoparticles is probably, one of the ideal modalities since MRI provides the sharpest spatial resolution among the imaging methods, with excellent dynamic information and anatomical contrast. The SPIO nanoparticles are not only safe, but they have a long half-life and biodegrading, the quality that is suitable for longitudinal tracking of adaptively transferred DCs. In

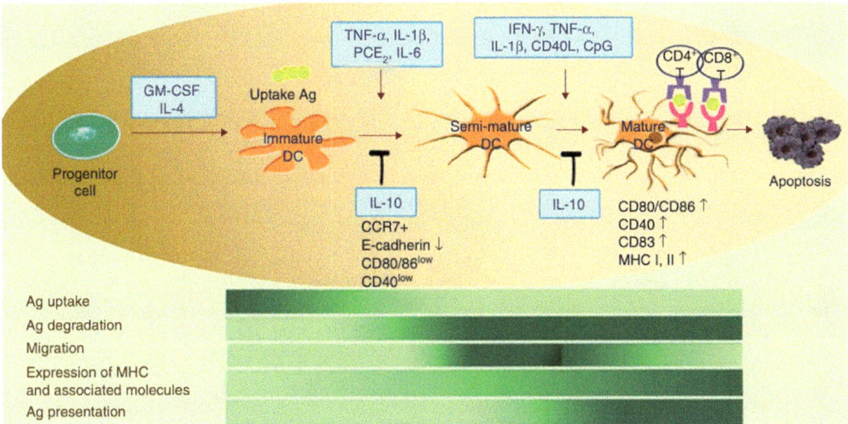

Fig. 6.16 Mechanism of DCs uptake of antigens/exogenous biologics and subsequent ontogeny. Data obtained from Pham et al. (2009)

addition, imaging DCs using T2 negative contrast provides a clear contrast to the background of soft tissue, such as the lymph nodes. In a clinical work using an FDA-approved dextran-coated SPIO (Feridex-USA, Endorem-Europe), it has been demonstrated that DCs phagocyte the nanoparticle substantially within a brief exposure of 200 μg/mL SPIO. Notably, this uptake has no interference with the process of cellular maturation (Fig. 6.17a). With a clinical 3-T MRI scanner, the injected SPIO-labeled DCs could be detected as a result of decreased signal intensity at the lymph node (Fig. 6.17b). The work also demonstrated MR imaging offers excellent resolution, and thus, detailed cellular distribution to the lymph nodes can be accounted appropriate. While imaging DCs via scintigraphy using ^{111}In-oxine-labeled DCs missed some lymph nodes due to poor resolution.

This work demonstrated the power of cellular MRI in clinical immune-cell therapy. It helps to monitor the accuracy of the injection and migratory capacity of DCs. With the assistance of MRI, physicians can track and know whether there is a likelihood the therapy would work, and correction can be implemented promptly to improve the procedure. For instance, MRI data showed that only 50% of SPIO-labeled DCs were correctly injected into the lymph nodes. Overall, this multidisciplinary approach toward DC-based therapy is feasible. As demonstrated in another study, which found that both dextran-coated SPIO and dextran alone all are safe to use. There is no registered toxicity 12 h after immature bone marrow-derived DCs were incubated with high concentration of dextran-coated SPIO or dextran (Fig. 6.18a,c). The nanoparticles do not induce cell maturation except at very high concentrations; a recognizable expression of CD40 was observed, but this happened only with extended incubation time (Fig. 6.18b). And this upregulation of CD40 seems to be attributed to the SPIO, not dextran (Fig. 6.18c,d)

To develop SPIO for labeling targeted ligands, the coating materials can serve this purpose fine. The weak nucleophiles of hydroxyl groups on dextran can be converted

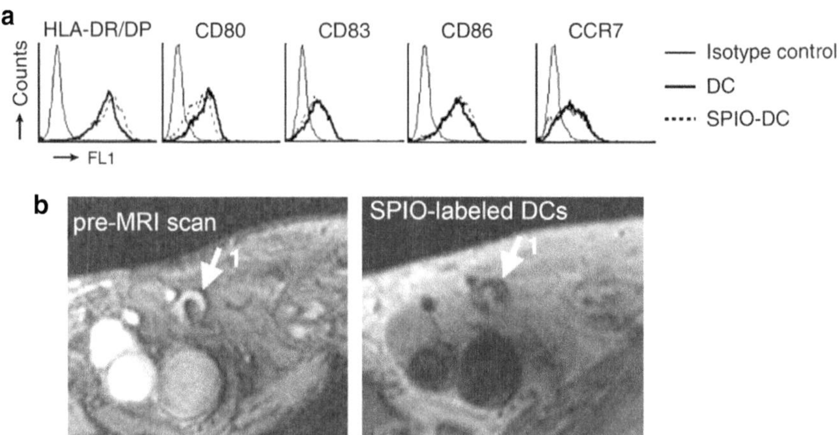

Fig. 6.17 **a** Cell sorting analysis to assess cell phenotypical expression of major histocompatibility complex class II (HLA-DR/DP), costimulatory molecules (CD80, CD86), DC maturation marker CD83 and chemokine receptor CCR7; **b** Tracking the delivery of SPIO-labeled DCs before (pre-MRI scan) and after intranodal injection in a patient. Data obtained from Vries et al. (2005) with permission from the author and Springer Nature

to amines using functionalized linkers. At present, the amine group appears to be an ideal chemical moiety for the functionalization of SPIO nanoparticles; this is due to its strong nucleophilicity and compatibility with a wide range of available amine-bearing biological materials. On the other hand, the functionalization of nanoparticles with carboxylic acid groups enables the conjugation of native peptides, proteins, or antibodies.

One notable work reported the conversion of oleylamine-coated SPIO into amine groups using dopamine as an anchoring analog (Young et al. 2009). The pyrocatechol moiety on dopamine can displace oleylamine and serve as a robust anchor to immobilize functional groups on the iron oxide shell of magnetic nanoparticles. Based on the spectroscopic analysis, it is suggested that the pyrocatechol behaves like a chelating agent, where it forms a tight bonding to the iron oxides by converting the under-coordinated cationic iron surface sites to bulk-like lattice structure (Xu et al. 2004).

Since dextran-coated SPIO nanoparticles are commercially available, developing chemistry methods to modify these products without affecting their physical property is of paramount importance since it will facilitate targeted imaging. The functionalized SPIO nanoparticles will also enable multimodal imaging, which we will discuss soon after. For direct conversion of weak nucleophilic hydroxyl groups from dextran-coated SPIO nanoparticles, versatile linkers with high reactivity of the epoxide ring. This highly strained ring can be opened in acid, basic, or sometimes, neutral reaction conditions. The synthesis of the epoxide linker started with Gabriel synthesis, which was discussed in great detail in Chap. 4 (Fig. 4.43).

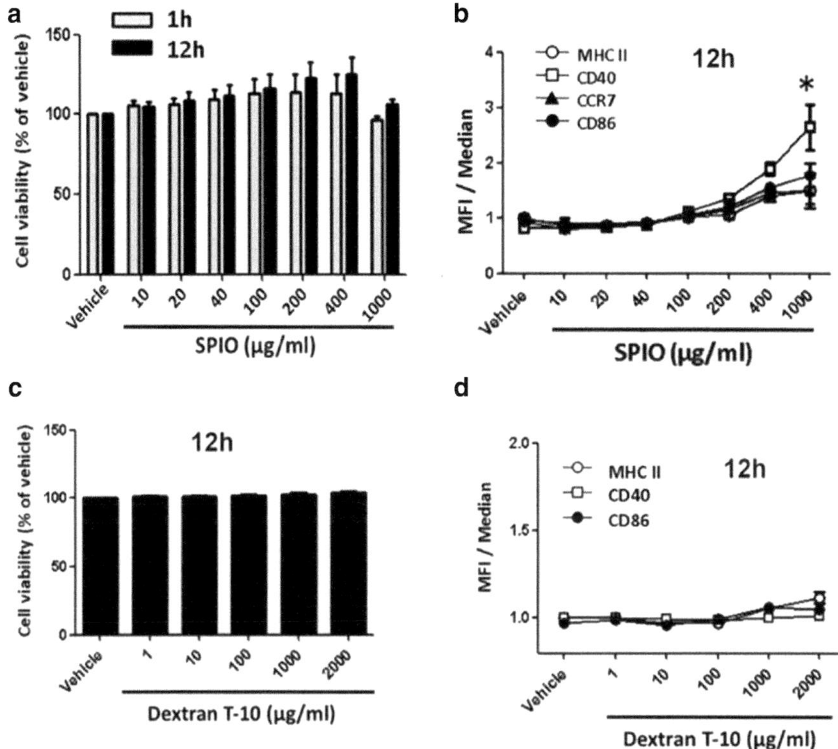

Fig. 6.18 Toxicity of immature bone marrow-derived DCs after incubation with dextran-coated SPIO and dextran T10; *, statistical significance P < 0.05, compared to the vehicle. Data obtained from Toki et al. (2013)

Another approach, which was used widely to functionalize dextran-coated SPIO nanoparticles for conjugation with amine-reactive bioligands, involved the oxidation of the unreactive hydroxyls of dextran into aldehydes using sodium periodate (Dunford and Hsuanyu 1999). This simple yet robust chemistry offers a wide spectrum of applications. For example, in an effort to develop a hybrid magneto- optical probe for tracking the migration of DCs. The advantage of the wet-lab chemistry for making this dual probe is that it used premade dextran-coated SPIO (SPIO-DX) and glutathione (GSH)-associated quantum dot (QD). The assembling occurred via a covalent linking of the oxidized dextran shell of magnetic particles to the glutathione ligands of QDs. The mechanism of periodate oxidation of the pendent diol groups of dextran has been reported somewhere in the past (Hermanson 2008). In principle, sodium periodate cleaves the carbon–carbon bond that possesses adjacent hydroxyls, converting the hydroxyl groups into highly active aldehyde moieties (Fig. 6.19). This simple oxidation reaction can be achieved effort-free; it occurred briefly in 1 h at room temperature. Usually, the size of desired products accumulates; thus, they can be separated from the starting materials via dialysis or using cut-off Centricon spin filters with an appropriate retentate membrane.

With several aldehyde groups coated on the surface of SPIO nanoparticles, exposing GSH-associated QDs would lead to the fabrication of smaller-sized QDs surrounding the SPIO via Schiff-based reaction. The resulting imine groups were further reduced to an amine using sodium borohydride. The conjugation of QDs onto the surface of SPIO can be confirmed by dynamic light scattering, where the particle size increases from 30-nm SPIO nanoparticles to 44 nm, assumingly the QD-SPIO complex. While no noticeable size change was observed when comparing the dextran-coated SPIO nanoparticles to the periodate oxidation product. The presence of the SPIO nanoparticles in the composite nanoparticle was exploited to facilitate purification using a magnetic column. The luminescence of this magneto-optical

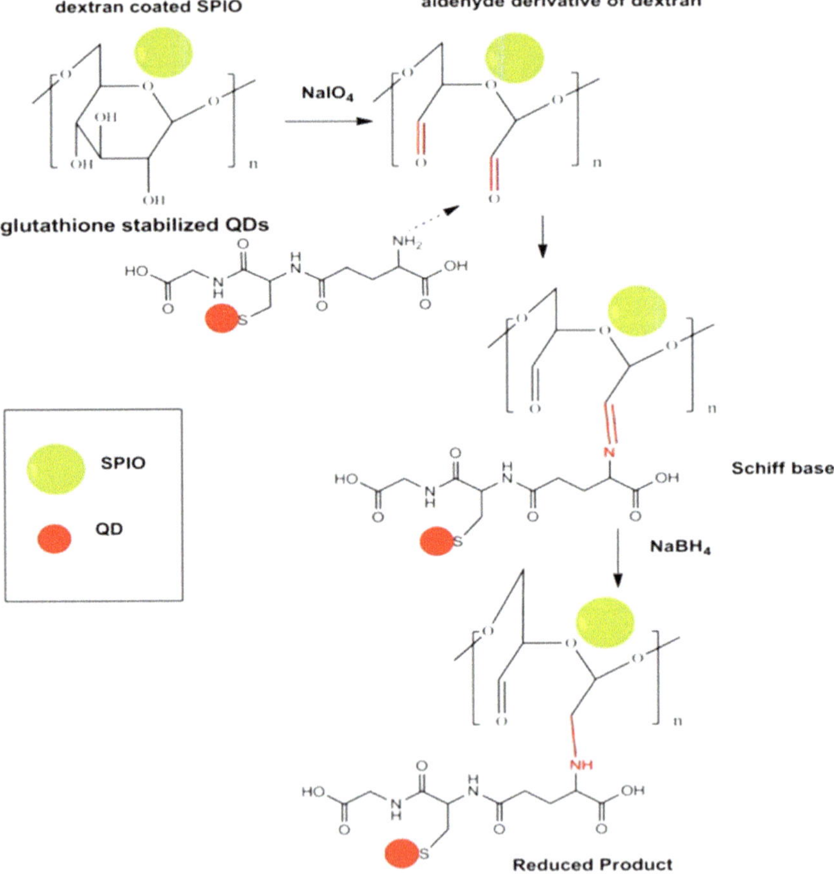

Fig. 6.19 A mechanism for developing a dual magneto-optical probe via the oxidation of hydroxyls on dextran-coated SPIO. Data obtained from Koktysh et al. (2011)

probe is strong, and it can be detected with a regular UV lamp, suggesting the dextran-GSH may serve as a spacer that helps to prevent fluorescent signal quenching by the paramagnetic core (Corr et al. 2008). Further, the composite nanoparticles can be concentrated by magnetic decantation using a magnetic separation stand, proving the complex's strong magnetism.

Another innovative chemistry has been developed to generate multimodal probes using a simple wet-lab operation. What is most noteworthy about this approach is that it enables the generation of sophisticated probes with a "plug-and-play" design, the process is experimentally simple, albeit appears to have enormous scope. First, the overall negative charge attributed by the dextran's hydroxyl groups was transformed into positive charge with the use of polyallylamine chloride (PAH) under acidic conditions. Successful surface modification with this layer alone doubles the size of the precursor SPIO-DX nanoparticles. Next, the SPIO-DX-PAH product was further combined with preformulated 3-mercaptopropionic acid (MPA)-stabilized (CdTe/CdS/ZnS) using as visible probes or (CdHgTe/CdS/ZnS) QDs for use as near-infrared probes. In this design, the charged polyelectrolyte PAH has two roles. First, it creates a net positive charge on the surface (Mackay et al. 2011) in order to fabricate the QDs via electrostatic interaction with the negative charge available from MPA-stabilized QDs. Second, it serves as a buffer zone to minimize the fluorescence quenching effect due to the proximity of the QDs to the iron core of the SPIO. Finally, the hybridized magneto-optical probe was coated with the transfecting reagent polylysine, which facilitates probe delivery across the cellular membrane. In every step of modification, the desired materials were purified using size-exclusion Centricon filters and characterized by quartz crystal microbalance (QCM).

The overall size of the polylysine-coated magneto-optical probe was about 123 nm, an ideal target for phagocytes, such as DCs. Particularly, as a transfecting reagent, polylysine also enhances DC uptake. To demonstrate the utility of this cell-based probe, the cells were injected into the footpad of animals, then, using non-invasive optical and MR imaging to track the distribution and homing of DCs in the draining lymph nodes. The MR imaging revealed the swelling of popliteal lymph node and an apparent signal reduction compared to the control lymph node due to the presence of T2-associated SPIO nanoparticles. The hybrid probe enables multi-faceted confirmation of the labeled cells using fluorescence microscopy of the excised lymph nodes (Fig. 6.20). Usually, after imaging, selected animals were sacrificed, and the excised lymph nodes were prepared for immunohistochemistry. Consecutive slides of the immunohistochemistry data indicated that these labeled DCs became mature (CD11c) while migrating to the draining lymph nodes. This information corroborates with MR imaging as it detected swelling lymph nodes in the live animals.

6.4 Imaging with Activatable Probes

Among the many molecular probes discussed so far, activatable contrast agents represent one of the most exciting behemoths in molecular imaging. If anything can tip

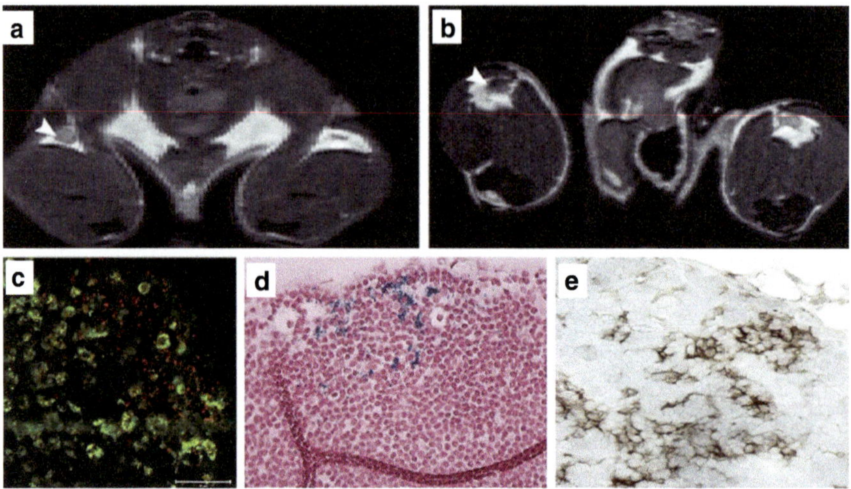

Fig. 6.20 Magneto-optical probe for imaging the migration of DCs. **a** Prescan of the right popliteal lymph node (arrowhead) using 9.4 T MRI scanner; **b** 24 h post-scan of the same lymph node (arrowhead). (C,D,E) Immunohistochemical staining of the popliteal lymph node; with **c** optical imaging via QDs; **d** Perls's staining for iron; **e** staining for DCs using anti-CD$_{11c}$ antibody. Data obtained from Mackay et al. (2011)

the balance toward a breakthrough in precision molecular imaging, the contribution of activatable probes would not be ignored. Yet still immature, "smart" activatable probes remain in the realm of the soon-to-be conquered domain, with enormous capability, implications, and opportunities. With the emergence of nanoscale molecular motors, the vision of self-propelled and autonomous surveillance incorporated with activatable probes for personalized imaging is not unimaginable. The advantage of this family of probes compared to the targeted probes discussed above is that they offer low background signals and high specificity.

The biological system is an overwhelmingly complex and dynamic environment. This network is not only responsible for the production and maintenance of the components that contribute to cellular functioning and homeostasis but also compensates for any irregular changes caused by internal and external perturbations (Bich et al. 2016). To detect the difference between normal conditions from pathological situations through these biological outputs requires understanding that this is not a yes/no algorithm like in a binary system where only two digits, 0 or 1, govern the whole matrix. But rather, the biological system is more complex, and thus probes should navigate in a fine line, reporting small outputs, fluctuating through a myriad of events, in between "more or less" gating (i.e., desirably detected intensity). And certainly, cross-activation of the probes (i.e., specificity) is highly possible in a complex biological milieu. So the signals that emerge as the result of biological action are a sum of more than a single event, but hopefully, or at least, the probe must guarantee the intended target contributes the most among others. In fact, some protein targets can recognize the probes with exquisite specificity, while it is not the case for others. For instance, some proteases can recognize and process multiple peptide substrates in a promiscuous manner. Given these foreseeable issues, aside from developing stable, reliable, reproducible probes, the associated design of a quantitative algorithm to threshold the noise during the imaging process is necessary.

Several biological targets can serve as a "logic switch" to activate the probes, and the most prominent family of such entities must include the proteases. These cleavable enzymes function like an executioner scissor. They catalyze specific proteolytic reactions, thereby producing new protein products contributing to cell development and homeostasis, including cell repair, proliferation, differentiation, wound repair, tissue remodeling/morphogenesis, neurogenesis, inflammation, autophagy, necrosis, apoptosis and more (Lopez-Otin and Bond 2008; Neurath and Walsh 1976). Depending on the mechanism of catalysis and the disease model for targeted imaging, proteases are classified into six distinct families, including aspartic, glutamic, metalloproteases, cysteine, serine, and threonine proteases, albeit glutamic proteases have not been identified in mammals up to date (Lopez-Otin and Bond 2008).

Other biological domains could serve for probe activation via protein topology. The three-dimensional structures arranged to generate different conformations, such as α-helix or β-sheet, either can influence the probe's physical property with switchable enhanced optical outputs. Several dyes, naturally or chemically developed, which will be discussed later, have such an ability.

The biological systems have many other dynamic environments that could be exploited for logic switches. The active pumps of the cellular membrane generating the gradient, caused by the electrochemical potential of ions across the cellular membrane, is a great mechanism that could be exploited to activate the probes. Other cellular motions could serve as molecular motors, such as myosins, kinesins, and dyneins responsible for cellular activities for muscle contraction and intracellular vesicle transport (Iino et al. 2020).

1. **Topologically activable probes**

Typically, any fluorescent dyes with donor and acceptor ends connected with or without a π-linker would experience a noticeable conformational degree of freedom. Particularly, this rotation will quench the fluorescent signal when the dye is unbound. However, upon binding to protein with favorable conformation, which will sterically and electronically stabilize the dye's ground state electronic distribution, reducing non-radiative relaxation pathways and increasing fluorescence efficiency (Wolfe et al. 2010). For example, the presence of two rotational degrees of freedom of pPZPQ (p-phenothiazine phenylquinoline) resulted in quenching of the delayed fluorescence of the quasi-equatorial conformers (Fig. 6.21). While the PZPQ analog, that has only one axis of rotation, exhibited short fluorescence of the quasi-axial conformer and thermally activated delayed fluorescence of the quasi-equatorial form (Sych et al. 2021).

Thioflavin T has the same characters as PZPQ. When in the free form, thioflavin T emits a weak fluorescent signal at a lambda max of approximately 440–445 nm because after it absorbs a photon, it proceeds to a state with a rotation angle around the axis (Wolfe et al. 2010). However, when bound to protein with a cross-β structure,

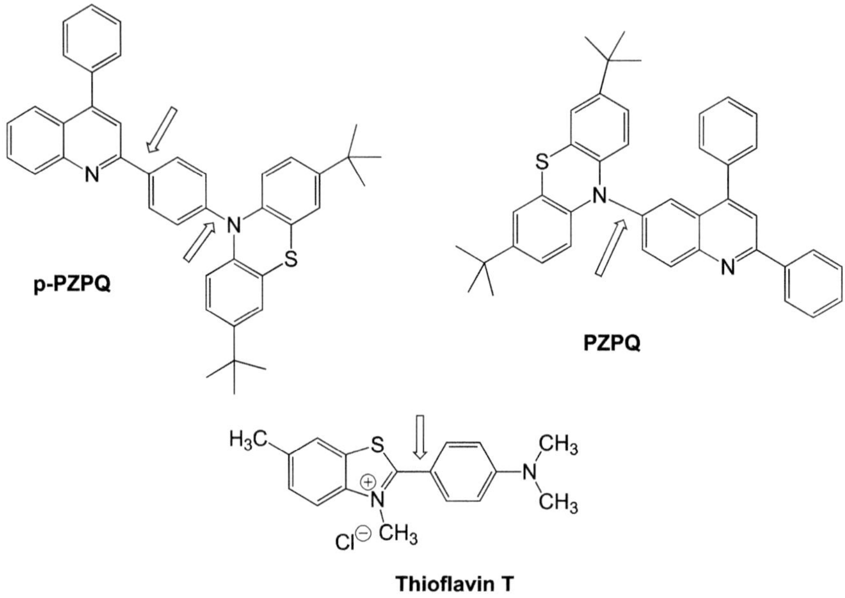

p-PZPQ

PZPQ

Thioflavin T

Fig. 6.21 Typical dyes would quench in free form due to molecular rotation along the axis linked between the end rings. Part of the data were deduced from Sych et al. (2021)

the internal molecular rotation of the dye is reduced, thus displaying a characteristic shift of the excitation lambda max, from 385 to 450 nm, and the emission lambda maximum, from 445 to 482 nm (LeVine 1993; Naiki et al. 1989), with approximately 1000-fold signal enhancement (Dyrager et al. 2017). This reporter exemplifies a model molecular switch capable of sensing protein topology. Thioflavin T serves as a "gold standard" reporter in numerous in vitro assays to detect protein aggregation with β-sheet-rich fiber, specifically amyloid-β. Since no crystal structure is available for amyloid-β, a detailed mechanism of thioflavin T binding to amyloid fibrils is unavailable. However, using a crystal structure of amyloid-like oligomer (β-2 microglobulin), it has been reported that when thioflavin T was cocrystalized with aggregated β-2 microglobulin, the fluorescence intensity of the resulted complex enhanced significantly compared to thioflavin T associated non-aggregated version of β-2 microglobulin (Wolfe et al. 2010). The detailed crystal structure of the binding complex showed that thioflavin T binds to aggregated β-2 microglobulin in a direction orthogonal to the β- sheet of the protein by stacking with the aromatic amino acids, leading to stabilization of the fluorophore's electronic distribution, decreasing radiation decay rate in both ground and excited states. Consequently, this phenomenon contributes to the observed increase in the fluorescence quantum yield. This binding is very specific to aggregated protein since no such observation in non-aggregated protein, making thioflavin T a unique logic reporter of amyloid aggregation. In general terms, topologically activatable dyes can be perceived in Fig. 6.22.

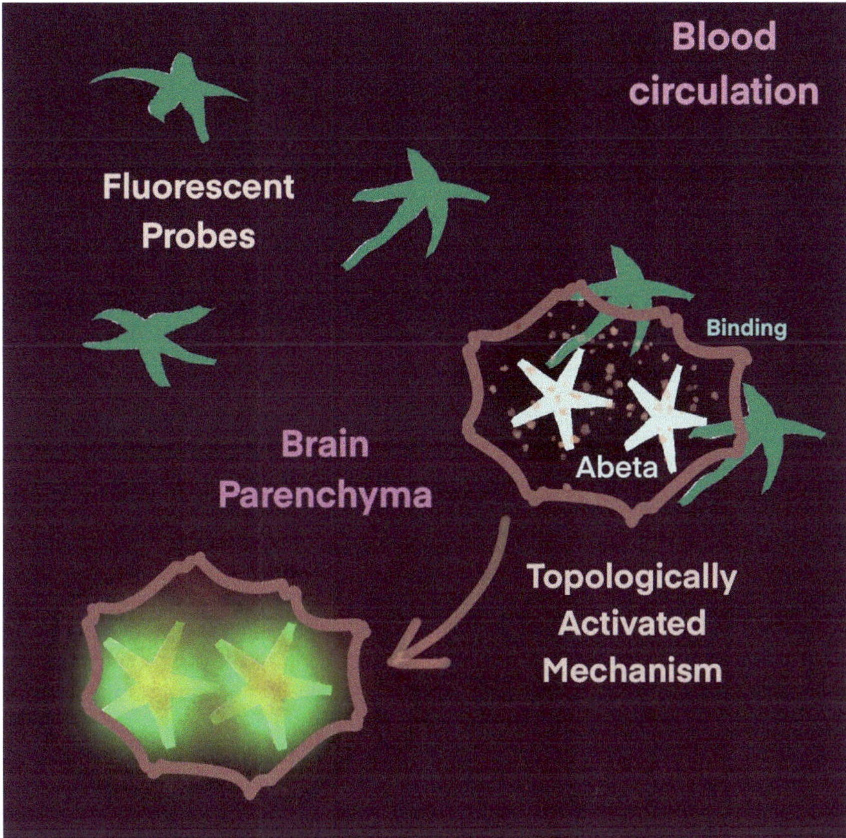

Fig. 6.22 Binding mechanism of topologically activatable probes

Aside from histological analysis of amyloid in tissue staining, thioflavin T serves as excellent fluorescent readouts for real-time monitoring the fibrillary kinetics (Naiki et al. 1989; Khurana et al. 2005). Other applications include high-throughput screening assays to identify amyloid inhibitors (McClure et al. 2019). This work postulates that at equilibrium, the binding of thioflavin T with amyloidal lysates can be displaced by a competitive inhibitor, resulting in low fluorescence quantum yield due to the conversion to the free dye in the reaction solution. The more competitive an inhibitor, the less the fluorescent signal will be recorded as more quenched fluorophores emerge in the solution.

For advanced application, a neutral version of thioflavin T (BTA-1) has been reported (Wu et al. 2008) to facilitate in vivo application since cationic moiety from the parent compound may impede the distribution across the blood–brain barrier. The neutral compound has a lipophilicity of 600-fold compared to the cationic counterpart. This may also be a reason why this dye exhibits better fluorescence compared to thioflavin. However, the electron-vibrational coupling between the BTA-1 and the solvent is reduced and thus an enhanced fluorescence emission. With the availability of this neutral compound, further binding mechanism employing the model system of the protofibril $A\beta_{16-22}$ (KLVFFAE), the aggregated oligomer of the Alzheimer's amyloid-β. The data showed that the positive charge on the benzothiazole ring has no role in the binding motif since there is no association of thioflavin to any negatively charged amino acid from the oligomer, but thioflavin T rather binds to hydrophobic and aromatic residues, confirming other reports suggesting it binds through aromatic via π-π stacking/hydrophobic interaction (Wolfe et al. 2010). BTA-1 has the biding affinity Ki of 20 nM compared to that of thioflavin T, 890 nM.

Aside from creating neutral dyes for in vivo imaging of amyloid-β, increasing the hydrophobicity also has some benefits. As shown in Fig. 6.23, modified thioflavin T analogs were developed to enhance affinity for amyloid fibrils, and improve blood–brain barrier penetration. This one-step chemical reaction is versatile and could be used for generating a large library of thioflavin T analogs if the yield improves. The condensation reaction between 4-amino-2-methoxybenzoic acid and 2-aminophenol or 2-aminothiophenol afforded the desired product, where the arylamine can be further alkylated as mono- or dimethylamine. A proof-of-principal study showed that even with a small collection of six compounds, one exhibited promising physical property merits further investigation. The 2-(2'-methoxy-4'-methylamino phenyl)benzoxazole has an affinity to amyloid-β fibrils ($K_D = 3.27\,\mu M$ vs. $4.12\,\mu M$) and fluorescence responsiveness ($F_{A\beta} = 36.1$ vs. 33.1) better than thioflavin T.

Once the design pattern of topologically activated probes is generalized, such as the heterogenous donor–acceptor entities linked through an axis of rotation acting as a molecular rotor, as described in a more engineering term (Aliyan et al. 2019), to quench fluorescence in the solution, more advanced analogs emerged with improved amyloid-β detection capability. For example, a neutral thienoquinoxaline has been demonstrated to serve as a great reporter for aggregated amyloid-β (1–40). It is not surprising that this chemical backbone has no or weak fluorescence as a free dye in the aqueous solution. When bound to amyloid-β the fluorescence emission

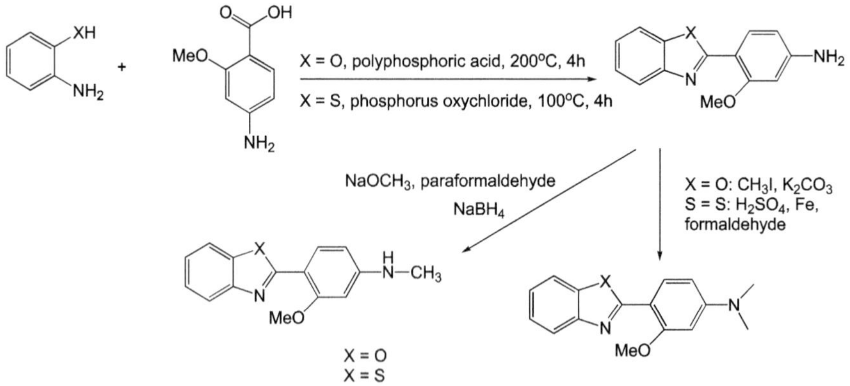

Fig. 6.23 Thioflavin T analogs for in vivo applications. Data deduced from Jung et al. (2013)

enhanced remarkably (Benzeid et al. 2012). It is worth noting that the Stokes shift of the dye extends over 100 nm, suitable for resolute imaging data without little signal interference.

The synthesis started with the generation of quinoxaline via the condensation reaction between *o*-phenylenediamine and ethyl pyruvate in refluxing ethanol (Fig. 6.24). Then, a condensation of quinoxaline with 4-N,N'-dimethylbenzaldehyde in neat condition at 200 °C for 2 h to afford a styryl-quinoxaline. This intermediate also has the ability to serve as a molecular switch to report aggregated amyloid-β with a long emission wavelength, given the extended unsaturated bond. Finally, the thieno-quinoxaline was achieved by refluxing styryl-quinoxaline with Lawesson's reagent in toluene.

It is worth mentioning another excellent chemistry for developing activatable probe mimetic of thioflavin T. Although the generated 1,2,3-triazole-linked benzoth-iazole and benzothiadiazole dyes display less superior to thioflavin T regarding signal intensity and other physical property to make these dyes in the road for in vivo applications. However, the chemistry itself and the unique structures of these benzothiazole-triazole backbones can serve as a template for future improvement and diversification of the chemical genetics along with these motifs for future and better amyloid-β-binding switcher dyes. In particularly, conversion of these amyloid-β binding molecules into PET radioligands is another possibility to exploit their full potential.

At the first impression, one would think the direct coupling between benzothiazole with triazole ring would accomplish the desired product (Fig. 6.25). However, the lack of suitable starting materials impeded this approach. Further, this idea lacks the capability for product diversification. Instead, in a more clever and straightforward approach, the benzothiazole ring was modified to facilitate an azide-alkyne Huisgen cycloaddition reaction to make the extended triazole ring. Based on this construct, the dipolarophile was generated available on the benzothiazole ring, which will react with the 1,3-dipolar moiety through cycloaddition reaction. Using the commercially available 2-amino-6-methoxy-benzothiazole, the iodo derivative can be easily achieved via diazotization-Sandmeyer iodination (Dyrager et al. 2017). Then, the

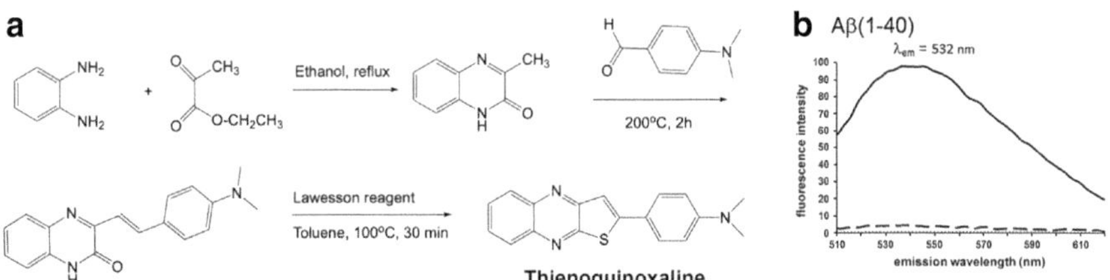

Fig. 6.24 **a** Chemical design and synthesis of a thioflavin T-derived topologically activated thieno-quinoxaline for the detection of aggregated amyloid-β (1–40); **b** the fluorescence signal intensity of thienoquinoxaline the absence (dotted line) and presence (continuous line) of aggregated Abeta. Data obtained from Benzeid et al. (2012)

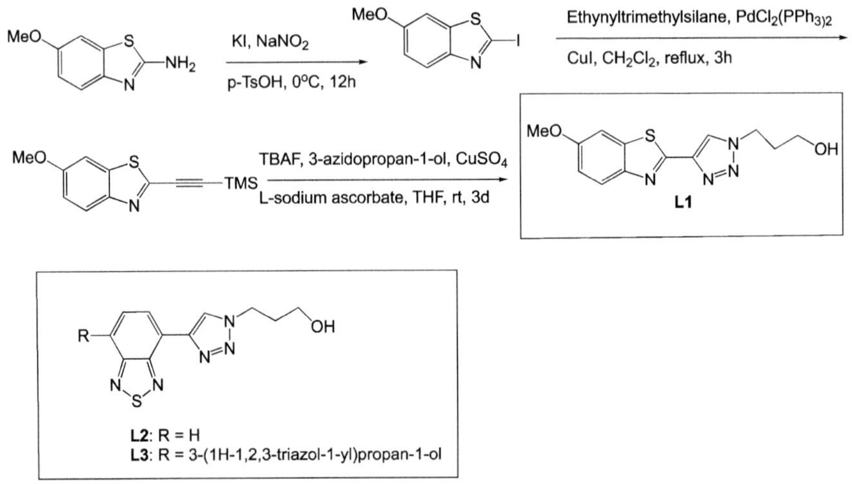

Fig. 6.25 Design and synthesis of benzothiazole-triazole activatable dye. Data obtained from Dyrager et al. (2017)

dipolarophile acetylene protected with TMS was introduced in the place of iodine using a palladium-mediated Sonogashira reaction.

After deprotection of the TMS group, Huisgen cycloaddition reaction between the dipolarophile alkyne with 3-azidopropan-1-ol to furnish the final product. Using similar chemistry, **L2** and **L3** analogs were also synthesized.

Histology staining of mouse brain sections obtained from APP/PS1 mice showed that three compounds bind to Amyloid-β with excellent specificity. The data corroborate with immunohistochemical staining using 6E10 antibodies.

Congo red dye is another notable activatable probe to some extent with a similar property to thioflavin T. The unique chemical structure of Congo red, including the peculiar elongated and planar backbone help to align along the axis of the amyloid, and thus revealing amyloid structure (Fig. 6.26). The induced anisotropy of Congo red, displaying a green color, is the hallmark of amyloids, and is therefore used to diagnose amyloidosis (Linke 2006). Congo red, along with thioflavin T, are commonly used for histological analysis of specimens. However, staining with Congo red requires stringent conditions, depending on the tissue thickness and orientation of the amyloid deposits (Nilsson 2004).

Chrysamine G has a chemical structure similar to Congo red (Fig. 6.26), except the lack of polar sulfonic acid moieties may actually help the molecule traverse to cross the blood–brain barrier more effectively. Both dyes exhibit good binding to amyloid-β with a similar mechanism as determined on postmortem human brain specimens. It has been proposed that the negative charge on sulfonate or carboxylate on Congo red or Chrysamine G, respectively, interact with the positive charges on the sidechain of the amino acid residues of the amyloid-β (Klunk et al. 1994, 1989); albeit Chrysamine G displays less signal intensity compared to Congo red under

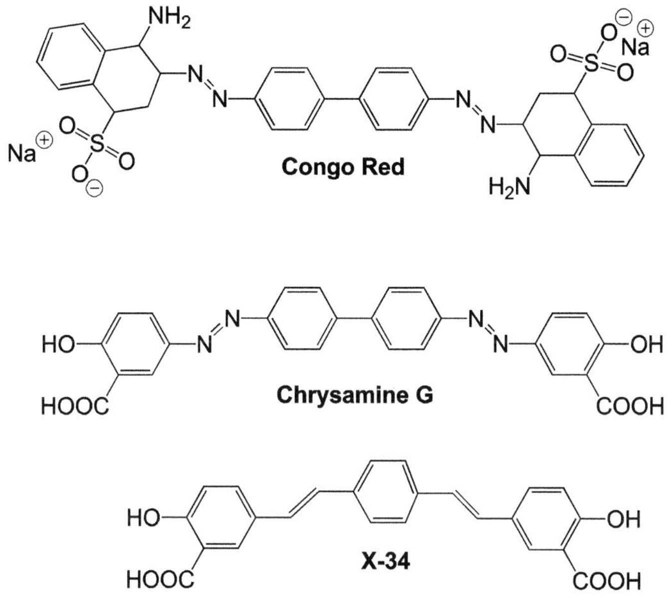

Fig. 6.26 Topologically activatable dyes (Styren et al. 2000)

both visible and fluorescent conditions. In stark contrast to Congo red, Chrysamine G shows no birefringence under polarized light (Klunk et al. 1995).

The combined limitations of Congo red and Chrysamine G are behind the development of an improved analog called X-34 (Fig. 6.27). Specifically, both the polar groups of sulfonate and azo structures on these dyes prevent the compound from penetrating the blood–brain barrier and thus limit in vivo applications. The chemical development of X-34 was directly evolved from Chrysamine G with the azo groups $(N = N)$ replaced by unsaturated alkene $(C = C)$ (Fig. 6.27). This one-step reaction via the Horner–Wadsworth–Emmons mechanism is ideal for the generation of an olefin with great selectivity using the aldehyde available from 5-formylsalicylic acid with phosphorus ylides generated from *p*-xylylenediphosphonic acid tetraethyl ester in the presence of a strong and excess base, such as sodium hydride or potassium tert-butoxide at room temperature. Given the simple reaction condition and the ease of obtaining large-scale material purely by recrystallization, this chemistry illustrates an impressive design, which serves as a model for future work. In ex vivo staining on postmortem human brain slides showed that X-34 stained neuritic and diffuse plaques, neurofibrillary tangles, neuropil threads, and cerebrovascular amyloid (Styren et al. 2000). It is specific for β-sheet structures in the same manner found in thioflavin S. As a matter of fact, both dye staining were abolished if the brain slides were pretreated with formic acid.

A number of analogs similar to X-34 have been reported as activatable fluorescent probes for the detection of amyloid-β (1–40) and amyloid-β (1–42) fibrils.

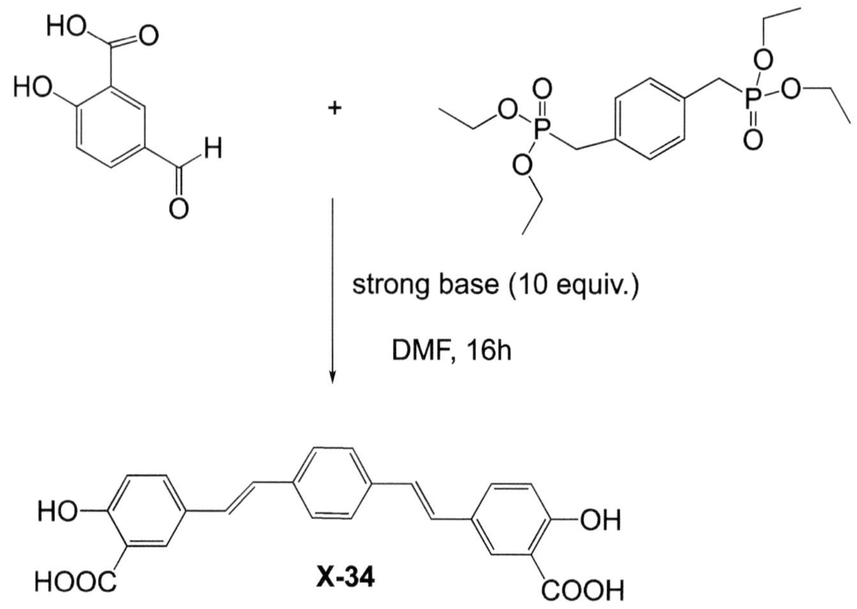

Fig. 6.27 Synthesis of X-34, an analog of Congo red. Data derived from Styren et al. (2000)

Compounds 2C40 and 2E10 are styryl-based probes that were synthesized via solid-phase chemistry, the procedure of which was described in Chap. 5 (Fig. 6.28). These activatable probes can sense amyloid-β (1–40) fibrils with remarkably enhanced fluorescent signal ranging from 27-fold (2C40) to 30-fold (2E10), and a 20-fold (2C40) and 16-fold (2E10) in case of amyloid-β (1–42) fibrils.

Similar analogs derived from 2C40 and 2E10 with neutral chemical structure enable in vivo application have also been reported (Li et al. 2007).

As mentioned in Chap. 2 that BODIPY dyes have an excellent optical property and photostability suitable for in vivo imaging. The dyes can be tuned into redshift

Fig. 6.28 Chemical structures of amyloid-β activated 2C40 and 2E10 reporters (Li et al. 2004)

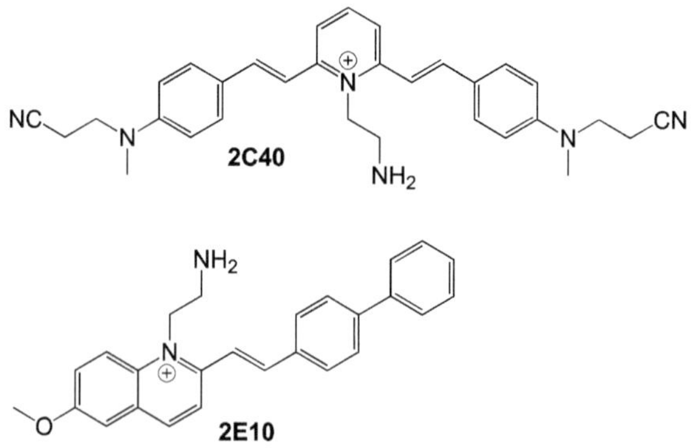

by simply changing the substituents on the parent compound. Further, the exceptionally high quantum yield and molar extinction coefficient of this family of dyes in aqueous condition make them ideal for developing activatable probes. BODIPY-based probes are versatile for numerous imaging applications, but most are based on targeted imaging applications. BODIPY's chemical backbone is hydrophobic, and it has the proclivity to form planarized π-stacking platform similar to thioflavin T. The question is how to convert this great physical attribute into an on–off switcher dye? One of the hypotheses focuses on deploying an electron-rich moiety, which serves as an electron donor to quench the fluorescence of BODIPY via photoinduced electron transfer (PeT) (Boens et al. 2012). The electron transfer phenomenon was known in biology and chemistry quite a long time ago, but not much had not been exploited for probe design until only recently. Rationally, any aliphatic or aromatic amine is ideal for serving as an electron-donating component for the design. Substituted amino groups are also great. In a solution phase, it has been demonstrated that treating aromatic amine derivatives with the acceptor dyes resulted in the former quenched the fluorescence from the latter. Notably, the linear and reverse correlation between the acceptor's fluorescence as the concentration of aromatic amines increased. No matter how high the concentration of the amines was used, the fluorescence spectra decreased but remained the same shape, suggesting no exciplex formation was involved in the quenching mechanism (Singh et al. 2000). This quenching mechanism is amenable to the Stern–Volmer (SV) linear relationship:

$$\frac{I_0}{I} = 1 + K_{SV}[Q] = 1 + k_q \tau_0 [Q]$$

where I_0 and I represent fluorescent intensities in the absence and presence of the quencher. K_{sv} is the Stern–Volmer constant, which is the slope of the linear equation. [Q] is the quencher concentration. K_q is the biomolecular quenching constant, and τ_0 is the singlet state lifetime of the acceptor in the absence of the quencher.

Equipped with this knowledge, one of the first "smart" activatable probes was developed for sensing amyloid-β via intramolecular PeT between aromatic amines and BODIPY dye (Ren et al. 2016). The construct of the probe is purely a simple modification of BODIPY acid through a 3-step synthesis (Fig. 6.29). First, the carboxylic acid group was converted to alcohol using borane in THF, followed by oxidation of an alcohol to the aldehyde using pyridine sulfur trioxide in DMSO/dichloromethane. Next, Borch reduction of aldehyde in the presence of N-methylaniline provided the desired probe. This chemistry enables the synthesis of several derivatives for assay to screen for binding to amyloid-β.

2. FRET Probes

This family of probes works based on an activation mechanism called fluorescence resonance energy transfer (FRET). FRET was discovered by Theodor Forster in the late 1940s (Energiewanderung 1946). The first biological application of this phenomenon started in the 1970s, focusing primarily on protein research (Didenko 2001). This non-radiative energy transfer process has numerous useful applications in

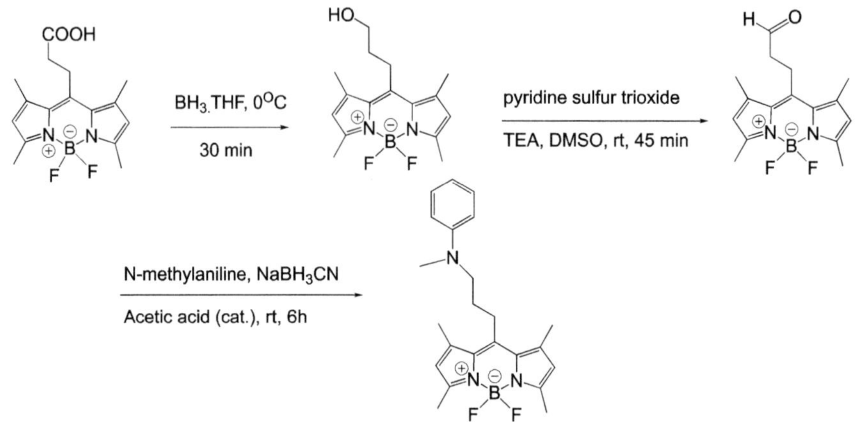

Fig. 6.29 BODIPY-based activatable probe for imaging amyloid-β. Data obtained from Ren et al. (2016) with permission from Elsevier

cellular physiology (sensing temperature, pH conditions, membrane potential biological assays (detection of cell's secondary messengers, metabolic inorganic metals, and protein–protein interactions), and recently in vivo molecular imaging. This book will cover only the chemical development of FRET probes using organic dyes.

The detection of enzyme activity in the intracellular targets, extracellular matrix, or DNA sequence with FRET components would alleviate unspecific background signals. Since FRET detection is dependent on the targeted enzyme, the detected signal outputs are not only specific but enhanced and thus foster higher resolution imaging. For the activatable probe design, when the fluorescent dyes cluster on a biological or synthetic scaffold in the vicinity of each other, FRET will occur. In this conformation, the energy of the donor dye in its electronic excited state transfers energy to the acceptor dye in the vicinity (it could be the same (homo-dyes) or different dyes (hetero-dyes) through non-radiative dipole–dipole coupling. In this restrained environment, the fluorescence of the donor dye will be intercepted and quenched by the quencher (acceptor), which is located in the vicinity (Fig. 6.30). Upon exposure to the targeted protease, the enzyme would cleave the peptide/protein scaffold, leading to the separation of the FRET pairs and resulting in an enhanced fluorescent signal.

Optimal probe design must take an account of the FRET efficiency as described (Rowland et al. 2015):

$$E = \frac{R_0^6}{R_0^6 + R^6}$$

where R_0 represents the Forster distance; in practice, this is the critical distance at which the energy transfer efficiency is 50%, while R stands for the donor–acceptor separation distance, both expressed in Angstrom. The Forster distance R_0 depends

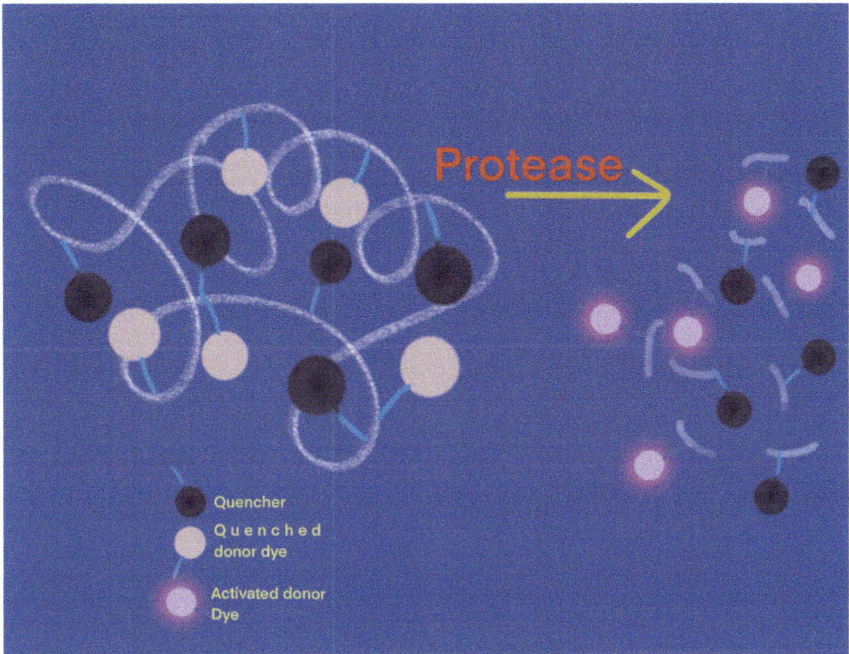

Fig. 6.30 Mechanism of FRET probe for the detection of proteases

on the donor's quantum yield (Q_D), the integral of the spectral overlap between the donor emission and acceptor absorbance (I), the transition dipole orientations of the donor and acceptor (κ_p), and the refractive index of the solution by the time of measurement (n_D). While N_A is the Avogadro constant.

$$R_0 = \left[\frac{Q_D I [9000(\ln 10)\kappa_p^2]}{128\, N_A \pi^5 n_D^4} \right]^{1/6}$$

Developing an effective FRET probe for in vivo application requires mastery of strategy and tactics. From a practical point-of-view, it is sometimes challenging to obtain a precise R_0 value, so it is intuitive to expect that this is a relative value. The FRET efficiency can be fine-tuned by optimizing the relative distance between donor and acceptor. FRET usually occurs over a distance amenable to the dimensions of biological scaffolds, such as proteins, peptides, and DNAs, and a distance between 10 to 100 Å, whether linear or angular, should be effective for FRET to operate (Sapsford et al. 2006). Further, the degree of spectral overlap between donor and acceptor can be exploited to enhance the efficiency of the activated fluorescence signal. Thus, the design would first obtain information about the FRET pair to ensure the maximal spectral overlap between the donor emission and acceptor absorbance. The FRET donor–acceptor pair can be the same (homo-dyes) or different dyes (hetero-dyes) as far as they satisfy the aforementioned criteria. However, the best design can be achieved if an emissive donor can couple with a "dark-hole" quenching acceptor. Currently, there are only a handful number of non-fluorescent quenchers available for FRET probe design. More creative chemistry for quenchers will give more impetus to this field.

However, acquiring concrete dipole orientation requires extended effort since κ_p can be obtained appropriately only in a very rare case when a crystal structure of the donor/acceptor is available. The dipole orientation of the fluorophores can be ranging from 0 for perpendicular to 4 for collinear/parallel scenarios (Sapsford et al. 2006). The probability of having the highest energy transfer is when the participating fluorophores align in a parallel orientation (Didenko 2001). In most cases, the random orientations of the FRET pair have a default value of 2/3, which is accepted as a relative value.

One of the best probes suitable for analysis in this lecture is the DNA hair-pin oligonucleotide-based FRET agent. If the distance between donor–acceptor rules in FRET, the conformal change of a single-strand of DNA with dynamic orientation shifting from a single-stranded loop to the hybridized linear form as the DNA anneals to the target serves as a perfect model for FRET design (Fig. 6.31). This type of probe has so many implications during the post-genomic and proteomic era when the demand for biomolecular recognition probes with high sensitivity and selectivity to process gigantic genomic and proteomic information keeps increasing (Tan et al. 2004). The exceptional specificity and thermodynamic feature of this probe are impeccable thanks to the DNA complementary and heat resistance, respectively. Basically, this type of probe could report a target that differs from others by as little as a single nucleotide (Manganelli et al. 2001; Wang et al. 2002). The principle design of the probe relies on the single-stranded oligonucleotides comprising a loop and stem structures (Tyagi and Kramer 1996). The former is a probe sequence, which is complimentary and thus will hybridize to the target nucleic acid sequence. The latter is a double-stranded oligonucleotide carrying the donor and acceptor via their respective linkers. The stem brings the dyes near each other at a distance that is amenable to the FRET mechanism for quenching the fluorescence signal. By deployment of the "dark-hole" quencher (4-(4'-dimethylaminobenzeneazo)benzoic acid (DABCYL dye)), the FRET signal was quenched robustly when the probe is in an inactivated form (Fig. 6.31). The energy transfer from the emission photons would be intercepted by the quencher and dissipated non-radiative through heat. However, in the presence of a perfectly matched target sequence, the linearization of the loop structure due to nucleotide hybridization disrupted the FRET transfer and thus resulted in the unleased fluorescent signal. Approximately, a 25-fold increase was observed in the fluorescence of the probe upon hybridization to their targets. For better hybridization, the length of the loop and stem need further optimization depending on the targets. Usually, the length of the stem should be somewhere between 5–8 bases. And the single-stranded portion makeup of the loop should have at least 15 nucleotides, or even better, 25−35.

DABCYL dye, which has a broad absorption spectrum between 360–560 nm, has proven to be a very effective quencher for visible dye for acquiring genomic and

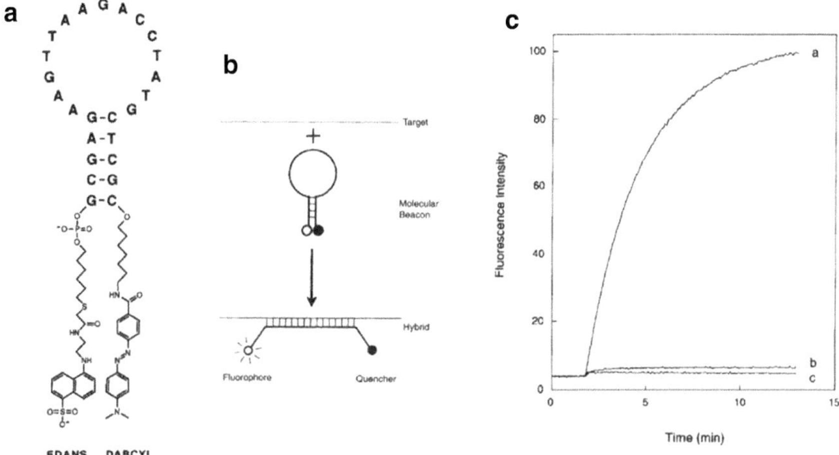

Fig. 6.31 **a** The chemical design of the molecular beacon; **b** activation of the probe via FRET mechanism; **c** specificity of the probe is perfectly complementary target (trace a), similar target with a single nucleotide mismatch (trace b), and target containing a single nucleotide deletion (trace c). Data obtained from Tyagi and Kramer (1996) with permission from Springer Nature

proteomic data from bioassays. Yet, there is a need to develop near-infrared (NIR) versions compatible with the NIR fluorochromes for in vivo application. Now our discussion will lead to the chemical design of a family of azulene for use as a NIR "dark-hole" quencher.

Azulene derivatives, including guaiazulene and chamazulene, are natural products found in many plants and mushrooms. Due to its unique physicochemical properties, azulene analogs have been used in a number of different high technology, including for the development of optoelectronic devices (Bakun et al. 2021). Further, in the past hundreds of years, these blueish compounds, as the name derived from, have been used in ancient medicine for various applications, including anti-allergy, antibacterial, anti-viral, antifungal, anti-diabetic, and anti-inflammation therapies. As mentioned in Chap. 2, azulene molecule is a fused bicyclic structure of the 5- and 7-member rings. The cyclopentadienide has an overall negative charge, while the tropylium is a cationic moiety. This peculiar electronic arrangement renders azulene suitable for numerous chemical modifications. For example, the electrophilic aromatic substitution reactions can be achieved at the 1- and 3-positions; while nucleophilic addition happens at the 4-, 6-, and 8-positions. For molecular probe design, azulene analogs have something to offer. These 10-π electron non-benzenoid aromatic hydrocarbons have an impressive absorbance, ranging from 524–560 nm. So the rationale at this stage is that if the azulene rings could be linked by an unsaturated carbon chain, the chance to fine-tune the dye to the NIR window is possible. As mentioned earlier, the azulene's versatile structure also enables the complex chemical design to extend the dye to the near-infrared spectrum, and also include functional groups for bioconjugation. Considering all of these information and requirements, the design focuses on stable dye given that the development of FRET probe for protease sensing requires dye conjugation with a peptide substrate. The chemistry started with 2-methyl azulene, which was obtained through a 3-step reaction started with tropolone (Pham et al. 2003). The preparation of the functional group for bioconjugation started with methylation on the 7-membered ring with methyllithium; the process involves the formation of azulenate ions of the Meisenheimer-type intermediate as the color changed from deep blue to pale yellow suspension (Fig. 6.32). Addition of methanol at $-70\ °C$, followed by dehydrogenation with p-chloranil to provide 2,4-dimethyl azulene. Treating this 2,4-dimethyl azulene with n-BuLi and bromoacetic acid in the presence of diisopropylamine at $-40\ °C$ provided the carboxylic acid product. From here, the polymethine bridge was designed as a cyclic structure to enhance stability through a simple condensation of azulene with squaric acid in refluxing butanol. With the absorbance lambda max at around 700 nm, this quencher is suitable to serve as an effective FRET acceptor for any fluorochrome, which has an emission lambda max from 600–750 nm.

To extend the spectral quenching range, the same chemistry was performed on guaiazulene. It is likely that the electronic donating isopropyl moiety on guaiazulene contributes to the bathochromic shift of the dyes in the NIR window ($\lambda_{max} = 750$ nm) suitable for paring with any commercial NIR dyes for a FRET construct. This robust chemistry facilitated a large-scaled dye synthesis with simple purification operation using flash column chromatography. The stability of the azulene family of dyes comes as no surprise that they survive harsh reaction conditions in solid-phase peptide chemistry.

To demonstrate the utility of NIR quencher for the FRET detection of caspase-3, guaiazulenyl squaraine dye was conjugated onto a peptide backbone while the peptide was still anchored onto the MBHA resin (Fig. 6.33). Then, the resin and other acidic-labile protecting groups were removed in TFA in the presence of the

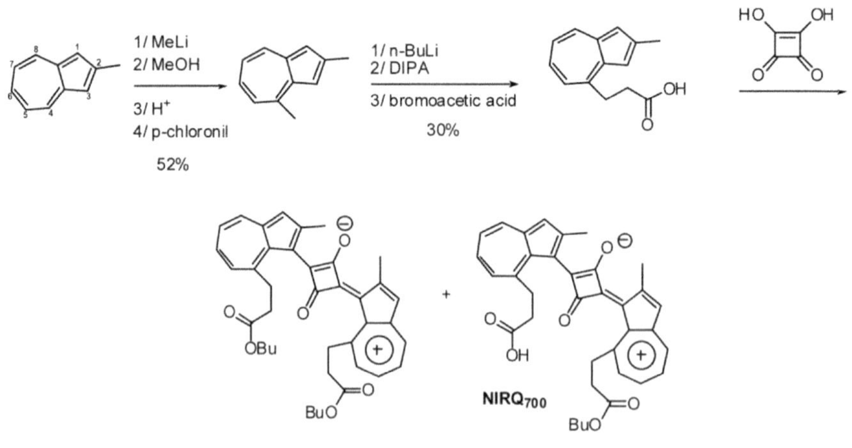

Fig. 6.32 Design a mono-functionalized NIR azulenyl squaraine quencher dyes. Data obtained from Pham et al. (2003)

scavengers. The unstable and scarce Alexa Fluor 680 dyes were attached to the peptide when it was cleaved off from the resin. The main rationale here is to prevent Alexa decomposition should it is exposed to a strong acidic condition. In one batch of peptide synthesis, several hundred milligrams of the guaiazulenyl-labeled peptide can be isolated and stored for future labeling with the donor dye, which usually needs only a few milligrams for the work. In vitro analysis of the specificity of the probe showed a fourfold activated signal in the presence of the caspase-3 enzyme. No registered activation signal when the caspase-3 specific inhibitor was present. When replacing the guaiazulenyl squaryl quencher with a visible analog of DABCYL, the activated fluorescence signal was 2.4-fold when the probe was exposed to caspase-3, indicating that FRET efficiency depends on the degree of the signal overlap between the compatible donor and acceptor (Pham et al. 2002).

The IRDye QC-1 (Fig. 6.34) is another versatile and stable NIR quencher capable of quenching the donor dyes in a FRET construct (Peng et al. 2009). This quencher has the chemical backbone of the cyclic heptamethine cyanine dye with NIR absorbance, thus capable of intercepting Cy7 fluorophores like ICG and others. In this clever design, incorporating the amino substitution on the indole scaffold resulted in the quenching of the fluorescent signal. Further, amino substitution also resulted in a wide absorbance spectrum, which enables the quencher to intercept NIR fluorophores and those in the visible regions. The art of this objective design is to conflate multiple

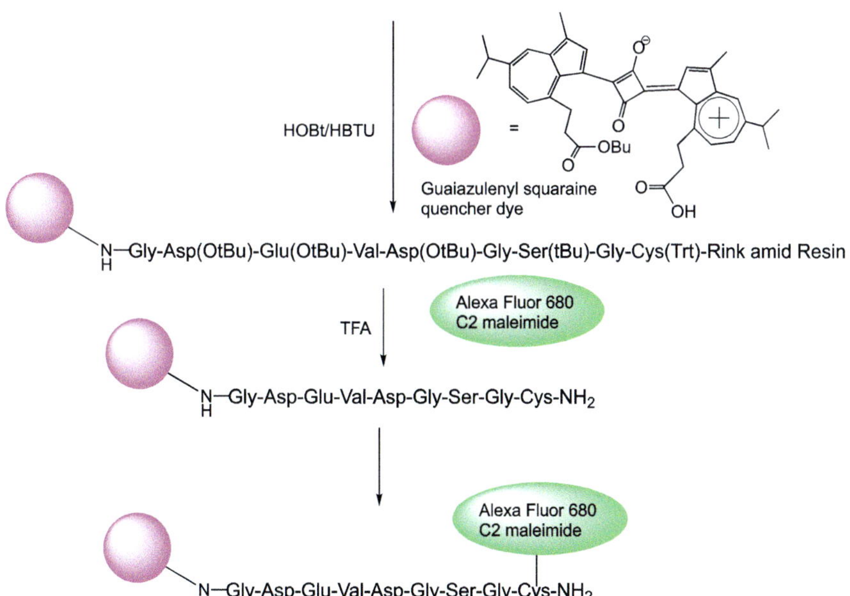

Fig. 6.33 A near-infrared FRET probe for the detection of caspase-3. Data derived from Pham et al. (2002)

features into one product that is impeccable for FRET study not only for bioassay but also in vivo application. Aside from a NIR quencher, the compound is activated with succinimide ester for labeling purposes. What stands out is the inclusion of four sulfonate groups to render the quencher water-soluble, suitable for working in biological systems.

To demonstrate the versatile application and stability of the quencher for FRET construct, the dye was labeled directly onto the N terminal of a caspase-3 substrate via solid-phase chemistry (Fig. 6.35). Then, the labeled peptide was treated with TFA and scavengers to remove the protection groups, and cleavage of the peptide from the resin. The product was isolated by filtration and ether precipitation. The labeled peptide was purified by HPLC before labeling the fluorophore onto the lysine sidechain to afford the FRET probe.

Recently, a novel class of NIR "dark-hole" quenchers was reported, offering more enthusiasm to the FRET work (Fig. 6.36). The underlying mechanism that regulates quenching property surrounding the conformation degree of freedom between the xanthene backbond and the heterocyclic attach to it via a single bond. While extending the bathochromic shift of the rhodamine dye by displacing the oxygen atom at position 10 by a silicon moiety (Myochin et al. 2015). A more detailed explanation of using the silicon-substituted xanthene scaffold for the NIR feature was described in Chap. 2 (Grimm et al. 2017).

The synthesis started with palladium-catalyzed deprotection of N-allyl groups using barbituric acid derivative at mild temperature. Then, using Sandmeyer reaction to covert the amine group of the primary aromatic amine into aryl halide via the aryl diazonium salt. The iodo-xanthone product is the key intermediate where the aryl halide was converted into an arylamine at the 3,6-positions using anisidine

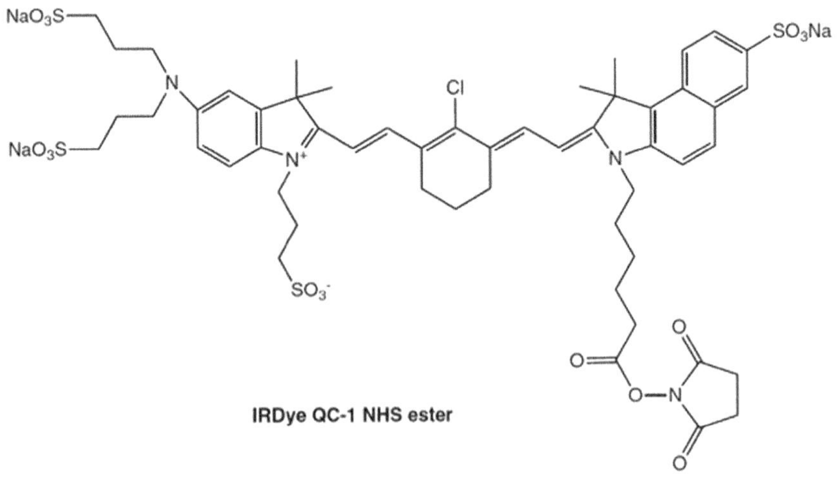

Fig. 6.34 Water-soluble heptamethinyl cyanine quencher

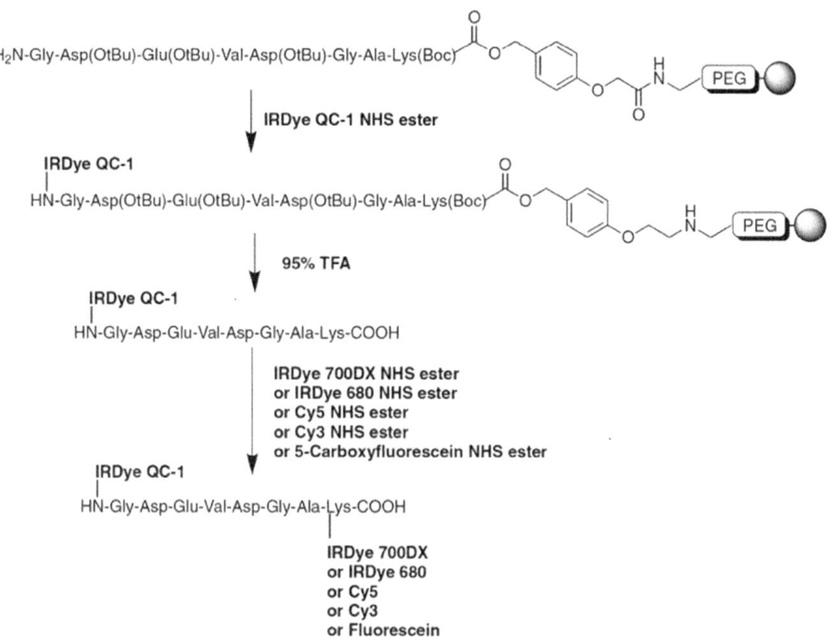

Fig. 6.35 Solid-phase peptide synthesis of a caspase-3 FRET probe. Data obtained from Peng et al. (2009) with permission from Elsevier

or indoline via the Buchwald–Hartwig amination reaction. Finally, treating the *o*-tolylmagnesium bromide to these intermediates afford the quenchers SiNQ660 and SiNQ780, which have the absorbance lambda max of 660 and 779 nm, respectively, without recorded fluorescence. This convergent synthetic strategy using reliable procedure facilitates combinatorial development of a library of quenchers in the future with diversifying chemical genetics of NIR quenchers.

One of the water-soluble quencher versions developed using this approach has been employed for the generation of an activatable NIR probe. In this FRET construct, peptide sequences were derived from the endopeptidase substrate for matrix metalloproteinases 1 (MMP-1). MMPs are a family of zinc-dependent endopeptidases that implicate extracellular matrix degradation. MMPs are also involved in embryogenesis, wound healing, inflammation, arthritis, cardiovascular disease, and cancer (Chakraborti et al. 2003; Egeblad and Werb 2002; Shiomi et al. 2010). Among over 20 types of MMPs, MMP-2, 9, and membrane-type I MMP (MT1-MMP) have more attention due to their prolific activity implicated in tumor invasion, metastasis, and angiogenesis (Itoh and Seiki 2006; Hinsbergh et al. 2006). The holy grail in cancer imaging is the ability of the probe to report these activities during the progress of cancer; it is anticipated that these versatile quenchers will contribute greatly to this endeavor. In this design, the water-soluble quencher swSiNQ780 was assembled along with the DY730 fluorophore on a cognate oligopeptide, which MMP-2, 9 and

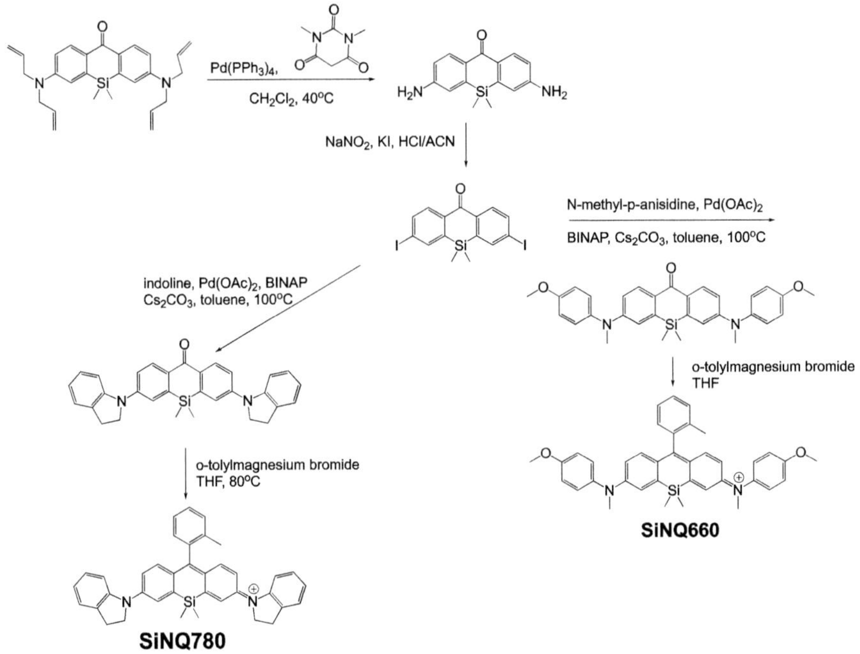

Fig. 6.36 Design of NIR quencher based on Si-rhodamine backbone. Data obtained from Myochin et al. (2015) with permission from the American Chemical Society

MT1-MMP can recognize (Fig. 6.37). The C terminal of the peptide was conjugated to a polyethylene glycol (PEG) chain to enhance in vivo circulation time. During in vitro analysis, the probe was confirmed for being activated by MMP-9 or MT1-MMP, resulting in over 20-fold fluorescence enhancement. In contrast, the intact form before exposure to enzyme yielded no signal. The specificity was confirmed when an identical probe with only D-amino acids, instead of L-amino acids showed no activatable fluorescent signal in the presence of MT1-MMP.

Usually, proteases are popular when it comes to designing specific cleavage sequences for FRET imaging. But that is not always the case, in the following work, which exploits a unique spectral change of a quencher called QSY-21. This quencher has 2 dynamic absorbance wavelengths that can interchange like an on/off switch activated by special physiological inputs. For example, under hypoxia, this quencher's absorbance at 660 nm decreases with the enhanced signal at 470 nm concomitantly.

While in the same solution, it exhibited only one lambda max at 660 nm under normoxia conditions. The underlying design of the probe implies that if QSY-21 is paired with a NIR fluorochrome whose emission spectrum will be intercepted by QSY-21's absorbance spectrum at 660 nm. However, in the presence of hypoxia, the lambda max at 470 nm dominates, and thus FRET lost the effect, unleashing enhanced fluorescent signals. This concept leads to the production of the Reversible Hypoxia Cy5 Probe (RHyCy5) (Fig. 6.38). Further data analysis of this probe showed that

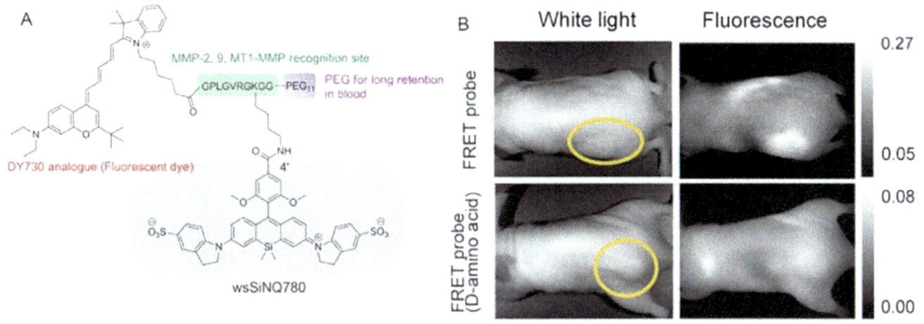

Fig. 6.37 **a** Design of a FRET probe for imaging MMPs in vivo; **b** testing the probe in HT-1080 tumor-bearing mice. Data obtained from Myochin et al. (2015) with permission from the American Chemical Society

QSY-21 was reversibly reduced under hypoxia (Takahashi et al. 2012), where QSY-21 became a radical and resulted in strong fluorescence. However, this emitted fluorescence quickly attenuated when the probe underwent from hypoxia to normoxia, certainly attributed to the instant oxidation of RHyCy5 radical in the air. Further, the probe could detect the fluctuation of hypoxia in human adenocarcinomas of lung cancer cells (A549).

The integration of nanotechnology with FRET improves the physicochemical property of the construct. The multivalency afforded by nanoparticles like quantum dots is impeccable that a single dye platform has no match. Further, these semiconductor nanocrystals, such as quantum dots (QDs) are packed with hundreds if not thousands of excitons, capable of illuminating signals with exceptionally high quantum yield. Other remarkable physical outputs of QDs include tunable emission

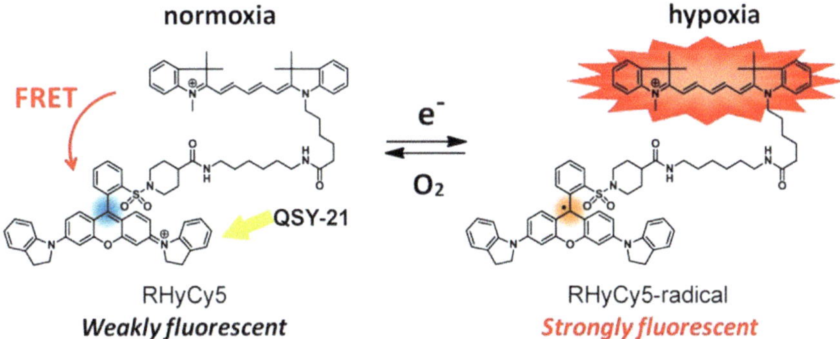

Fig. 6.38 Peculiarly dual absorbance spectra of QYS-21 quencher make it a good candidate for developing a FRET probe for the detection of hypoxia. The proposed "on/off" detection of hypoxia via the radical form of RHyCy5. Data obtained from Takahashi et al. (2012) with permission from the American Chemical Society

spectra with nearly negligible photobleaching and a long fluorescence lifetime (Sapsford et al. 2006; Medintz and Mattoussi 2009). These give the impetus for in vivo translation.

Aside from QDs, other nanoparticles like iron oxide and gold nanoparticles all share the same advantage for FRET design via multivalency for multimodal imaging. For example, the cross-linked iron oxide (CLIO) nanoparticle enables the conjugation of homogenous copies of Cy5.5 dyes to afford a FRET probe for MRI and optical imaging. Obviously, there is no FRET intercept between the nanoparticles and the dyes based on the spectral analysis between CLIO (lambda max at the ultraviolet region of the spectrum) and Cy5.5 (lambda max at 680 nm). However, the quenching occurred between the dyes, thanks to the CLIO particle multivalency, offering a platform to derivatize the dyes in the vicinity to each other (Fig. 6.39). The synthesis of this probe started first with the development of the arginine oligomer peptide, where the C terminal was ended with a cysteine. The use of arginine peptide is twofold here; first, it serves as a spacer. Second, the positive charge on its sidechain would interact with the negative charge from the phospholipid cellular membrane via electrostatic interaction and other mechanisms help to shuttle the probe inside the cells. This mechanism is one of the tentative explanations. However, it is not everything that makes arginine peptides a special delivery module. Obviously, the guanidine head group also has some special properties for this phenomenon, aside from electrostatic interaction (Wender et al. 2000). While cysteine provides a handle to couple with SPDP linker or SIA via the disulfide or thioether bond, respectively. The succinimide ester available on the other end of the bifunctional linker can react with the aminated CLIO nanoparticles. Finally, Cy5.5 succinimide ester can be conjugated to the N terminal of the peptide.

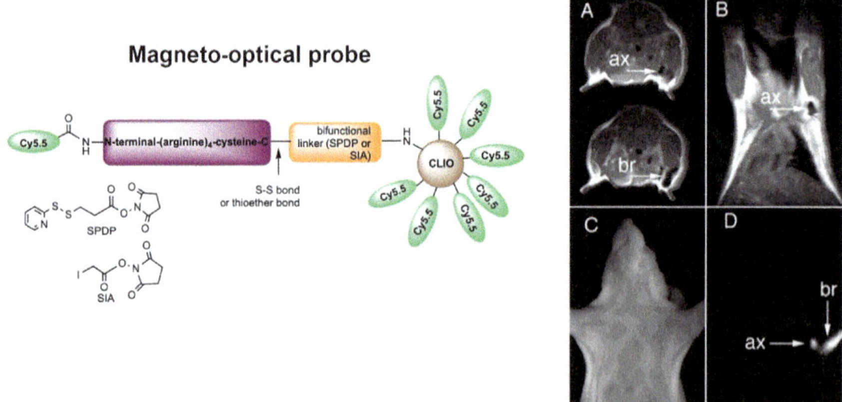

Fig. 6.39 Design of an optical-magneto FRET probe for in vivo application. MR imaging of the probe, **a** homing of the probe in the axillary (ax) and brachial (br) lymph nodes; **b** coronal view of the axillary lymph nodes; **c** white light imaging; **d** NIR fluorescence imaging of the lymph nodes. Data derived from Josephson et al. (2002) with permission from the American Chemical Society

When treated with trypsin, the enzyme cleaved the peptide backbone, the dyes separated from the nanoparticle and from each other, resulting in 27–36-fold fluorescent signal enhancement, while no signal detection if the peptide is made of D-amino acids. The specificity was also confirmed, where DTT can activate only the probe assembled via the disulfide bond, not the thioether counterpart. To demonstrate the utility of this probe for non-invasive imaging; particularly, for the potential tracking of macrophages, the probe was administered to mice via subcutaneous injection and homing of the nanoparticles to the draining lymph nodes, where macrophages can activate the probe. T2-mediated MR imaging is an excellent tool for the detection of CLIO nanoparticles in the lymph nodes. The data show that the probe homed in many draining lymph nodes resulted in reduced signal intensity. The data corroborated with optical imaging (Fig. 6.39). This work proved a practical implication of optical imaging for in vivo work. Deep tissue imaging is still a challenge for optical techniques, but surface imaging is its mantra.

Up to this point, we have not truly discussed whether nanoparticles could perform as an active partner in either donor or acceptor role in FRET. One of the questions one would ask is whether the nanoparticles are way too large compared to small organic molecules to be amenable to the Forster distance theory. To answer this question, a prototypical FRET construct was built using CdSe-ZnS core–shell QDs tethered with Cy3 dye via a "rod-like" peptide comprising of repeated YEHK sequence (Medintz and Mattoussi 2009). To analyze whether the FRET efficiency generated from the nanoparticles depends on the donor-to-acceptor distance with an inversed 6th power rule due to dipole–dipole interaction mechanism, different repeated YEHK peptide chains were used, including 1, 3, 5, 7, 14 and 21 repeats to control the calculated separation distance (r) between Cy3 and QDs. With this construct, the FRET efficiencies (E) were derived from the QD emission signal using Forster distance R_0, which was obtained from the experimental values, including spectral overlap integral between QD and Cy3, the QD quantum yield, and normalized to the configuration one-to-one donor–acceptor pair. Overall, the data confirmed that the Forster dipole–dipole distance accurately reflects FRET transfer mechanism between QD and small organic dyes.

More work has demonstrated that the QD-dye system is not only compatible with FRET technology, but the FRET quenching efficiency is remarkably high. With the ability to affiliate ligands through surface fabrication, QDs or nanoparticles in generals enable multivalency and facilitate biomolecular assembly in a "plug-and-play" mode, just limited by our own imagination. In the next construct, a FRET activatable probe was designed for imaging the activity of β-lactamase (Bla), a family of bacterial enzymes that use penicillin and cephalosporins as cleavage substrates (Xu et al. 2006). This probe design is unique because no peptide substrate is needed to activate the FRET signal. But rather a Bla-specific lactam chemical scaffold with a cyanine acceptor dye at one terminus, while biotin was tethered through a spacer on the other terminus. This construct can make self-assembling onto the 605-nm streptavidin-QDs (Invitrogen) with remarkable affinity (Kd $= 10^{-14}$ M) (Fig. 6.40). The data from this study showed that distance between the substrate and the QDs along with the density of the substrates are crucial to providing optimal

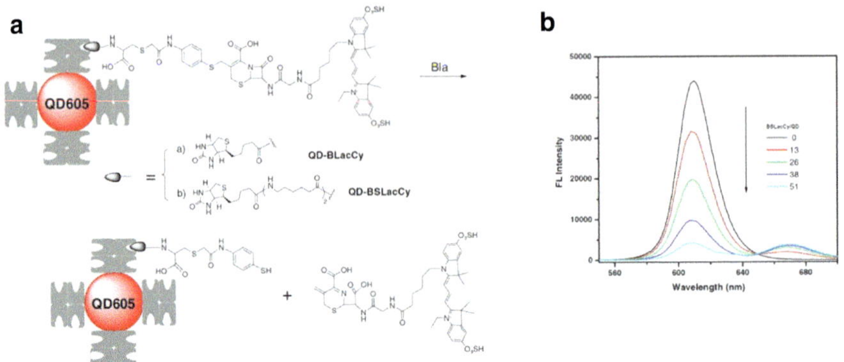

Fig. 6.40 a Self-assembled QD-based FRET probe for imaging β-lactamase using non-peptide scaffold. In the presence of Bla, the FRET pair will be separated due to the opening of the lactam ring and restoring the QD emitted signal; **b** Increasing ratio of BSLac-cy:QDs resulting in quenching of QD signal. Data obtained from Xu et al. (2006) with permission from Elsevier

quenching efficiency. It is apparent that the incorporation of a longer spacer reduced steric hindrance and allowed the enzyme to access the binding site. The fluorescent signal from the QDs can be quenched by the corresponding cyanine dye as much as 95%. Notably, this self-assembled probe opens up more thought for future design, bringing this field a step closer to molecular automation. More information about QD-based FRET design for sensing enzyme activity could be found in references herein (Medintz et al. 2006; Shi et al. 2006, 2007).

3. "Smart" activatable probes for MR imaging

This class of molecular probes, as the name suggested, is solely based on the magnetic iron oxide nanoparticles, albeit the ones that would be "switched on" by the detection targets. Several biological processes and entities, such as protein–protein interaction, DNA-DNA hybridization or enzyme activities, can regulate this magnetic nanoparticle activation. In one of the pioneering works, where the term magnetic relaxation switches (MRS) was first introduced to the imaging field; the underlying mechanism behind this technique showed that biological-mediated assembly of iron nanoparticles resulted in an apparent lower T_2 relaxation time of neighboring water molecules and vice versa (Perez et al. 2002). Depending on the targets, whether they assemble the iron oxide nanoparticles together or dissociate the agglomeration into its constituent subsets, the T_2 relaxation time will decrease or increase, respectively (Fig. 6.41a). This technique is predicted to have widespread implications in molecular imaging, high-throughput, and functional imaging assays to screen new drug development. These nanosensors can discern the difference between oligonucleotides with a single mismatched nucleotide. In this construct, the aminated dextran-coated iron oxide nanoparticles with 45 nm size were labeled with complementary and thiolated oligonucleotide via the SPDP bifunctional linker. To generate MRS, for every targeted DNA sequence, the 2 portions of nanoparticles, each derivatized of the adjacent complementary oligonucleotide. With this design, upon DNA hybridization, the targeted DNA will trigger nanoparticle self-assembly into clusters, evidenced by increasing size up to 140 nm (Fig. 6.41b, c). This nanomaterial union resulted in a significant reduction in T_2 relaxation time (Fig. 6.41d). Any single nucleotide mismatched or extra presence of irrelevant nucleotide will abolish the MRS. Notably, the method can perform in a high-throughput manner using multiple-well plates with robust signal readout capability (Fig. 6.41e). Quantitatively, the observed reduction in T_2-weighted MR imaging inversely corresponds to the increasing concentration of added oligonucleotides.

Another activatable probe for MR imaging was described recently using the FRET type of approach termed two-way magnetic resonance tuning (TMRET) (Wang et al. 2020). In this design, a pheophorbide a-paramagnetic Mn^{2+} chelate (p-Mn) and SPIO nanoparticles were encapsulated inside a disulfide cross-linked micelle (DCM). In this tightly constrained domain within the micelle core, the T_1 and T_2 MRI signals are "switched" off, albeit they will turn on only when the increased distance between the SPIO and Mn^{2+} is made possible by the biological-mediated degradation of the micelles (Fig. 6.42).

It is unclear what exactly quenched the signal, but it is certain that the SPIO aggregation did not induce T_2 quenching. There is a possibility that the spin fluctuation of the T_1 contrast agent is slow, combined with the weakening of the T_2 contrast agent, and further compounded by the diamagnetic field; altogether reducing the effective dipole field (Fig. 6.42).

Glutathione (GSH) serves as the main biological component, which activates this probe. GSH plays an important role in cellular defense mechanisms and metabolic processes, including cell differentiation, proliferation, and apoptosis. Disrupting GSH homeostasis is involved in the etiology and progression of many human

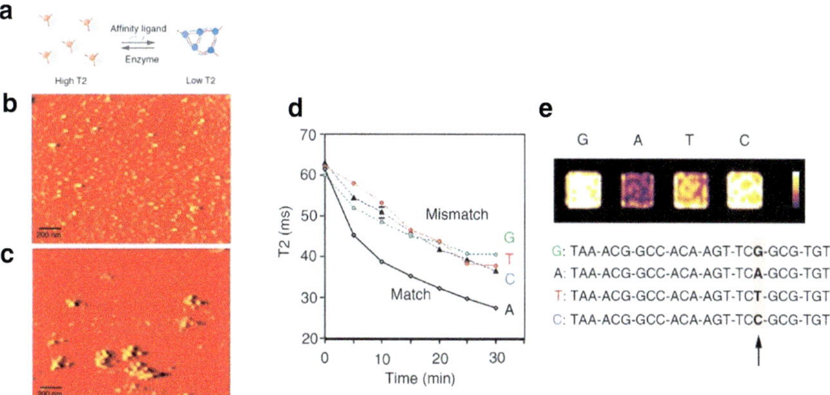

Fig. 6.41 Design and applications of magnetic relaxation switches for sensing molecular interactions. Data obtained from Perez et al. (2002) with permission from Springer Nature

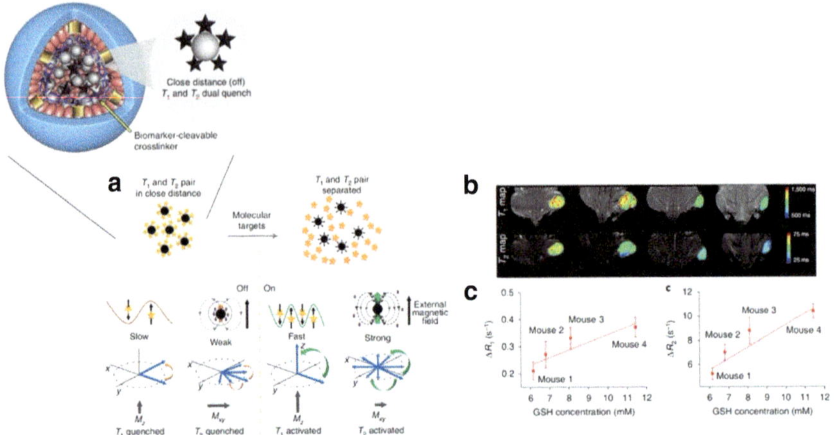

Fig. 6.42 **a** Mechanism of activation of TMRET nanosensor; **b** A linear relationship between MRI relaxation rates and the concentration of GSH in tumor. Data obtained from Wang et al. (2020) with permission from Springer Nature

Fig. 6.43 **a** ROS oxidation to activate hydroethidine; **b** a tentative mechanism to explain how radicals oxidize hydroethidine. Data adapted from Zielonka et al. (2008)

diseases, including cancer (Traverso et al. 2013). Aside from GSH, the notorious acidic tumor pH condition also contribute to the degradation of the disulfide micelles, as a result increasing the detected T_1 and T_2 MRI signal.

To evaluate this new imaging platform for in vivo quantification of the targets, in this case, GSH in tumors, xenograft nude mice with human prostate cancer cells (PS-3) were implanted on the flank of the animals. After an intravenous dose of DCM@P-Mn-SPIO as a TMRET nanosensor, the GSH from tumors was detected using a 7.0 T MRI scanner. The data show that the R_1 and R_2 mapped images corroborate with the GSH levels (6.15, 6.80, 8.10 and 11.44 mM) in the tumors as they are quantified using ThiolTracker Violet (Fig. 6.42 b,c). Further, the values of ΔR_1 and ΔR_2 reported by the TMRET nanosensor were found to increase linearly with the intratumoral GSH concentrations.

4. ROS-Activatable Probes

Reactive oxygen species (ROS) are a term to describe the products of oxygen molecules called free oxygen radicals after reaction with electrons. So, when discussing ROS, it could be one of these products or all (Phaniendra et al. 2015), including hydroxyl radical ($^{\bullet}OH$), peroxide ($^{\bullet}O_2^{2-}$), superoxide anion ($^{\bullet}O_2^{-}$), nitric oxide ($NO^{\bullet}$), peroxyl radicals ($ROO^{\bullet}$), alkoxyl radicals ($RO^{\bullet}$), sulfonyl radicals ($ROS^{\bullet}$), and thiol peroxyl radicals ($RSOO^{\bullet}$). Non-radical ROS comprises hydrogen peroxide (H_2O_2), hypochlorous acid (HOCl), peroxynitrite (ONO^{-}), nitrosoperoxycarbonate anion ($O = NOOCO_2^{-}$), singlet oxygen ($^1O_2$), and hydroxyl ion (OH^{-}). Mitochondria mainly, but not the only source, produce ROS. ROS plays an important roles in normal physiology, mostly through redox regulation of protein phosphorylation, ion channels, and cell signaling cascade pathways. Given their vital role in cellular functions, a lack of ROS or increasing levels of these reactive substances implicate multiple diseases. For example, a lack of ROS leads to chronic granulomatous disease and certain autoimmune disorders, while a surplus of ROS results in cardiovascular and neurodegenerative diseases (Brieger et al. 2012). Elevated ROS activity has also been detected in many stages of tumor development and progression (Liou and Storz 2010).

Cellular metabolic activity that leads to an imbalance of ROS, particularly oxidative states, is referred to as oxidative stress. To keep ROS activity in control, cells possess effective antioxidant defensive mechanisms. Enzymes, such as superoxide dismutase (SOD), catalase (CAT), and glutathione peroxidase (GPx), are responsible for protecting from cellular damage (Pizzino et al. 2017). Notably, more knowledge emerges about GPx, these enzymes, which contain selenium, involve the degradation of hydrogen peroxide and other peroxides into alcohols. For non-enzyme components, glutathione is one of the most potent radical scavengers that cells use to abrogate the deleterious effect of ROS. This is a tripeptide Glu-Cys-Gly where the Glu residue forms an amide bond to Cys N terminal with the δ-COOH group. Glutathione reduces hydroxyl radical ($^{\bullet}OH$) and superoxide anion ($^{\bullet}O_2^{-}$) to water; in turn, it is oxidized to form a disulfide product, which is a dimer of the tripeptide linked by an S–S bond.

Given the vast involvement of ROS in pathological diseases, the development of molecular probes to report this activity is of paramount biomedical importance. Most of the contrast agents designed for the detection of ROS are functional imaging probes because they do not bind or, in any case, associate with the physical interaction with the targets. Instead, ROS activity would oxidize these probes and switch them on as they undergo from a reduced to an oxidized form. A hydroethidine (HE) family, also known as dihydroethidium dyes, exemplifies this category. Hydroethidine dyes were one of the first systems reported as a ROS indicator. In a reduced form, the hydrogen-containing dye does not fluoresce. However, upon interaction with ROS products, the radicals convert HE to 2-OH-E$^+$ to establish an effective conjugation network and resulting in strong fluorescence (Fig. 6.43). The presence of a hydroxyl moiety on the aromatic ring of the final product 2-OH-E$^+$, not E$^+$ (Zielonka et al. 2005), suggests aromatic electronic rearrangement followed by oxygen radical insertion to the position ortho to the amine. To investigate this mechanism, Fremy's salt (nitrosodisulfonate (NDS) radical dianion) is a great candidate to serve as a free radical due to its long-live to explore whether the oxygen radical involves in the generation of 2-hydroxyethidium (2-OH-E$^+$). Based on the observation, with the excess of the oxygen radical, the reduced form of hydroethidine is oxidized to the corresponding radical cation (HE$^{•+}$) or the amine radical (HE($^•$NH)). These radical intermediates may interact with NDS radical to form the product 2-OH-E$^+$ (Fig. 6.43). It is worth mentioning that no E$^+$ was formed during the process; as a matter of fact, if taking E$^+$ reacting with NDS, the formation of 2-OH-E$^+$ was not observed. At least up to now, it is known that superoxide radical anion (O$_2^{•-}$) and NDS can serve as specific oxidative reagents for the conversion of HE into 2-OH-E$^+$. While it has been reported that other biologically active oxidizing species, such as hydroxyl radical, peroxyl radicals, peroxynitrite, hydrogen peroxide and peroxidase, react with HE to form different fluorescent products (Zielonka et al. 2005).

This probe has tremendous implications for imaging ROS in virtue of its unique distribution and specific retention in the brain, where ROS is prominent in several neurological disorders, including inflammation, cancer, Parkinson's disease, Alzheimer's, and cerebrovascular disease, including stroke. For in vivo application, HE was formulated with a mixture of DMSO and saline prior to injection to mice via intraperitoneal (IP) injection. As judged from the chemical structure, HE is an amphiphilic molecule, meaning that it can traverse in aqueous and cross the lipophilic site like the blood–brain barrier. And thus, HE is not entirely soluble in buffer or organic solvent, but both. Since a small amount of organic solvent was present in the formulation, it is intuitive to distribute the probe via the IP route. In fact, the study showed that no registered toxicity occurred with this approach. The data showed that the probe penetrates to the brain parenchyma as early as 10 min post-injection (Fig. 6.44). However, accumulation and retention need a longer time to ensure washout is completed before imaging. Further, animal hair was removed before scanning using a 470 nm excitation and 590 nm long-pass emission filters to facilitate deep tissue imaging. Another worth noting point about this work is to overcome the limitation of optical imaging in deep tissue by using fluorescence lifetime to unmix the fluorescence intensity depicted from tissue autofluorescence (AF) in

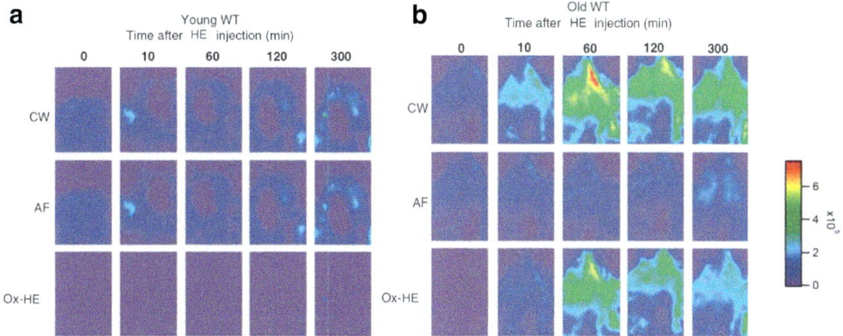

Fig. 6.44 Imaging ROS activity in aged mice using HE probe employing time-domain unmixing method. The dye is distributed in the brains, but the retention is ROS-dependent; otherwise, it would be washed out from the brain. CW (continuous wave), AF (autofluorescence), Ox-DHE (oxidized dihydroethidium). Data obtained from Hall et al. (2012) with permission from Sage Publishing

live animals. With this technique, every detected pixel is a collection of fluorescence intensity (continuous wave) and lifetime, thus removing the confounding tissue AF from the fluorophore-specific signal (Hall et al. 2012).

Another convincing evidence demonstrates the capability to image ROS in the brain. This work showed novel chemistry to modify HE for labeling with tritium (Abe et al. 2014). The [^{3}H]hydromethidine was developed by protecting the aromatic amine of the phenanthridine with the Boc groups (Fig. 6.45). Then, the protected compound was treated with [methyl-^{3}H]methyl nosylate in refluxing acetonitrile for 5 h. Usually, in the radioisotope-labeling reaction, the tritium is a limiting reagent; therefore, large excess of protected phenanthridine should be removed before moving on to the next reaction. Fortunately, the alkylation has distinct mobility compared to the starting material, and the purification was performed using preparative thin-layer chromatography. In the next step, the di-Boc methidium 4-nitrobenzenesulfonate intermediate was reduced by sodium borohydride, followed by Boc deprotection to provide [^{3}H]hydromethidine product. In vitro study showed that [^{3}H]hydromethidine interacts with $^\bullet O_2^-$ and $^\bullet OH$ but not with H_2O_2. The autoradiographic imaging of treated mice using this probe demonstrate unequivocal specificity of the agent for detection of ROS (Fig. 6.46). The mouse brain was first injected one-half of the hemisphere with sodium nitroprusside (SNP), which can produce $^\bullet O_2^-$ and $^\bullet OH$ in vivo. The probe was quickly distributed into major organs, including the brains after intravenous injection, followed by prompt clearance in healthy tissue. The accumulation of the probe in the brain is ROS-dependent. In the first few minutes post-injection, the signal was detected high in the striatum where NSP was injected, albeit significant uptake was found in other subregions of the brain. Twenty minutes later, after a non-specific washout, the retention was specific and lasted over an hour.

HE is a half-dye of the cyanine version, which can serve as a consummate model to extend the emission wavelength for deep tissue imaging; presumably, a similar ROS activation mechanism can be achieved starting with hydrocyanine dyes. Chemically,

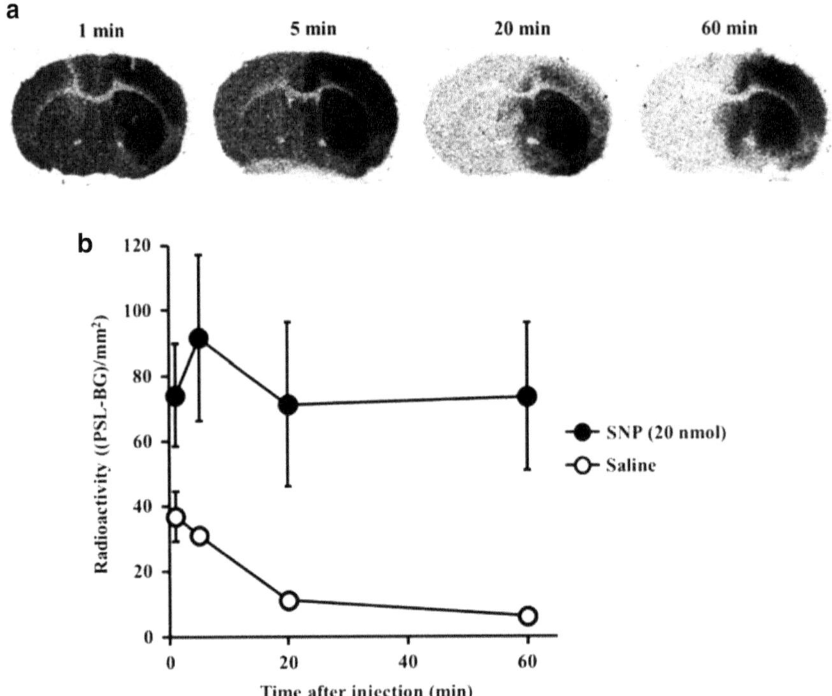

Fig. 6.45 Design of [³H]Hydroethidine probe for in vivo applications. Data obtained from Abe et al. (2014) with permission from Sage Publishing

Fig. 6.46 **a** Autoradiographic imaging of ROS activity in a mouse brain, in which half was treated with sodium nitroprusside (NSP) using [³H]hydromethidine; **b** radioactivity profiles in SNP-injected triatum (•) and saline-injected striatum (o). Data obtained from Abe et al. (2014) with permission from Sage Publishing

the non-fluorescent hydrocyanine dyes (Fig. 6.47) can be readily obtained using any chemistry methods discussed in Chap. 2, then the iminium cations of the cyanine dyes can be reduced to hydrocyanine using sodium borohydride. The powerful implication of this approach is that any commercial cyanine dye can also be converted to ROS probe using sodium borohydride chemistry and thus open up an extensive repertoire of dyes for in vivo applications. The conversion of a colored dye with water-soluble features into a reduced form has two advantages. First, the color (purple, blue, bluish-green and green for Cy3, Cy5, Cy5.5 and Cy7, respectively) will change to colorless molecule, so the progress of the reaction can be monitored easily. Usually, with one equivalent or less sodium borohydride, the reaction can finish from 10–30 min. Second, the reduced dyes are more hydrophobic than the cationic starting materials, and thus after the reaction, the addition of water and organic solvents like methylene chloride or ethyl acetate into the reaction would facilitate purification. In this two-phase extraction, the reduced product can be extracted in the organic phase, while the starting material is in the aqueous phase. The hydrocyanine dyes have higher stability, sensitivity, and reactivity than the HE counterpart, and thus they are suitable for imaging ROS in biological samples and in vivo studies.

To investigate ROS activities in vivo using hydrocyanine dyes, wild-type mice were treated with lipopolysaccharide (LPS) to generate ROS via activated macrophages and neutrophils. In this LPS-induced acute inflammation model, it is worth mentioning that the LPS injection route is crucial. Studies in the past show that intravenous injection of LPS resulted in high toxicity that can cause lethality to animals even at a dose as low as 3 mg/Kg, and even with a dose of 0.01 mg/Kg, animals experienced nearly 10% body weight loss 24 h post-injection (Barton et al. 2019). Most of cases, the highest dose of LPS can be injected into mice via IP is about 5 mg/Kg, to secure animal survival, although the intense immune response can cause animal severe and noticeable weight loss. Using the hydro-Cy7 probe, in vivo imaging of the mouse model showed a significantly greater fluorescence intensity

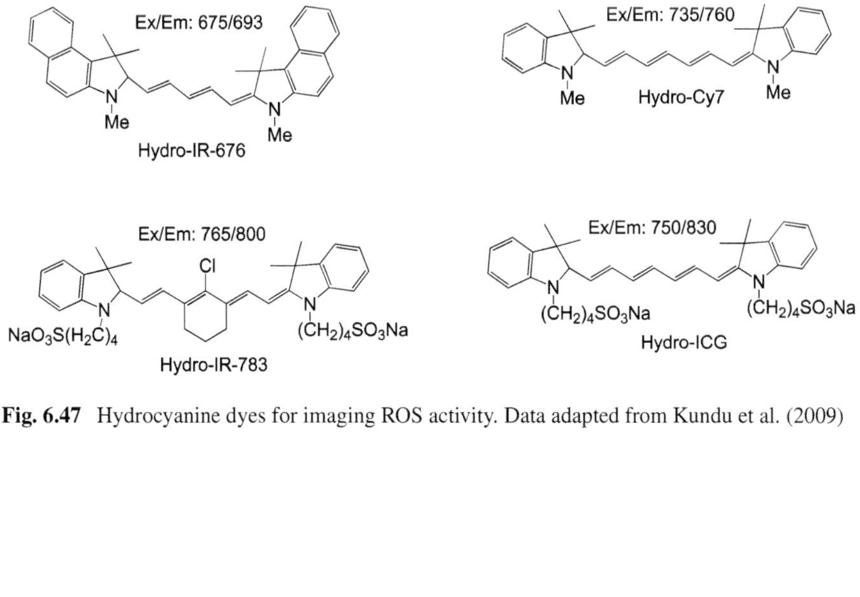

Fig. 6.47 Hydrocyanine dyes for imaging ROS activity. Data adapted from Kundu et al. (2009)

in the abdominal region. On average, the signal depicted from LPS-treated mice is twofold compared to saline-treated animals.

With the exceptional specificity and the ability to cross the blood–brain barrier for delineating ROS-prone tissues in the brain, these activatable probes have the potential and the gravitas for human application. But that is possible if integrating this technology with an imaging modality that can penetrate deep tissues. In this context, the clinical PET and MR imaging fit the purpose by virtue of being unencumbered by tissue penetration. However, one important aspect needs further evaluation in the design, particularly for neuroimaging. Even for an ideal probe for the brain, only a few percent of the injected dose could destine for the target. PET imaging seems to be a great candidate when we put together all these aspects as a recipe for successful human translation.

Up to date, only a few attempts have been reported for the development of ROS PET agents for the brain. Although HE can cross the BBB, after modification of the backbone for [^{18}F] labeling, the ROS PET radioligand called (Wee et al. 2018) FDMT loses the ability to penetrate the BBB (Chu et al. 2014). The attempt to improve this issue resulted in the design of [^{18}F]ROStrace (Hou et al. 2018). With the availability of this probe, non-invasive, repeated, and in vivo imaging of ROS was realized recently. The synthesis started with hydroxyethyl insertion into the hydroxyl moiety on di-*tert*-butyl [6-(4-hydroxyphenyl)phenanthridine-3,8-diyl] dicarbamate (Fig. 6.48). The alcohol product was treated with iodomethane to methylate the aromatic amine, followed by a reduction with sodium borohydride. The ROS PET precursor was synthesized with the tosylate group to serve for aliphatic [^{18}F]F$^-$ radiolabeling. The [^{18}F]labeling procedure was performed with [^{18}F]F$^-$ using routine conditions. Besides the PET agent, this work also reported a practical method for developing the standard, sometimes called the "cold" compound, which is a must in every new probe development. The "cold" compound assisted with the authentication of the labeling material by providing comparative HPLC and TLC profiles. In addition, this compound enabled the assessment of the efficacy of this new probe in scavenging ROS products. The data showed that superoxide radicals selectively oxidize this probe in a manner similar to those observed in HE. The oxidation of the probe was completely abrogated in the presence of superoxide dismutase (SOD).

In vivo imaging data demonstrated that [^{18}F]ROStrace probe quickly entered the brains of control and LPS-treated mice. However, the probe was noted with fast washout in control mice. In contrast, [^{18}F]ROStrace showed a favorable biodistribution profile, with rapid and prolong uptake noted in the brain of the LPS-treated mice (Fig. 6.49a). It is apparent that retention in the brain was ROS-dependent. Overall, the in vivo imaging data showed that the probe was distributed across many subregions of the brain and resided in the brain with a high concentration in LPS-treated mice for up to 20 min and started to plateau at 40 min (Fig. 6.49 b–c).

Another PET radioligand was developed recently for imaging ROS in Alzheimer's disease. This probe derives from a rare amino acid called L-ergothioneine (ERGO). ERGO is a potent antioxidant synthesized by actinomycetes, cyanobacteria, methylbacteria, and some fungi, like mushrooms (Cumming et al. 2018; Halliwell et al. 2016). The underlying mechanism of this probe is based on the hypothesis that a

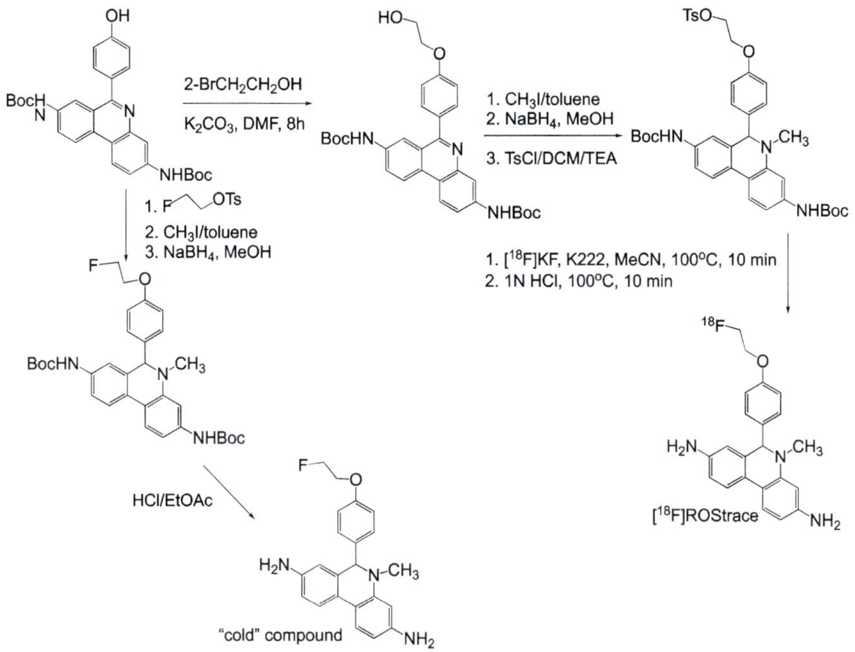

Fig. 6.48 Design of a ROS probe for neuroimaging using PET. Data obtained from Hou et al. (2018)

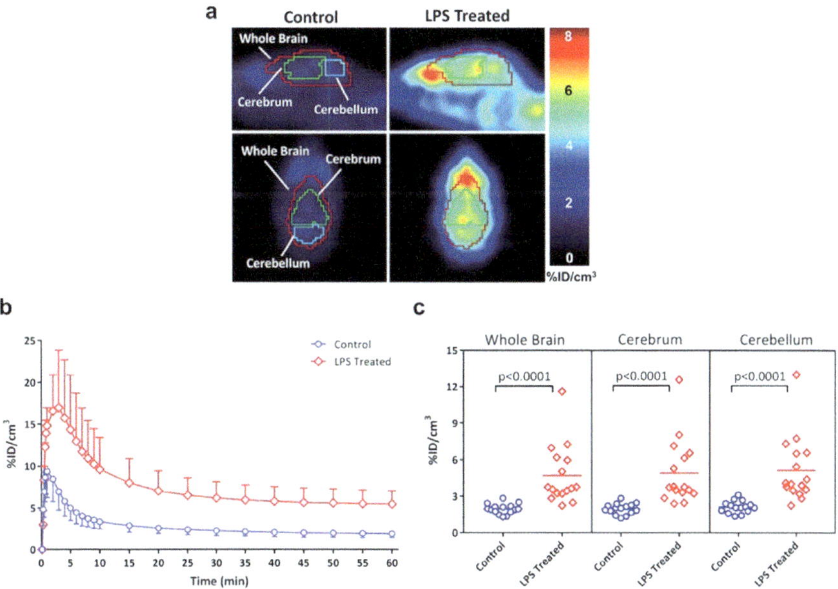

Fig. 6.49 In vivo neuroimaging of ROS using [^{18}F]ROStrace. Data obtained from Hou et al. (2018)

high concentration of ROS radicals and transition metals in the brain of Alzheimer's disease subjects would trap the probe and result in enhanced PET signals (Behof et al. 2022).

Since ERGO is an amino acid with a chemical structure similar to histidine, the construction of the probe using histidine as a template seems to be a reasonable design. Furthermore, this approach also reserves the same configuration since amino acids are stereoisomers. As said, there are many challenges associated with this chemistry given the small molecule, but there are four active functional groups. Orthogonal protection and deprotection operations are intensive in this chemistry. These protection groups also help leverage the molecules into organic solvents since ERGO is a very polar compound.

The synthesis started with protected L-histidine (Fig. 6.50). The carboxylic group was first capped with a methyl ester using EDC and a catalytic amount of DMAP. Next, the acid-labile Boc group was removed using TFA. Next. Reductive methylation of amine with aldehyde and $NaBH_3CN$ to produce a dimethylated intermediate, followed by introducing the thioketone into the imidazole ring and protecting other active functional groups to provide a precursor. Direct methylation of the precursor was achieved using $[^{11}C]MeOTf$, followed by removing the protecting groups to afford the $[^{11}C]ERGO$ radioligand.

In vivo imaging using dynamic PET acquisition showed that $[^{11}C]ERGO$ PET radioligand was distributed to the brain via a transporter called OCTN1. As shown in Fig. 6.51, the retention of the probe in the brains of 5XFAD mice is more profound than observed in WT counterparts. The data suggested that oxidative stress in Alzheimer's disease might trap the probe, resulting in probe accumulation in the

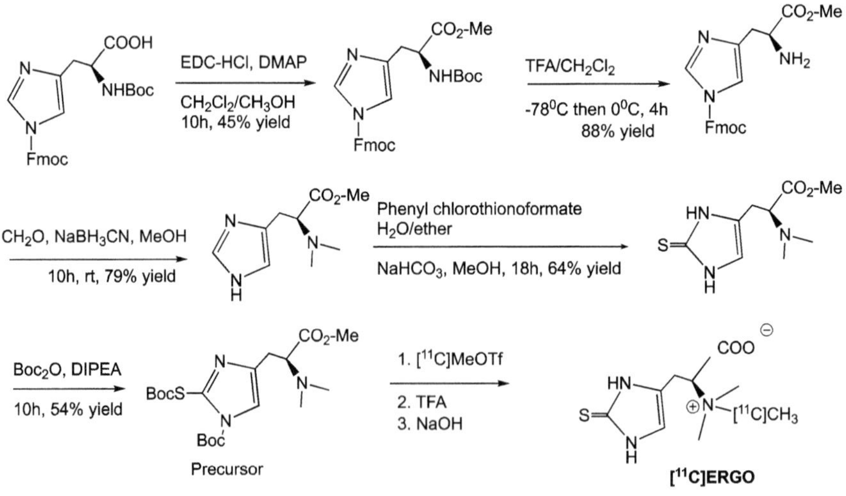

Fig. 6.50 Synthesis of a $[^{11}C]ERGO$ PET radioligand for imaging ROS. Data obtained from Behof et al. (2022)

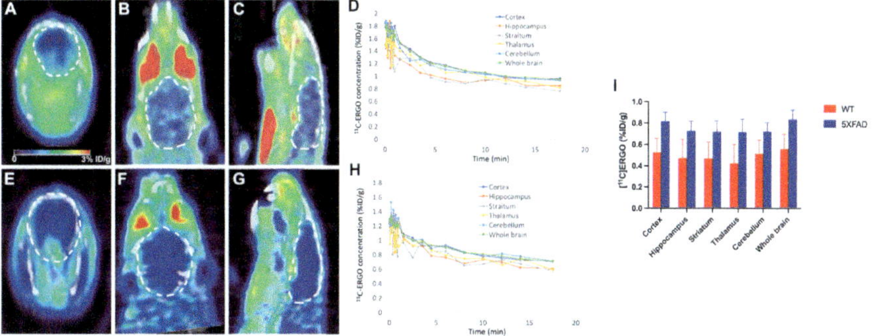

Fig. 6.51 Dynamic PET scans to image the distribution of the [^{11}C]ERGO radioligand on the brains of 5XFAD mice (ABCD) versus WT control (EFGH). The uptake was quantified and compared between cohorts (I). Data obtained from Behof et al. (2022)

brain as visualized in PET images with enhanced spatial and temporal resolution compared to control mice.

Another dye designed for imaging hydrogen peroxide belongs to the family of dihydroxyphenoxazone, called Amplex Red (10-acetyl-3,7-dihydroxyphenoxazine). This tricyclic iminoquinone probe has been widely used for developing assays to detect hydrogen peroxide in biological systems. The probe activation relies on peroxidase, such as horseradish peroxidase to catalyze the oxidation of this non-fluorescent compound to form Resorufin, which has an emission at 585 nm (Fig. 6.52).

This activatable probe has been used for the detection of cholesterol and cholesteryl ester, which are hydrolyzed by cholesterol esterase into cholesterol. Total cholesterol in the sample is further oxidized into a ketone product by cholesterol oxidase and releasing H_2O_2. Cholesterol has been shown to mediate Ca^{2+} influx into the cells and increase cellular vulnerability (Bastiaanse et al. 1994). Most importantly, the cholesterol assays have contributed greatly to the quantification of cholesterol to facilitate a greater understanding of its implications in cellular functions related to protein–protein and protein–lipid interactions. Given the existence of large lipid components in the brain, the role of cholesterol in neurodegenerative diseases is

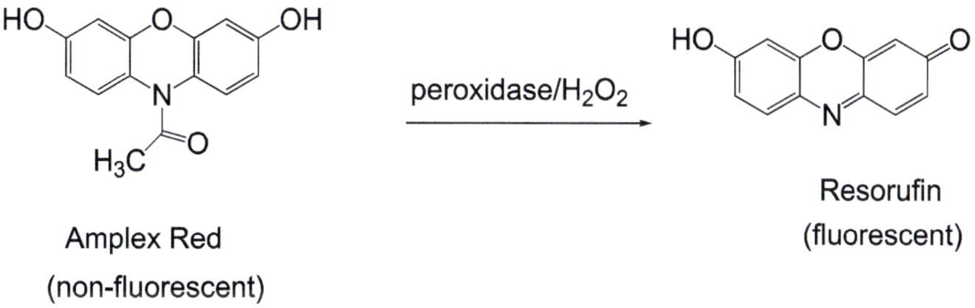

Fig. 6.52 Activatable amplex red dye

a large topic of research. Particularly, many reports demonstrated the relation of cholesterol to Alzheimer's disease. The levels of cholesterol change in the aging. At young age, approximately 87% of the cholesterol is localized in the inner layer of the brain plasma membrane. However, the percentage of cholesterol increases in the outer layer during aging, offsetting the initial transmembrane asymmetry to the point where amyloid precursor protein (APP) can insert into the membrane. This starts to perturb the biophysical integrity of the membrane and the activity of several transmembrane or associated-membrane proteins (Fabiani and Antollini 2019). Using Amplex Red in the assay to detect cholesterol, the work demonstrated that mature neuron membrane cholesterol levels were significantly higher than those detected in younger counterparts (Nicholson and Ferreira 2009). The work also suggested that this event is reversible, and it could be a target for therapy. For example, reducing membrane cholesterol in mature neurons reduced their susceptibility to Abeta-dependent calpain activation.

With the impressive backdrop of the Amplex Red application for imaging, it is instructive to delve into the chemical design of this dye; from there, further modifications can be achieved and motivated for future deep tissue imaging and for in vivo applications. The synthesis started with the per-acylation process of Resazurin in the presence of a strong reducing reagent (SnCl$_2$) and acylation agent. Next, the transformation into Amplex Red could be achieved by selectively deprotection of O-acyl groups using sodium sulfite in a mixture of dioxane in water (Fig. 6.53).

Given the promise of this family of dyes, it is thoughtful for further modification to fine-tune the emission wavelength to the red or near-infrared spectrum. One can use electronic effects to achieve such an approach, as discussed in Chap. 2. For example, the absorbance of azulene-based dyes shifts from 700 to 750 nm after manipulating the electronic effect. The other potential approach for fine-tune the dyes for bathochromic shift involved displacing the oxygen atom on the phenoxazine ring for selenium. Toward this direction, a family of phenoselenazinequinone and related structures of phenoselenazone derivatives have been developed and reported with excellent near-infrared capability. With a chemical scaffold similar to those of

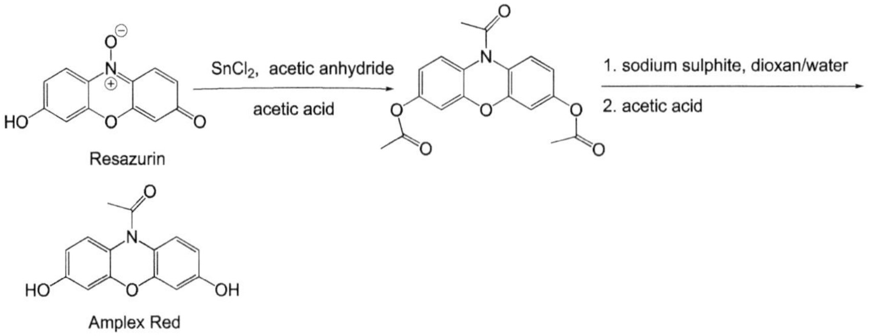

Fig. 6.53 Synthesis of Amplex Red. Data obtained from Eltz et al. (1991)

Amplex Red, the absorbance of these selenium-based dyes was recorded up above 750 nm (Ishida et al. 1989).

6.5 Conclusion

It is not easy to cover every chemical design and application of molecular probes in this book. Nevertheless, the systematic categorization of different imaging strategies into groups enables the distinction of each approach's similarities and differences so one can generalize the characteristic of the probes and predict the outcome. Aside from providing a didactic platform, the information collected from this book provides an essential foundation to motivate more advanced development of molecular probes in the future. If biomarker imaging contributes to the realization of precision and personalized medicine, the development of innovative molecular probes remains a public healthcare imperative.

Bonus: tips on intravenous injection

Although this book describes only preclinical validations of molecular probes in small animal models, the data should be obtained and reported appropriately and accordingly if the clinical translation is the next goal. The biodistribution, pharmacokinetics, pharmacodynamics, and toxicity of treated exogenous compounds in animal experiments are paramount. They are directly related to future translation to humans, and these data link directly to how it is introduced to the host. Thus, each injection route will result in a very different outcome. The injection route dictates the biodistribution and pharmacokinetics of the probes in vivo. The most significant advantage of intravenous injection is that outcome obtained from this work can be extrapolated to humans. While the method is simple when performed on large animal models, it is a challenge in mice.

A few things need to know before we discuss the methods in detail. Intravenous injection can be performed on either side of a mouse's tail's lateral veins (positions 3 and 9). As a rule of thumb, one would start the injection from the near tip of the tail and continue to move up in the event initial attempts fail, or in some cases, more than one injection is necessary. It is important to treat this work as a small surgical procedure, meaning that only serious preparation and handling of the work with precision can guarantee successful intravenous injections. Theoretically, an alcohol pad can be used to sanitize the tail before injection. However, in practice, alcohol can induce vasoconstriction of the veins, thus preventing good injection. Another note worth mentioning that refrain from marking the tail with Sharpie and any kind of marker which will obscure the view. Next, intravenous injection is quite different from retroorbital injection. The latter should not be reported as an intravenous injection. The biodistribution and pharmacokinetics data generated from retroorbital methods cannot be extrapolated to humans. Lastly, there is a myth that intravenous injection of nude mice is easier than performing the same work on C57BL6 or BALB/c mice. The truth is that even the veins of nude mice are visible; they are very small compared to those of BL6 or BALB/c mice. Further, they are also very shallow; altogether, intravenous injection of nude mice can be a full day of frustration if the performer does not consider this information.

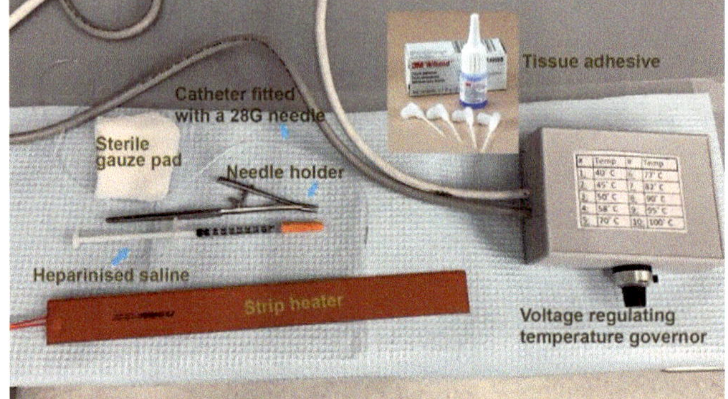

As shown in the figure above, the supplies needed for injection are comprised of a strip heater, a voltage regulating temperature governor, a needle holder, and the catheter-needle line. This setting enables precise temperature during the process. Usually, the strip heater can be replaced by a balloon of hot water. However, the limitation of this operation is that one would have to continue replacing the balloon with the desired temperature. Further, extra care requires to ensure water is not around the injection area to prevent accidental needle sticks due to slippery.

Animals should be sedated for at least 5 min with isoflurane (2%, with oxygen) before dilating the tail veins with a strip heater. The injection is performed with a needle holder using the catheter-needle line prefilled with heparinized saline. Since the mouse tail veins are very shallow, the needle insertion at an angle of 5–8 degrees would suffice. When blood is visible in the catheter, the needle plane is now adjusted parallel to the surface of the tail while continuing to push ahead about 1–2 mm to make sure the needle remains in place firmly. Finally, the probe can be injected from the other end of the catheter. If the probe needs to be injected while the animal stays inside a scanner, a drop of tissue adhesive solution would be applied at the insertion point to secure the needle catheter before transferring the animal to the imaging scanner. In general, for a small mouse, the ideal volume of each injection is approximately 100 µL, and maximal volume is 200 µL. After injection, the animal should be removed from isoflurane immediately, and the bleeding should be controlled by providing some pressure on the wound using a sterilized gauge pad.

References

M.S. Attene-Ramos, C.P. Austin, M. Xia, High throughput screening. Encyclopedia of Toxicol. **2**, 916–917 (2014)

R. McClure, R. Redha, P. Vinson, W. Pham, A robust and scalable high-throughput compatible assay for screening amyloid-beta-binding compounds. J. Alzheimers Dis. **70**, 187–197 (2019)

R. McClure, D. Yanagisawa, D. Stec, D. Abdollahian, D. Koktysh, D. Xhillari, R. Jaeger, G. Stanwood, E. Chekmenev, I. Tooyama, J.C. Gore, W. Pham, Inhalable curcumin: offering the potential for translation to imaging and treatment of Alzheimer's disease. J. Alzheimers Dis. **44**, 283–295 (2015)

D. Yanagisawa, T. Amatsubo, S. Morikawa, H. Taguchi, M. Urushitani, N. Shirai, K. Hirao, A. Shiino, T. Inubushi, I. Tooyama, In vivo detection of amyloid beta deposition using (1)(9)F magnetic resonance imaging with a (1)(9)F-containing curcumin derivative in a mouse model of Alzheimer's disease. Neuroscience **184**, 120–127 (2011)

D. Yanagisawa, N.F. Ibrahim, H. Taguchi, S. Morikawa, K. Hirao, N. Shirai, T. Sogabe, I. Tooyama, Curcumin derivative with the substitution at C-4 position, but not curcumin, is effective against amyloid pathology in APP/PS1 mice. Neurobiol. Aging **36**, 201–210 (2015)

D. Yanagisawa, N.F. Ibrahim, H. Taguchi, S. Morikawa, T. Tomiyama, I. Tooyama, Fluorine-19 magnetic resonance imaging for detection of amyloid beta oligomers using a keto form of curcumin derivative in a mouse model of Alzheimer's disease. Molecules **26** (2021)

D. Yanagisawa, N. Shirai, T. Amatsubo, H. Taguchi, K. Hirao, M. Urushitani, S. Morikawa, T. Inubushi, M. Kato, F. Kato, K. Morino, H. Kimura, I. Nakano, C. Yoshida, T. Okada, M. Sano, Y. Wada, K.N. Wada, A. Yamamoto, I. Tooyama, Relationship between the tautomeric structures of curcumin derivatives and their Abeta-binding activities in the context of therapies for Alzheimer's disease. Biomaterials **31**, 4179–4185 (2010)

M.D. Seemann, Human PET/CT scanners: feasibility for oncological in vivo imaging in mice. Eur. J. Med. Res. **9**, 468–472 (2004)

W.W. Moses, Fundamental limits of spatial resolution in PET. Nucl. Instrum. Methods Phys. Res. A **648**(Supplement 1), S236–S240 (2011)

N.M. deSouza, E. Achten, A. Alberich-Bayarri, F. Bamberg, R. Boellaard, O. Clement, L. Fournier, F. Gallagher, X. Golay, C.P. Heussel, E.F. Jackson, R. Manniesing, M.E. Mayerhofer, E. Neri, J. O'Connor, K.K. Oguz, A. Persson, M. Smits, E.J.R. van Beek, C.J. Zech, R. European Society of, Validated imaging biomarkers as decision-making tools in clinical trials and routine practice: current status and recommendations from the EIBALL* subcommittee of the European Society of Radiology (ESR). Insights Imaging **10**, 87 (2019)

A. Pyo, D.Y. Kim, H. Kim, D. Lim, S.Y. Kwon, S.R. Kang, H.S. Kim, H.S. Bom, J.J. Min, Ultrasensitive detection of malignant melanoma using PET molecular imaging probes. Proc. Natl. Acad. Sci. USA **117**, 12991–12999 (2020)

H. Liu, S. Liu, Z. Miao, Z. Deng, B. Shen, X. Hong, Z. Cheng, Development of 18F-labeled picolinamide probes for PET imaging of malignant melanoma. J. Med. Chem. **56**, 895–901 (2013)

A. Zhu, D. Lee, H. Shim, Metabolic PET imaging in cancer detection and therapy response. Semin. Oncol. **38**, 55–69 (2011)

B. Krug, R. Crott, M. Lonneux, J.F. Baurain, A.S. Pirson, T. Vander Borght, Role of PET in the initial staging of cutaneous malignant melanoma: systematic review. Radiology **249**, 836–844 (2008)

J.D. Wagner, D.S. Schauwecker, D. Davidson, S. Wenck, S.H. Jung, G. Hutchins, FDG-PET sensitivity for melanoma lymph node metastases is dependent on tumor volume. J. Surg. Oncol. **77**, 237–242 (2001)

J. Rouanet, M. Quintana, P. Auzeloux, F. Cachin, F. Degoul, Benzamide derivative radiotracers targeting melanin for melanoma imaging and therapy: preclinical/clinical development and combination with other treatments. Pharmacol. Ther. **224**, 107829 (2021)

S.E. Koch, J.R. Lange, Amelanotic melanoma: the great masquerader. J. Am. Acad. Dermatol. **42**, 731–734 (2000)

E. Wee, R. Wolfe, C. McLean, J.W. Kelly, Y. Pan, Clinically amelanotic or hypomelanotic melanoma: Anatomic distribution, risk factors, and survival. J. Am. Acad. Dermatol. **79**, 645–651 e644 (2018)

B.J. Fueger, J. Czernin, I. Hildebrandt, C. Tran, B.S. Halpern, D. Stout, M.E. Phelps, W.A. Weber, Impact of animal handling on the results of 18F-FDG PET studies in mice. J. Nucl. Med. **47**, 999–1006 (2006)

O. Kelada, J.C. Tseng, J. Peterson, Best practices for preclinical 18F-FDG PET imaging. J. Nucl. Med. **59**, 1155 (2018)

D. Delbeke, W.H. Martin, Metabolic imaging with FDG: a primer. Cancer J. **10**, 201–213 (2004)

J. Morales, L. Li, F.J. Fattah, Y. Dong, E.A. Bey, M. Patel, J. Gao, D.A. Boothman, Review of poly (ADP-ribose) polymerase (PARP) mechanisms of action and rationale for targeting in cancer and other diseases. Crit. Rev. Eukaryot. Gene. Expr. **24**, 15–28 (2014)

A.A.E. Ali, G. Timinszky, R. Arribas-Bosacoma, M. Kozlowski, P.O. Hassa, M. Hassler, A.G. Ladurner, L.H. Pearl, A.W. Oliver, The zinc-finger domains of PARP1 cooperate to recognize DNA strand breaks. Nat. Struct. Mol. Biol. **19**, 685–692 (2012)

V. Ossovskaya, I.C. Koo, E.P. Kaldjian, C. Alvares, B.M. Sherman, Upregulation of Poly (ADP-Ribose) Polymerase-1 (PARP1) in triple-negative breast cancer and other primary human tumor types, genes. Cancer **1**, 812–821 (2010)

J. Tang, D. Salloum, B. Carney, C. Brand, S. Kossatz, A. Sadique, J.S. Lewis, W.A. Weber, H.G. Wendel, T. Reiner, Targeted PET imaging strategy to differentiate malignant from inflamed lymph nodes in diffuse large B-cell lymphoma. Proc. Natl. Acad. Sci. USA **114**, E7441–E7449 (2017)

M.E. Wielgos, R. Rajbhandari, T.S. Cooper, S. Wei, S. Nozell, E.S. Yang, Let-7 status is crucial for PARP1 expression in HER2-overexpressing breast tumors. Mol. Cancer Res. **15**, 340–347 (2017)

C.P. Irwin, Y. Portorreal, C. Brand, Y. Zhang, P. Desai, B. Salinas, W.A. Weber, T. Reiner, PARPi-FL–a fluorescent PARP1 inhibitor for glioblastoma imaging. Neoplasia **16**, 432–440 (2014)

T. Reiner, J. Lacy, E.J. Keliher, K.S. Yang, A. Ullal, R.H. Kohler, C. Vinegoni, R. Weissleder, Imaging therapeutic PARP inhibition in vivo through bioorthogonally developed companion imaging agents. Neoplasia **14**, 169–177 (2012)

S.M. Aukema, R. Siebert, E. Schuuring, G.W. van Imhoff, H.C. Kluin-Nelemans, E.J. Boerma, P.M. Kluin, Double-hit B-cell lymphomas. Blood **117**, 2319–2331 (2011)

S. Hu, Z.Y. Xu-Monette, A. Tzankov, T. Green, L. Wu, A. Balasubramanyam, W.M. Liu, C. Visco, Y. Li, R.N. Miranda, S. Montes-Moreno, K. Dybkaer, A. Chiu, A. Orazi, Y. Zu, G. Bhagat, K.L. Richards, E.D. Hsi, W.W. Choi, X. Zhao, J.H. van Krieken, Q. Huang, J. Huh, W. Ai, M. Ponzoni, A.J. Ferreri, F. Zhou, G.W. Slack, R.D. Gascoyne, M. Tu, D. Variakojis, W. Chen, R.S. Go, M.A. Piris, M.B. Moller, L.J. Medeiros, K.H. Young, MYC/BCL2 protein coexpression contributes to the inferior survival of activated B-cell subtype of diffuse large B-cell lymphoma and demonstrates high-risk gene expression signatures: a report from the international DLBCL Rituximab-CHOP consortium program. Blood **21**, 4021–4031 (2013); quiz 4250

K.J. Savage, N.A. Johnson, S. Ben-Neriah, J.M. Connors, L.H. Sehn, P. Farinha, D.E. Horsman, R.D. Gascoyne, MYC gene rearrangements are associated with a poor prognosis in diffuse large B-cell lymphoma patients treated with R-CHOP chemotherapy. Blood **114**, 3533–3537 (2009)

W. Pham, S. Kobukai, C. Hotta, J.C. Gore, Dendritic cells: therapy and imaging. Expert Opin. Biol. Ther. **9**, 539–564 (2009)

M. Bajbouj, M. Vieth, T. Rosch, S. Miehlke, V. Becker, M. Anders, H. Pohl, A. Madisch, T. Schuster, R.M. Schmid, A. Meining, Probe-based confocal laser endomicroscopy compared with standard four-quadrant biopsy for evaluation of neoplasia in Barrett's esophagus. Endoscopy **42**, 435–440 (2010)

J. Luo, M.D. Smith, D.A. Lantrip, S. Wang, P.L. Fuchs, Efficient syntheses of pyrofolic acid and pteroyl azide, reagents for the production of carboxyl-differentiated derivatives of folic acid. JACS **119**, 10004–10013 (1997)

S. Wang, R.J. Lee, C.J. Mathias, M.A. Green, P.S. Low, Synthesis, purification, and tumor cell uptake of 67Ga-deferoxamine–folate, a potential radiopharmaceutical for tumor imaging. Bioconjug. Chem. **7**, 56–62 (1996)

T. Ishikawa, Guanidine chemistry. Chem. Pharm. Bull. (Tokyo) **58**, 1555–1564 (2010)

M.D. Kennedy, K.N. Jallad, D.H. Thompson, D. Ben-Amotz, P.S. Low, Optical imaging of metastatic tumors using a folate-targeted fluorescent probe. J. Biomed. Opt. **8**, 636–641 (2003)

L.M. Randall, R.M. Wenham, P.S. Low, S.C. Dowdy, J.L. Tanyi, A phase II, multicenter, open-label trial of OTL38 injection for the intra-operative imaging of folate receptor-alpha positive ovarian cancer. Gynecol. Oncol. **155**, 63–68 (2019)

W.B. Pratt, D.O. Toft, Regulation of signaling protein function and trafficking by the hsp90/hsp70-based chaperone machinery. Exp. Biol. Med. (maywood) **228**, 111–133 (2003)

K. Richter, J. Buchner, Hsp90: chaperoning signal transduction. J. Cell Physiol. **188**, 281–290 (2001)

S. Sato, N. Fujita, T. Tsuruo, Modulation of Akt kinase activity by binding to Hsp90. Proc. Natl. Acad Sci. USA **97**, 10832–10837 (2000)

T.W. Schulte, M.V. Blagosklonny, L. Romanova, J.F. Mushinski, B.P. Monia, J.F. Johnston, P. Nguyen, J. Trepel, L.M. Neckers, Destabilization of Raf-1 by geldanamycin leads to disruption of the Raf-1-MEK-mitogen-activated protein kinase signalling pathway. Mol. Cell. Biol. **16**, 5839–5845 (1996)

P. Workman, Combinatorial attack on multistep oncogenesis by inhibiting the Hsp90 molecular chaperone. Cancer Lett. **206**, 149–157 (2004)

H. Zhang, F. Burrows, Targeting multiple signal transduction pathways through inhibition of Hsp90. J. Mol. Med. (berl) **82**, 488–499 (2004)

K.H. Huang, J.M. Veal, R.P. Fadden, J.W. Rice, J. Eaves, J.P. Strachan, A.F. Barabasz, B.E. Foley, T.E. Barta, W. Ma, M.A. Silinski, M. Hu, J.M. Partridge, A. Scott, L.G. DuBois, T. Freed, P.M. Steed, A.J. Ommen, E.D. Smith, P.F. Hughes, A.R. Woodward, G.J. Hanson, W.S. McCall, C.J. Markworth, L. Hinkley, M. Jenks, L. Geng, M. Lewis, J. Otto, B. Pronk, K. Verleysen, S.E. Hall, Discovery of novel 2-aminobenzamide inhibitors of heat shock protein 90 as potent, selective and orally active antitumor agents. J. Med. Chem. **52**, 4288–4305 (2009)

J.J. Barrott, P.F. Hughes, T. Osada, X.Y. Yang, Z.C. Hartman, D.R. Loiselle, N.L. Spector, L. Neckers, N. Rajaram, F. Hu, N. Ramanujam, G. Vaidyanathan, M.R. Zalutsky, H.K. Lyerly, T.A. Haystead, Optical and radioiodinated tethered Hsp90 inhibitors reveal selective internalization of ectopic Hsp90 in malignant breast tumor cells. Chem. Biol. **20**, 1187–1197 (2013)

U. Schmiedl, M. Ogan, H. Paajanen, M. Marotti, L.E. Crooks, A.C. Brito, R.C. Brasch, Albumin labeled with Gd-DTPA as an intravascular, blood pool-enhancing agent for MR imaging: biodistribution and imaging studies. Radiology **162**, 205–210 (1987)

A. Matsumura, Y. Shibata, K. Nakagawa, T. Nose, MRI contrast enhancement by Gd-DTPA-monoclonal antibody in 9L glioma rats. Acta Neurochir Suppl. (wien) **60**, 356–358 (1994)

D. Shahbazi-Gahrouei, M. Williams, S. Rizvi, B.J. Allen, In vivo studies of Gd-DTPA-monoclonal antibody and gd-porphyrins: potential magnetic resonance imaging contrast agents for melanoma. J. Magn. Reson. Imaging **14**, 169–174 (2001)

H. Kobayashi, N. Sato, A. Hiraga, T. Saga, Y. Nakamoto, H. Ueda, J. Konishi, K. Togashi, W. Brechbiel, 3D-Micro-MR angiography of mice using macromolecular MR contrast agents with polyamidoamine dendrimer core with reference to their pharmacokinetif properties. Magn. Reson. Med. **45**, 454–460 (2001)

S.D. Swanson, J.F. Kukowska-Latallo, A.K. Patri, C. Chen, S. Ge, Z. Cao, A. Kotlyar, A.T. East, J.R. Baker, Targeted gadolinium-loaded dendrimer nanoparticles for tumor-specific magnetic resonance contrast enhancement. Int. J. Nanomed. **3**, 201–210 (2008)

J. Schultz, K. Kaminker, Myeloperoxidase of the leukocyte of normal human blood. I. content and localization. Arch. Biochem. Biophys. **96**, 465–467 (1962)

A. Tobler, H.P. Koeffler, Myeloperoxidase: localization, structure, and function. in *Blood Cell Biochemistry*, ed. by J.R. Harris. vol 3 (Plenum Publishing Co., New York, 1991), pp. 255–288

H.B. Dunford, Y. Hsuanyu, Kinetics of oxidation of serotonin by myeloperoxidase compounds I and II. Biochem. Cell Biol. **77**, 449–457 (1999)

J.W. Chen, W. Pham, R. Weissleder, A. Bogdanov Jr., Human myeloperoxidase: a potential target for molecular MR imaging in atherosclerosis. Magn. Reson. Med. **52**, 1021–1028 (2004)

M. Cella, D. Scheidegger, K. Palmer-Lehmann, P. Lane, A. Lanzavecchia, G. Alber, Ligation of CD40 on dendritic cells triggers production of high levels of interleukin-12 and enhances T cell stimulatory capacity: T-T help via APC activation. J. Exp. Med. **184**, 747–752 (1996)

F. Koch, U. Stanzl, P. Jennewein, K. Janke, C. Heufler, E. Kampgen, N. Romani, G. Schuler, High level IL-12 production by murine dendritic cells: upregulation via MHC class II and CD40 molecules and downregulation by IL-4 and IL-10. J. Exp. Med. **184**, 741–746 (1996)

I.J. de Vries, W.J. Lesterhuis, J.O. Barentsz, P. Verdijk, J.H. van Krieken, O.C. Boerman, W.J. Oyen, J.J. Bonenkamp, J.B. Boezeman, G.J. Adema, J.W. Bulte, T.W. Scheenen, C.J. Punt, A. Heerschap, C.G. Figdor, Magnetic resonance tracking of dendritic cells in melanoma patients for monitoring of cellular therapy. Nat. Biotechnol. **23**, 1407–1413 (2005)

S. Toki, R.A. Omary, K. Wilson, J.C. Gore, R.S. Peebles Jr., W. Pham, A comprehensive analysis of transfection-assisted delivery of iron oxide nanoparticles to dendritic cells. Nanomedicine **9**, 1235–1244 (2013)

K.L. Young, C. Xu, J. Xie, S. Sun, Conjugating Methotrexate to magnetite (Fe(3)O(4)) nanoparticles via trichloro-s-triazine. J. Mater. Chem. **19**, 6400–6406 (2009)

C. Xu, K. Xu, H. Gu, R. Zheng, H. Liu, X. Zhang, Z. Guo, B. Xu, Dopamine as a robust anchor to immobilize functional molecules on the iron oxide shell of magnetic nanoparticles. J. Am. Chem. Soc. **126**, 9938–9939 (2004)

D. Koktysh, V. Bright, W. Pham, Fluorescent magnetic hybrid nanoprobe for multimodal bioimaging. Nanotechnology **22**, 275606 (2011)

G.T. Hermanson, in *Bioconjugate Techniques*, 2nd ed (2008). pp. 499

S.A. Corr, Y.P. Rakovich, Y.K. Gun'ko, Multifunctional magnetic-fluorescent nanocomposites for biomedical applications. Nanoscale Res. Lett. **3**, 87–104 (2008)

P.S. Mackay, G.J. Kremers, S. Kobukai, J.G. Cobb, A. Kuley, S.J. Rosenthal, D.S. Koktysh, J.C. Gore, W. Pham, Multimodal imaging of dendritic cells using a novel hybrid magneto-optical nanoprobe. Nanomedicine **7**, 489–496 (2011)

L. Bich, M. Mossio, K. Ruiz-Mirazo, A. Moreno, Biological regulation: controlling the system from within. Biol. Philos. **31**, 237–265 (2016)

C. Lopez-Otin, J.S. Bond, Proteases: multifunctional enzymes in life and disease. J. Biol. Chem. **283**, 30433–30437 (2008)

H. Neurath, K.A. Walsh, Role of proteolytic enzymes in biological regulation (a review). Proc. Natl. Acad. Sci. USA **73**, 3825–3832 (1976)

R. Iino, K. Kinbara, Z. Bryant, Introduction: molecular motors. Chem. Rev. **120**, 1–4 (2020)

G. Sych, R. Pashazadeh, Y. Danyliv, O. Bezvikonnyi, D. Volyniuk, A. Lazauskas, J.V. Grazulevicius, Reversibly switchable phase-dependent emission of quinoline and phenothiazine derivatives towards applications in optical sensing and information multicoding. Chemistry **27**, 2826–2836 (2021)

L.S. Wolfe, M.F. Calabrese, A. Nath, D.V. Blaho, A.D. Miranker, Y. Xiong, Protein-induced photophysical changes to the amyloid indicator dye thioflavin T. Proc. Natl. Acad. Sci. USA **107**, 16863–16868 (2010)

H. LeVine 3rd., Thioflavine T interaction with synthetic Alzheimer's disease beta-amyloid peptides: detection of amyloid aggregation in solution. Protein Sci. **2**, 404–410 (1993)

H. Naiki, K. Higuchi, M. Hosokawa, T. Takeda, Fluorometric determination of amyloid fibrils in vitro using the fluorescent dye, thioflavin T1. Anal. Biochem. **177**, 244–249 (1989)

C. Dyrager, R.P. Vieira, S. Nystrom, K.P.R. Nilsson, T. Storr, Synthesis and evaluation of benzothiazole-triazole and benzothiadiazole.-triazole scaffolds as potential molecular porbes for amyloid-beta aggregation. New J. Chem. **41**, 1566 (2017)

R. Khurana, C. Coleman, C. Ionescu-Zanetti, S.A. Carter, V. Krishna, R.K. Grover, R. Roy, S. Singh, Mechanism of thioflavin T binding to amyloid fibrils. J. Struct. Biol. **151**, 229–238 (2005)

C. Wu, Z. Wang, H. Lei, Y. Duan, M.T. Bowers, J.E. Shea, The binding of thioflavin T and its neutral analog BTA-1 to protofibrils of the Alzheimer's disease Abeta(16–22) peptide probed by molecular dynamics simulations. J. Mol. Biol. **384**, 718–729 (2008)

S.J. Jung, Y.D. Park, J.H. Park, S.D. Yang, M.G. Hur, K.H. Yu, Synthesis and evaluation of thioflavin-T analogs as potential imaging agents for amyloid plaques. Med. Chem. Res. **22**, 4263–4268 (2013)

A. Aliyan, N.P. Cook, A.A. Marti, Interrogating amyloid aggregates using fluorescent probes. Chem. Rev. **119**, 11819–11856 (2019)

H. Benzeid, E. Mothes, E.M. Essassi, P. Faller, G. Pratviel, A thienoquinoxaline and a styryl-quinoxaline as new fluorescent probes for amyloid-beta fibrils. Comptes. Redus. Chimie **15**, 79–85 (2012)

S.D. Styren, R.L. Hamilton, G.C. Styren, W.E. Klunk, X-34, a fluorescent derivative of Congo red: a novel histochemical stain for Alzheimer's disease pathology. J. Histochem. Cytochem. **48**, 1223–1232 (2000)

R.P. Linke, Congo red staining of amyloid: improvements and practical guide for a more precise diagnosis of amyloid and the different amyloidosis. in *Book: Protein Misfolding, Aggregation, and Conformational Diseases* (2006), pp. 239–276

M.R. Nilsson, Techniques to study amyloid fibril formation in vitro. Methods **34**, 151–160 (2004)

W.E. Klunk, M.L. Debnath, J.W. Pettegrew, Development of small molecule probes for the beta-amyloid protein of Alzheimer's disease. Neurobiol. Aging **15**, 691–698 (1994)

W.E. Klunk, J.W. Pettegrew, D.J. Abraham, Quantitative evaluation of congo red binding to amyloid-like proteins with a beta-pleated sheet conformation. J. Histochem. Cytochem. **37**, 1273–1281 (1989)

W.E. Klunk, M.L. Debnath, J.W. Pettegrew, Chrysamine-G binding to Alzheimer and control brain: autopsy study of a new amyloid probe. Neurobiol. Aging **16**, 541–548 (1995)

Q. Li, J.S. Lee, C. Ha, C.B. Park, G. Yang, W.B. Gan, Y.T. Chang, Solid-phase synthesis of styryl dyes and their application as amyloid sensors. Angew. Chem. Int. Ed. Engl. **43**, 6331–6335 (2004)

Q. Li, J. Min, Y.H. Ahn, J. Namm, E.M. Kim, R. Lui, H.Y. Kim, Y. Ji, H. Wu, T. Wisniewski, Y.T. Chang, Styryl-based compounds as potential in vivo imaging agents for beta-amyloid plaques. ChemBioChem **8**, 1679–1687 (2007)

N. Boens, V. Leen, W. Dehaen, Fluorescent indicators based on BODIPY. Chem. Soc. Rev. **41**, 1130 (2012)

M.K. Singh, H. Pal, A.V. Sapre, Interaction of the excited singlet state of neutral red with aromatic amines. Photochem. Photobiol. **71**, 300–306 (2000)

W. Ren, M. Xu, S.H. Liang, H. Xiang, L. Tang, M. Zhang, D. Ding, X. Li, H. Zhang, Y. Hu, Discovery of a novel fluorescent probe for the sensitive detection of beta-amyloid deposits. Biosens. Bioelectron. **75**, 136–141 (2016)

T. Forster, Energiewanderung und fluoreszenz Naturwissenschaften, **33**, 166–175 (1946)

V.V. Didenko, DNA probes using fluorescence resonance energy transfer (FRET): designs and applications. Biotechniques **31**, 1106–1121 (2001)

C.E. Rowland, C.W. Brown, I.L. Medintz, J.B. Delehanty, Intracellular FRET-based probes: a review. Methods Appl. Fluoresc. **3**, 042006 (2015)

K.E. Sapsford, L. Berti, I.L. Medintz, Materials for fluorescence resonance energy transfer analysis: beyond traditional donor-acceptor combinations. Angew Chem. Int. Ed Engl. **45**, 4562–4589 (2006)

S. Tyagi, F.R. Kramer, Molecular beacons: probes that fluoresce upon hybridization. Nat. Biotechnol. **14**, 303–308 (1996)

W. Tan, K. Wang, T.J. Drake, Molecular beacons. Curr. Opin. Chem. Biol. **8**, 547–553 (2004)

R. Manganelli, S. Tyagi, I. Smith, Real time PCR using molecular beacons : a new tool to identify point mutations and to analyze gene expression in mycobacterium tuberculosis. Methods Mol. Med. **54**, 295–310 (2001)

H. Wang, J. Li, H. Liu, Q. Liu, Q. Mei, Y. Wang, J. Zhu, N. He, Z. Lu, Label-free hybridization detection of a single nucleotide mismatch by immobilization of molecular beacons on an agarose film. Nucleic Acids Res. **30**, e61 (2002)

P. Bakun, B. Czarczynska-Goslinska, T. Goslinski, S. Lijewski, In vitro and in vivo biological activities of azulene derivatives with potential applications in medicine. Med. Chem. Res. 1–13 (2021)

W. Pham, R. Weissleder, C.-H. Tung, A practical approach for the preparation of monofunctional azulenyl squaraine dye. Tetrahedron Lett. **44**, 3975–3978 (2003)

W. Pham, R. Weissleder, C.-H. Tung, An azulene dimer as a near-infrared quencher. Angew. Chem. Int. Ed. **41**, 3659–3662 (2002)

X. Peng, H. Chen, D.R. Draney, W. Volcheck, A. Schutz-Geschwender, D.M. Olive, A nonfluorescent, broad-range quencher dye for forster resonance energy transfer assays. Anal. Biochem. **388**, 220–228 (2009)

T. Myochin, K. Hanaoka, S. Iwaki, T. Ueno, T. Komatsu, T. Terai, T. Nagano, Y. Urano, Development of a series of near-infrared dark quenchers based on Si-rhodamines and their application to fluorescent probes. J. Am. Chem. Soc. **137**, 4759–4765 (2015)

J.B. Grimm, T.A. Brown, A.N. Tkachuk, L.D. Lavis, General synthetic method for Si-Fluoresceins and Si-Rhodamines. ACS Cent. Sci. **3**, 975–985 (2017)

S. Chakraborti, M. Mandal, S. Das, A. Mandal, T. Chakraborti, Regulation of matrix metalloproteinases: an overview. Mol. Cell Biochem. **253**, 269–285 (2003)

M. Egeblad, Z. Werb, New functions for the matrix metalloproteinases in cancer progression. Nat. Rev. Cancer **2**, 161–174 (2002)

T. Shiomi, V. Lemaitre, J. D'Armiento, Y. Okada, Matrix metalloproteinases, a disintegrin and metalloproteinases, and a disintegrin and metalloproteinases with thrombospondin motifs in non-neoplastic diseases. Pathol. Int. **60**, 477–496 (2010)

Y. Itoh, M. Seiki, MT1-MMP: a potent modifier of pericellular microenvironment. J. Cell Physiol. **206**, 1–8 (2006)

V.W. van Hinsbergh, M.A. Engelse, P.H. Quax, Pericellular proteases in angiogenesis and vasculogenesis. Arterioscler Thromb. Vasc. Biol. **26**, 716–728 (2006)

S. Takahashi, W. Piao, Y. Matsumura, T. Komatsu, T. Ueno, T. Terai, T. Kamachi, M. Kohno, T. Nagano, K. Hanaoka, Reversible off-on fluorescence probe for hypoxia and imaging of hypoxia-normoxia cycles in live cells. J. Am. Chem. Soc. **134**, 19588–19591 (2012)

I.L. Medintz, H. Mattoussi, Quantum dot-based resonance energy transfer and its growing application in biology. Phys. Chem. Chem. Phys. **11**, 17–45 (2009)

L. Josephson, M.F. Kircher, U. Mahmood, Y. Tang, R. Weissleder, Near-infrared fluorescent nanoparticles as combined MR/optical imaging probes. Bioconjug. Chem. **13**, 554–560 (2002)

P.A. Wender, D.J. Mitchell, K. Pattabiraman, E.T. Pelkey, L. Steinman, J.B. Rothbard, The design, synthesis, and evaluation of molecules that enable or enhance cellular uptake: peptoid molecular transporters. Proc. Natl. Acad. Sci. USA **97**, 13003–13008 (2000)

C. Xu, B. Xing, J. Rao, A self-assembled quantum dot probe for detecting beta-lactamase activity. Biochem. Biophys. Res. Commun. **344**, 931–935 (2006)

I.L. Medintz, A.R. Clapp, F.M. Brunel, T. Tiefenbrunn, H.T. Uyeda, E.L. Chang, J.R. Deschamps, P.E. Dawson, H. Mattoussi, Proteolytic activity monitored by fluorescence resonance energy transfer through quantum-dot-peptide conjugates. Nat. Mater. **5**, 581–589 (2006)

L. Shi, V. De Paoli, N. Rosenzweig, Z. Rosenzweig, Synthesis and application of quantum dots FRET-based protease sensors. J. Am. Chem. Soc. **128**, 10378–10379 (2006)

L. Shi, N. Rosenzweig, Z. Rosenzweig, Luminescent quantum dots fluorescence resonance energy transfer-based probes for enzymatic activity and enzyme inhibitors. Anal. Chem. **79**, 208–214 (2007)

J.M. Perez, L. Josephson, T. O'Loughlin, D. Hogemann, R. Weissleder, Magnetic relaxation switches capable of sensing molecular interactions. Nat. Biotechnol. **20**, 816–820 (2002)

Z. Wang, X. Xue, H. Lu, Y. He, Z. Lu, Z. Chen, Y. Yuan, N. Tang, C.A. Dreyer, L. Quigley, N. Curro, K.S. Lam, J.H. Walton, T.Y. Lin, A.Y. Louie, D.A. Gilbert, K. Liu, K.W. Ferrara, Y. Li, Two-way magnetic resonance tuning and enhanced subtraction imaging for non-invasive and quantitative biological imaging. Nat. Nanotechnol. **15**, 482–490 (2020)

N. Traverso, R. Ricciarelli, M. Nitti, B. Marengo, A.L. Furfaro, M.A. Pronzato, U.M. Marinari, C. Domenicotti, Role of glutathione in cancer progression and chemoresistance. Oxid Med. Cell. Longev. **2013**, 972913 (2013)

A. Phaniendra, D.B. Jestadi, L. Periyasamy, Free radicals: properties, sources, targets, and their implication in various diseases, Indian. J. Clin. Biochem. **30**, 11–26 (2015)

K. Brieger, S. Schiavone, F.J. Miller Jr., K.H. Krause, Reactive oxygen species: from health to disease. Swiss. Med. Wkly. **142**, w13659 (2012)

G.Y. Liou, P. Storz, Reactive oxygen species in cancer. Free Radic. Res. **44**, 479–496 (2010)

G. Pizzino, N. Irrera, M. Cucinotta, G. Pallio, F. Mannino, V. Arcoraci, F. Squadrito, D. Altavilla, A. Bitto, Oxidative stress: harms and benefits for human health. Oxid Med. Cell Longev. **2017**, 8416763 (2017)

J. Zielonka, J. Vasquez-Vivar, B. Kalyanaraman, Detection of 2-hydroxyethidium in cellular systems: a unique marker product of superoxide and hydroethidine. Nat. Protoc. **3**, 8–21 (2008)

J. Zielonka, H. Zhao, Y. Xu, B. Kalyanaraman, Mechanistic similarities between oxidation of hydroethidine by Fremy's salt and superoxide: stopped-flow optical and EPR studies. Free Radic. Biol. Med. **39**, 853–863 (2005)

D.J. Hall, S.H. Han, A. Chepetan, E.G. Inui, M. Rogers, L.L. Dugan, Dynamic optical imaging of metabolic and NADPH oxidase-derived superoxide in live mouse brain using fluorescence lifetime unmixing. J. Cereb. Blood Flow Metab. **32**, 23–32 (2012)

K. Abe, N. Takai, K. Fukumoto, N. Imamoto, M. Tonomura, M. Ito, N. Kanegawa, K. Sakai, K. Morimoto, K. Todoroki, O. Inoue, In vivo imaging of reactive oxygen species in mouse brain by using [3H]hydromethidine as a potential radical trapping radiotracer. J. Cereb. Blood Flow Metab. **34**, 1907–1913 (2014)

K. Kundu, S.F. Knight, N. Willett, S. Lee, W.R. Taylor, N. Murthy, Hydrocyanines: a class of fluorescent sensors that can image reactive oxygen species in cell culture, tissue, and in vivo. Angew. Chem. Int. Ed Engl. **48**, 299–303 (2009)

S.M. Barton, V.A. Janve, R. McClure, A. Anderson, J.A. Matsubara, J.C. Gore, W. Pham, Lipopolysaccharide induced opening of the blood brain barrier on aging 5XFAD mouse model. J. Alzheimer Disease **67**, 503–513 (2019)

W. Chu, A. Chepetan, D. Zhou, K.I. Shoghi, J. Xu, L.L. Dugan, R.J. Gropler, M.A. Mintun, R.H. Mach, Development of a PET radiotracer for non-invasive imaging of the reactive oxygen species, superoxide, in vivo. Org. Biomol. Chem. **12**, 4421–4431 (2014)

C. Hou, C.J. Hsieh, S. Li, H. Lee, T.J. Graham, K. Xu, C.C. Weng, R.K. Doot, W. Chu, S.K. Chakraborty, L.L. Dugan, M.A. Mintun, R.H. Mach, Development of a positron emission tomography radiotracer for imaging elevated levels of superoxide in neuroinflammation. ACS Chem. Neurosci. **9**, 578–586 (2018)

B.M. Cumming, K.C. Chinta, V.P. Reddy, A.J.C. Steyn, Role of ergothioneine in microbial physiology and pathogenesis. Antioxid Redox Signal **28**, 431–444 (2018)

B. Halliwell, I.K. Cheah, C.L. Drum, Ergothioneine, an adaptive antioxidant for the protection of injured tissues? a hypothesis. Biochem. Biophys. Res. Commun. **470**, 245–250 (2016)

W.J. Behof, C.A. Whitmore, J.R. Haynes, A.J. Rosenberg, M.N. Tantawy, T.E. Peterson, F.E. Harrison, R.B. Beelman, P. Wijesinghe, J.A. Matsubara, W. Pham, Improved synthesis of an ergothioneine PET radioligand for imaging oxidative stress in Alzheimer's disease. FEBS Lett. Online Ahead of Print (2022)

E.M. Bastiaanse, D.E. Atsma, M.M. Kuijpers, A. Van der Laarse, The effect of sarcolemmal cholesterol content on intracellular calcium ion concentration in cultured cardiomyocytes. Arch. Biochem. Biophys. **313**, 58–63 (1994)

C. Fabiani, S.S. Antollini, Alzheimer's disease as a membrane disorder: spatial cross-talk among beta-amyloid peptides Nicotinic Acetylcholine Receptors and Lipid Rafts. Front Cell Neurosci. **13**, 309 (2019)

A.M. Nicholson, A. Ferreira, Increased membrane cholesterol might render mature hippocampal neurons more susceptible to beta-amyloid-induced calpain activation and tau toxicity. J. Neurosci. **29**, 4640–4651 (2009)

H. von der Eltz, G. Hans-Joachim, M. Klaus, Hydrolase substrates, U.S. Patent, 5,035,998 (1991)

T. Ishida, M. Sakakibara, Y. Ueno, Synthesis of 4H-Benzo[a]phenoselenazin-4-ones as near-IR dyes. J. Heterocyclic Chem. **26**, 785–788 (1989)

Index